AGROFORESTRY AND CLIMATE CHANGE MITIGATION

Harnessing Multitrophic Interactions for Sustainable Plant Protection and Ecosystem Resilience

AGROFORESTRY AND CLIMATE CHANGE MITIGATION

Harnessing Multitrophic Interactions for Sustainable Plant Protection and Ecosystem Resilience

Hanuman Singh Jatav, PhD

Vishnu D. Rajput, PhD

Ipsita Samal, PhD

Tanmaya Kumar Bhoi, PhD

Editors

AAP | APPLE ACADEMIC PRESS

First edition published 2026

Apple Academic Press Inc.
1265 Goldenrod Circle, NE,
Palm Bay, FL 32905 USA

760 Laurentian Drive, Unit 19,
Burlington, ON L7N 0A4, Canada

CRC Press
2385 NW Executive Center Drive,
Suite 320, Boca Raton FL 33431

4 Park Square, Milton Park,
Abingdon, Oxon, OX14 4RN, UK

Apple Academic Press exclusively co-publishes with CRC Press, an imprint of Taylor & Francis Group, LLC

Library and Archives Canada Cataloguing in Publication

CIP data on file with Canada Library and Archives

Library of Congress Cataloging-in-Publication Data

CIP data on file with US Library of Congress

ISBN: 978-1-77964-138-0 (hbk)
ISBN: 978-1-77964-139-7 (ebk)
DOI: 10.1201/9781779641397

About the Editors

Hanuman Singh Jatav, PhD
Assistant Professor, Soil Science and Agricultural Chemistry, Sri Karan Narendra Agriculture University, Jobner, Jaipur, Rajasthan, India

Hanuman Singh Jatav, PhD, is an Assistant Professor of Soil Science and Agricultural Chemistry at Sri Karan Narendra Agriculture University in Jobner, Jaipur, Rajasthan, India. He specializes in sewage sludge, biochar, micronutrients, heavy metals, and soil fertility management, having taught for over 8 years and conducted research for 12 years. He earned his BSc (Hons.) in Agriculture from Maharana Pratap University of Agriculture and Technology in Udaipur, Rajasthan, India, and his MSc (Agriculture) and PhD in Soil Science and Agricultural Chemistry from the Institute of Agricultural Sciences at Banaras Hindu University in Varanasi, Uttar Pradesh, India, under the supervision of Prof. Satish Kumar Singh (FISSS, FISAC). Dr. Jatav is dedicated to advancing environmental sustainability through his research, having published over 120 papers including research papers, review papers, books, book chapters, and technical papers in various national and international journals and publishers. He serves as an Academic Editor and member of subject expertise in many internationally acclaimed journals, and is an active member of more than 20 national and international scientific societies. Dr. Jatav has been recognized for his contributions, receiving a Certificate of Gratitude as an Ambassador by IUCN: International Union for Conservation of Nature-Gland, Switzerland, at the World Conservation Congress during the Global Youth Summit. He has also been honored as the Outstanding Assistant Professor of Sri Karan Narendra Agriculture University, Jobner, Jaipur, Rajasthan, India

Besides this Dr. Jatav has been invited as Member to nominate "Blue Planet Prize" from The Asahi Glass Foundation, Tokyo, Japan. Dr. Jatav has served as a Visiting Professor at Southern Federal University, Russia. In recognition of his academic excellence and contributions, Dr. Jatav was honored by District administration, Government of Rajasthan.

Vishnu D. Rajput. PhD
Leading Researcher (Associate Professor), Head of International Laboratory of Nanobiotechnology, Southern Federal University

Vishnu D. Rajput, PhD, is a Leading Researcher (Associate Professor) and Head of the "International Lab on Nanobiotechnology & Soil Health" at Southern Federal University, Rostov-on-Don, Russia. His current research focuses on the toxic effects of bulk and nanoforms of metals, including bioaccumulation, bio/geo-transformations, uptake, translocation, and their impact on plant physiology, morphology, anatomy, and the ultrastructure of cellular and subcellular organelles. Dr. Rajput's work also involves investigating cytomorphometric modifications and DNA damage induced by these metal forms. Dr. Rajput has made significant contributions to the field, particularly in understanding how nanoparticles and heavy metals interact with plants, soil, microbial communities, and the broader environment. He has explored remediation approaches using biochar/nano-biochar-based sorbents and nanotechnological interventions. With extensive experience and experimental work, he has provided comprehensive insights into the state of research in environmental science related to these interactions and remediation technologies. He has an impressive publication record, with a total of 351 scientific publications, including 204 peer-reviewed full-length articles, 18 books, 70 chapters (Scopus indexed), and 44 conference articles. Dr. Rajput serves on the editorial boards of various high-impact journals and leads several grants, including Mega-Grant and BRICS, and participates as a performer/co-PI in numerous national and international projects. Dr. Rajput has been recognized for his outstanding contributions to academic, creative research, and publication activities. He has received the "Certificate for Appreciation 2019," "Certificate of Honor 2020," Diploma Award 2021, and Letter of Gratitude 2022 and 2023 from Southern Federal University, Russia. Additionally, he has been granted "Highly Qualified Specialist" status by the Russian government and is listed among the top 2% of Russian scientists.

Ipsita Samal, PhD

Scientist (Entomology), Indian Council of Agricultural Research—National Research Centre on Litchi, Mushahari, Muzaffarpur, Bihar, India

Ipsita Samal, PhD, is currently working as Scientist (Entomology) at ICAR-National Research Centre on Litchi, Mushahari, Muzaffarpur, Bihar, India. She has completed her PhD (Entomology) from Indian Agricultural Research Institute, New Delhi. She has obtained her MSc degree from *Indian Agricultural Research Institute* (IARI), New Delhi, and did her graduation from the College of Agriculture, Bhawanipatna, OUAT, Odisha. Her scientific interests are in field of entomology, host plant resistance, integrated pest management aspects of litchi, and understanding biochemical-molecular basis of resistance in insect pests. She is an awardee of Dr. S. Pradhan memorial award (From ICAR-IARI, New Delhi), ICAR-JRF, ICAR-NET, and IARI-PhD in entomology. She has published few articles, research papers and book chapters and presented her work in various national and international conferences.

Tanmaya Kumar Bhoi

Scientist, Arid Forest Research Institute, Jodhpur, Rajasthan, India

Tanmaya Kumar Bhoi is a distinguished scientist with a specialization in the intricate realm of insect science. His extensive expertise encompasses research in the areas of host plant resistance, biological control, and integrated insect pest management. With a dedicated focus on advancing agricultural practices, his contributions have significantly impacted the field of entomology. His research delves into the mechanisms of host-plant resistance, aiming to enhance crop resilience against insect pests through innovative strategies. Additionally, his work in biological control emphasizes sustainable and environmentally friendly approaches to manage insect populations, mitigating the reliance on chemical interventions. His commitment to advancing knowledge is underscored by a prolific publication record, with numerous scholarly contributions both at the national and international levels. These publications reflect not only the depth of his research but also its broader impact on the scientific community and agricultural practices. He is multifaceted expertise positions him as a key figure in the ongoing efforts to develop effective and sustainable solutions for insect pest management, contributing significantly to the advancement of entomological science and its practical applications in forestry.

Contents

Contributors

Souvik Acharya
Faculty Centre for Integrated Rural Development and Management, Ramakrishna Mission Vivekananda Educational and Research Institute, Kolkata, India

Mohsin Ali
School of Chemistry and Chemical Engineering, Hubei University, Wuhan, PR China

Feby Atee
Department of Entomology, Tamil Nadu Agricultural University, Lawley Road, Coimbatore, Tamil Nadu, India

Koushik Banerjee
ICAR-Mahatma Gandhi Integrated Farming Research Institute, Motihari, Bihar, India

Jagadeesh Bathula
Department of Natural Resource Management and Conservation, Forest College and Research Institute, Mulugu, Telangana, India

P. Bhavana
Department of Entomology, Odisha University of Agriculture and Technology, Bhubaneswar

Mahadev Bera
School of Environment and Disaster Management, Ramakrishna Mission Vivekananda Educational and Research Institute, Narendrapur, Kolkata, India

Sunaullah Bhat
Crop Protection Division, ICAR-Vivekananda Parvatiya Krishi Anusandhan Sansthan, Almora, Uttarakhand, India

Tanmaya Kumar Bhoi
ICFRE-Arid Forest Research Institute (AFRI), Jodhpur, Rajasthan, India

Ankita Mukhopadhyay Biswas
Department of Agriculture, Brainware University, Barasat, Kolkata, West Bengal, India

Chandni
ICAR-Central Institute of Temperate Horticulture, Regional Station, Mukteshwar, Uttrakhand

Prashant Chauhan
Department of Plant Pathology, CCS Haryana Agricultural University, Hisar, Haryana

Sunil Choudhary
S.B.R.M Government College Nagaur, Rajasthan, India

Rajeswari Das
Department of Soil Science and Agricultural Chemistry, School of Agriculture, GIET University, Gunupur, Rayagada, Odisha, India

Sashanka S. Dash
Department of Entomology, College of Agriculture, OUAT, Bhubaneswar, Odisha, India

Sumanta Das
School of Environment and Disaster Management, Ramakrishna Mission Vivekananda Educational and Research Institute, Narendrapur, Kolkata; India School of Agriculture and Food Sustainability, The University of Queensland, St Lucia, Queensland, Australia

Suman Samilita Dash
Department of Entomology, Odisha University of Agriculture and Technology, Bhubaneswar, Odisha, India

Swarnali Duary
Department of Agriculture, Brainware University, Barasat, Kolkata, West Bengal, India

Vinod Kumar Dubey
Department of Entomology, Post Graduate College of Agriculture, Dr. Rajendra Prasad Central Agricultural University, Pusa (Samastipur), Bihar, India

Suman Dutta
School of Genetics and Plant Breeding, Ramakrishna Mission Vivekananda Educational and Research Institute, Narendrapur, Kolkata, India

Pabitra Kumar Ghosh
Department of Agriculture, Brainware University, Barasat, Kolkata, West Bengal, India

Ann Theresa Jose
Division of Soil Science and Agricultural Chemistry, ICAR-IARI, New Delhi

Kanchan Kadawla
Department of Entomology, College of Agriculture, Chaudhary Charan Singh Haryana Agricultural University, Hisar, Haryana, India

Nutan Karki
Crop Protection Division, ICAR-Vivekananda Parvatiya Krishi Anusandhan Sansthan, Almora, Uttarakhand, India

Prasun Karmakar
Forest Protection Division, Rain Forest Research Institute, Jorhat, Assam, India

Alagesan Keerthana
Department of Entomology, Tamil Nadu Agricultural University, Lawley Road, Coimbatore, Tamil Nadu, India

Ifrah Khan
ICFRE-Arid Forest Research Institute (AFRI), Jodhpur, Rajasthan, India

J. Komal
Basic Seed Multiplication and Training Centre, Central Silk Board, Kharaswan, Jharkhand, India

Sharat Kothari
ICFRE-Arid Forest Research Institute, Jodhpur, Rajasthan, India

Abhishek Kumar
Department of Agriculture, Maharishi Markandeshwar (Deemed to be University), Mullana, Ambala, Haryana, India

Anil Kumar
Sugarcane Research Institute, Department of Entomology, Post Graduate College of Agriculture, Dr. Rajendra Prasad Central Agricultural University, Pusa (Samastipur), Bihar, India

Naveen Kumar
Division of Soil Science and Agricultural Chemistry, ICAR-IARI, New Delhi, India

Golmei Langangmeilu
Department of Agronomy, School of Agriculture, GIET University, Gunupur, Rayagada, Odisha, India

K. Logeswaran
Division of Entomology and Nematology, Indian Institute of Horticultural Research, Bengaluru, Karnataka, India

Deepak Kumar Mahanta
Forest Entomology Discipline, Forest Protection Division, Indian Council of Forestry Research, and Education (ICFRE)-Forest Research Institute (ICFRE-FRI), Dehradun, Uttarakhand, India

Subash C. Mahapatra
ICRP on Agroforestry, OUAT, Bhubaneswar, Odisha, India

Mita Meher
Department of Agricultural Extension, School of Agriculture, GIET University, Gunupur, Rayagada, Odisha, India

S. N. Murthy Midde
Indian Council of Forestry Research and Education (ICFRE)-Arid Forest Research Institute (ICFRE-AFRI), Jodhpur, Rajasthan, India

Akshay Mishra
Forest Protection Division, Rain Forest Research Institute, Jorhat, Assam, India

Jasti Sri Vishnu Murthy
Department of Agricultural Entomology, College of Agriculture, Vellanikkara, Kerala Agricultural University, Thrissur, Kerala, India

Alapati Nymisha
Division of Soil Science and Agricultural Chemistry, ICAR-IARI, New Delhi, India

Soumita Pal
Department of Agriculture, Government of West Bengal, Santipur, West Bengal, India

Aradhana Panda
Division of Entomology, Sher-e-Kashmir University of Agricultural Sciences and Technology, Kashmir, India

Sucharita Panda
Department of Agroforestry, School of Agriculture, GIET University, Gunupur, Rayagada, Odisha, India

Sivaraj Paramasivam
Department of Agricultural Extension, Amrita School of Agricultural Sciences, Amrita Vishwa Vidyapeetham, Coimbatore, India

Amit Umesh Paschapur
Crop Protection Division, ICAR-Vivekananda Parvatiya Krishi Anusandhan Sansthan, Almora, Uttarakhand, India

Anshuman Patel
Indian Council of Forestry Research and Education (ICFRE)-Rain Forest Research Institute, Jorhat, Assam, India

Mahesh Pathak
Department of Entomology, School of Crop Protection, College of Post Graduate Studies in Agricultural Sciences, Central Agricultural University (Imphal), Umiam, Meghalaya, India

Vishwa Prajapati
Indian Council of Forestry Research and Education (ICFRE)-Arid Forest Research Institute (ICFRE-AFRI), Jodhpur, Rajasthan, India

Pooja
Department of Plant Pathology, CCS Haryana Agricultural University, Hisar, Haryana, India

Sajiya. Quadri
Department of Entomology, College of Agriculture, OUAT, Bhubaneswar, Odisha, India

Bindu Gudi Ramakrishna
Department of Agricultural Entomology, College of Agriculture, Vellanikkara, Kerala Agricultural University, Thrissur, Kerala, India

Atiraj Rathi
Indian Council of Forestry Research and Education-Arid Forest Research Institute (AFRI), Jodhpur, Rajasthan, India

Chada A. Reddy
Department of Entomology, School of Crop Protection, College of Post Graduate Studies in Agricultural Sciences, Central Agricultural University (Imphal), Umiam, Meghalaya, India

Lellapalli Rithesh
Department of Plant Pathology, Kerala Agricultural University, Thiruvananthapuram, Kerala, India

Saleem Sajjad
College of Agriculture, University of Sargodha, Sargodha, Punjab, Pakistan

Ipsita Samal
ICAR-National Research Centre on Litchi, Mushahari, Ramna, Muzaffarpur, Bihar, India

Satyabrata Sarangi
Department of Entomology, Odisha University of Agriculture and Technology, Bhubaneswar, Odisha, India

Kasturi Sarmah
Department of Agricultural Entomology, Centre of Plant Protection Studies, Tamil Nadu Agricultural University, Coimbatore, Tamil Nadu, India

Parthasarathy Seethapathy
Department of Plant Pathology, Amrita School of Agricultural Sciences, Amrita Vishwa Vidyapeetham, Coimbatore, India

Sourav Sen
Department of Agricultural Entomology, Bidhan Chandra Krishi Viswavidyalaya, Mohanapur, Nadia, West Bengal, India

Kuldeep Sharma
Indian Council of Forestry Research and Education (ICFRE)-Arid Forest Research Institute (AFRI), Jodhpur, Rajasthan, India

Sangeeta Singh
Indian Council of Forestry Research and Education (ICFRE)-Arid Forest Research Institute (AFRI), Jodhpur, Rajasthan, India

Aditya Pratap Singh
Department of Plant Breeding and Genetics, School of Agriculture, GIET University, Gunupur, Rayagada, Odisha, India

Victor Thingujam
ICAR-Research Complex for Eastern Region, Regional station, Palandu, Ranchi, Jharkhand, India

Muhammad Danish Toor
Department of Soil Science and Plant Nutrition, Faculty of Agriculture, Ondokuz Mayis University, Samsun, Turkey; Department of Agro-Chemistry and Soil Science, Faculty of Agronomy, Agricultural University, Plovdiv, Bulgaria

Manoj K. Tripathy
Department of Entomology, College of Agriculture, OUAT, Bhubaneswar, Odisha, India

Vipula Vyas
Indian Council of Forestry Research and Education (ICFRE)-Arid Forest Research Institute (AFRI), Jodhpur, Rajasthan, India

Mohammad Abdul Waseem
Department of Basic and Social Sciences, Forest College and Research Institute, Mulugu, Telangana, India

Syeda Duaa Zahra
School of Chemistry and Chemical Engineering, Hubei University, Wuhan, PR China

Abbreviations

ABA	abscisic acid
ADS	Asian dust storms
AFE	agroforestry ecosystem
AFOLU	agricultural, forestry, and other land use
AFS	agroforestry systems
AM	arbuscular mycorrhizae
AMF	arbuscular mycorrhizal fungi
BGA	blue-green algae
CBB	coffee berry borer
CCS	carbon capture and storage
EIP	extrinsic incubation period
ELISA	enzyme-linked immunosorbent assays
EMF	ectomycorrhizal fungi
GHGs	greenhouse gases
IGP	intraguild predation
IPCC	Intergovernmental Panel on Climate Change
IPM	Integrated Pest Management
ISR	induced systematic resistance
JEV	Japanese encephalitis virus
MGCA	molecular gut content analysis
NAPA	National Adaptation Programme of Action
NTFPs	nontimber forest product
OAs	organic amendments
PGPRs	plant growth promoting rhizobacteria
PSB	phosphorus-solubilizing bacteria
PSO	phosphate-solubilizing organism
TTI	tritrophic interaction
UNFCCC	United Nations Framework Convention on Climate Change

Acknowledgment

Everyone who helps directly or indirectly in completing a task deserves a token of thanks. It gives me pleasure to mention and express my feelings to those who helped me by motivating me to compile this work. I, Dr. Hanuman Singh Jatav, as an author of this book, *Agroforestry and Climate Change Mitigation: Harnessing Multitrophic Interactions for Sustainable Plant Protection and Ecosystem Resilience,* give a token of thanks to Prof. Pushpendra Singh Chauhan, Hon'ble Vice-Chancellor, Sri Karan Narendra Agriculture University, Jobner-, Jaipur, for his motivation to compile such a work. I am deeply indebted to Dr. Amitava Rakshit, Professor, Banaras Hindu University, Varanasi, for his lifelong motivation in compiring this work. I also extend my thanks to Prof. K. K. Sharma (HOD-SSAC), and Dr. S. K. Dadhich for their direct and indirect support.

It is very difficult to acknowledge with just a token of words Prof. Harphool Singh, Prof. S.R. Dhaka, who not only supported but also guided me to compile this upshot for the students. I would also like to give thanks from the bottom of my heart to Dr. Sanjay K. Attar, Dr. Muddser A. Khan, Dr. K. C. Verma, Dr. Kailash Chandra, Dr. Mujahid Khan, Dr. Lokesh Kumar Jat, Dr. Champa L. Khatik, Dr. Shubita Kumawat, Dr. Ramu Meena Dr. Ajeet Singh, and Dr. Vishnu D. Rajput for their direct and indirect support in compiling this outcome. At the outset, being a student of the great institution, Banaras Hindu University, Varanasi, I am thankful to the Prof. A. K. Ghosh, Prof. Nirmal De, Prof. Surendra Singh (FISSS), Prof. Priyankar Raha, Prof. P. K. Sharma, Dr. Y. V. Singh, and Dr. Ashish Latare for their support and guidance during my studies. I am grateful to my school teacher Shri Surendra Sarthak Ji, Shri Ghanshyam Ji, and Shri Harkesh Ji, Shri Hanuman Ji. I am also delighted to offer thanks to my friends Dr. Sunil Kumar, Dr. Dinesh Jinger, Dr. Sonu Kumar, Dr. Surendra Singh Jatav, Dharamveer Verma, and Rajat. I am also thankful to Prof. Bijay Singh and Prof. Satish Kumar Singh for their regular efforts in compiling and critical evaluation of this book. I am thankful to Dr. Vishnu D. Rajput for his continued support to check the content to make it worthy and valuable.

I also extend my sincere gratitude to my family members who always support me in any difficult situation. *The graces of the Lord Hanuman Ji, Goddess Maa Sharde, along with my parents, Smt. Supedi Devi and Shri Babulal Jatav, have always blessed me and given me patience and power to overcome difficulties that came my way in the accomplishment of this endeavour. I cannot dare to say thanks but only pray to bless me always.*

— Dr. H. S. Jatav
Assistant Professor
Soil Science and Agricultural Chemistry,
COA, Shekhawati, Sikar, Fatehpur
SKNAU, Jobner, Rajasthan, India

Preface

Agroforestry, the integrated management of trees and crops or livestock on the same piece of land, is increasingly recognized as a sustainable agricultural practice that offers a range of ecological, economic, and social benefits. Among its many advantages, agroforestry systems have the potential to enhance biodiversity, improve soil fertility, increase resilience to climate change, and provide livelihood opportunities for farmers. One of the key aspects of agroforestry that has garnered significant attention in recent years is its role in promoting sustainable plant protection through multitrophic interactions. Multitrophic interactions refer to the complex relationships among organisms at different trophic levels, including plants, herbivores, predators, and parasitoids. These interactions can have profound effects on the dynamics of pest populations and the overall health of agroecosystems.

This book explores the fascinating world of multitrophic interactions in agroforestry systems and their implications for sustainable plant protection. It brings together a diverse range of perspectives, including ecological, agronomic, and economic viewpoints, to provide a comprehensive overview of the topic. The chapters in this book cover a wide array of topics, including the role of biodiversity in pest control, the impact of agroforestry practices on multitrophic interactions, the use of natural enemies for pest management, and the potential of agroforestry for enhancing ecosystem services.

We hope that this book will serve as a valuable resource for researchers, students, policymakers, and practitioners interested in understanding and harnessing the power of multitrophic interactions in agroforestry for sustainable plant protection. By exploring the intricate web of relationships among organisms in agroforestry systems, we can unlock new strategies for managing pests and promoting the health and resilience of agricultural landscapes. We would like to express our sincere gratitude to all the authors who contributed to this book, as well as to the reviewers who provided valuable feedback. We hope that this book will inspire further research and innovation in the field of agroforestry and contribute to the development of sustainable agricultural practices around the world.

CHAPTER 1

Agroforestry and Climate Change Mitigation: Strategies for Sustainable Plant Protection

IFRAH KHAN,* VIPULA VYAS, ATIRAJ RATHI, TANMAYA KUMAR BHOI, and SANGEETA SINGH

ICFRE-Arid Forest Research Institute (AFRI), Jodhpur, Rajasthan, India

*Corresponding author

ABSTRACT

Agroforestry is a sustainable land-use system that integrates herbaceous crops, livestock, and woody perennials such as trees, shrubs, palms, and bamboos on the same land unit, either spatially or temporally. This practice not only enhances crop production but also provides timber and nontimber products, thereby generating additional income for farmers. According to the Intergovernmental Panel on Climate Change, mitigation refers to corrective actions that reduce greenhouse gas emissions and environmental impacts, such as rising CO_2 levels. Climate change poses serious threats to human civilization, including reduced agricultural productivity, food insecurity, resource depletion, and increased incidence of pests and diseases. Agroforestry offers an effective mitigation strategy through carbon sequestration, whereby atmospheric CO_2 is absorbed and stored in tree biomass and soil. By combining land cultivation, afforestation, and changes in land use, agroforestry contributes significantly to climate change mitigation under mechanisms like the Clean Development Mechanism. The primary goal of mitigation is to lower greenhouse gas concentrations, thereby enhancing soil carbon storage, improving soil

Agroforestry and Climate Change Mitigation: Harnessing Multitrophic Interactions for Sustainable Plant Protection and Ecosystem Resilience. Hanuman Singh Jatav, Vishnu D. Rajput, Ipsita Samal, Tanmaya Kumar Bhoi (Eds.)

DOI: 10.1201/9781779641397-1

fertility, and sustaining long-term crop yields. Thus, agroforestry not only supports environmental sustainability but also strengthens food security and rural livelihoods in the face of climate change.

1.1 INTRODUCTION

Climate change is one of the most significant worldwide issues of the contemporary period, with its implications regarding the economy, society, politics, science, morality, and ethics. It has detrimental effects on human civilization, including reduced agricultural output, food insecurity, resource depletion, and the spread of diseases and pests [1]. A climate-dependent economic activity that satisfies the demands of human civilization is crop production. It is obvious that agriculture will undoubtedly be significantly impacted by climate changes. The main industry in low-income nations for rural development and livelihoods is agriculture. However, land pressure and climate change have a negative impact on agricultural systems in emerging nations, endangering food production. Even while rigorous farming techniques are supported all over the world, their relative success is exacerbated by reduced productivity brought on by land degradation. Localized negative response over climate, abundance and security of food, and income through farm are caused by the diverse environmental effects of agricultural intensification and food production, which have detrimental effects on biodiversity and soil [2, 3]. One of the most noticeable land-use practices in agro-ecological zones and landscapes is agroforestry. Agroforestry is becoming more and more popular due to its capacity to meet a variety of needs for on-farm adaptation and play a major part in agricultural, forestry, and other land use (AFOLU)-related mitigation pathways in the face of food shortages and growing climate change concerns. Agroforestry lowers human influence on natural forests and offers ecosystem services while producing benefits and revenue through wood energy and carbon, enhanced soil richness, and improved regional climate scenarios. The majority of these advantages support international efforts to reduce atmospheric greenhouse gas concentrations while also directly assisting with local adaptation. Extended rotation system for soil preservation, heterogeneity of crop, home gardens, border plants, crops flowing throughout the year, fence intercropping, live hedgegrows, enhanced fallow crops, and varied stratum agroforestry are just a few of the many land management techniques that make up agroforestry systems. In developing nations, combining adaptation and mitigation is recognized as a necessity, especially in the AFOLU area. It consists of several models in various parts of the world and is essential in reducing the adverse effects of climate change by expanding the diversity of tree crops, which increases the capacity for sequestering carbon dioxide more than if solitary agricultural crops were grown. In order to maximize the economic benefits to farmers, agroforestry systems should be managed to fulfill the needs for both production (fuelwood, nutritive fruits, food, timber, nontimber forest product [NTFPs], etc.) and protection (shade, effective nutrient cycling, watershed management, reducing soil erosion, improving soil health and fertility, etc.). In summary, by offering smallholder farmers numerous outputs and raising rural living standards, agro-forestry has enormous potential to improve farm productivity, income, and employment prospects as well as address the crises of food and nutritional uncertainties in the growing population. Therefore, agroforestry helps farmers both adapt to and mitigate the effects of climate change.

1.2 CLIMATE CHANGE

Earth's climate has remained stable for more than 12,000 years that in turn plays a major role for the survival of humankind's existence [4]. The average global temperature rose by over 1.5°F over the last century, and another 0.5–8.6°F is predicted to climb over the next 100 years. This is a serious issue, since even little variations in the average global temperature can have a big impact on the weather and climate [5].

We are living in a pivotal period as climate change is the defining problem of our time. The effects of climate change are widespread and unprecedented in magnitude, ranging from altered weather patterns that jeopardize food production to increasing sea levels that raise the possibility of catastrophic flooding. Future adaptation to these effects will be more expensive and time-consuming if immediate action is not taken now.

Climate change is a word used to describe long-term variations in temperature and weather patterns. These fluctuations may be caused by significant volcanic eruptions or variations in the sun's activity. However, human activity has been the primary cause of climate change since the

1800s, mostly as a result of the combustion of fossil fuels like coal, oil, and gas.

As a result of the burning of fossil fuels, greenhouse gases (GHGs) are released into the atmosphere, enveloping the earth like a blanket and trapping solar heat, raising global temperatures.

There are a few fundamental, reputable scientific connections:

Since the Industrial Revolution, the concentration of GHGs has been steadily rising and so have mean global temperatures.

Carbon dioxide (CO_2), the most prevalent GHG and the source of approximately two-thirds of all GHGs, is primarily produced by burning fossil fuels.

Since the 2015 signing of the Paris Agreement, there has been progress, according to the UN Environment Programme's latest Emissions Gas Report. When the agreement went into effect, it was estimated that policies in place would result in a 16% increase in greenhouse gas emissions in 2030. Currently, a 3% increase is anticipated. Predicted greenhouse gas emissions for 2030 still need to decrease by 28% for the 2°C pathway outlined in the Paris Agreement and 42% for the 1.5°C scenario.

The Intergovernmental Panel on Climate Change (IPCC) established by the World Meteorological Organization and UN Environment offers an unbiased scientific data source [6]. Regional variations in mean climate and extremes become increasingly broad and noticeable with each degree of global warming.

1.3 IPCC'S SIXTH ASSESSMENT REPORT

The Synthesis Report of the IPCC's Sixth Assessment Report states that there are numerous, workable, and practical solutions that can be implemented right now to cut greenhouse gas emissions and prepare for climate change brought on by humans. The IPCC suggests implementing "climate-resilient development" which combines steps to mitigate the effects of climate change with efforts to prevent or minimize greenhouse gas emissions in ways that yield additional advantages [7].

For example, having access to clean energy and technology can enhance health, particularly in the case of women and children. Walking, cycling, public transportation, and low-carbon electrification can all promote equity, better health, job prospects, and air quality. The financial gains from improving air quality alone would likely outweigh the costs associated with lowering or eliminating emissions in terms of health benefits.

Development that is climate resilient gets harder and harder with each degree of global warming. As a result, decisions made in the coming years will have a significant impact on how our world and future generations develop.

These decisions must be based on a variety of values, worldviews, and knowledge, such as scientific, indigenous, and local understanding, in order to be effective. This strategy will enable the construction of climate-resilient infrastructure and enable locally relevant, socially acceptable solutions.

Based on IPCC assessments, we know that:

- Without a question, human activity has contributed to the warming of the land, ocean, and atmosphere. The atmosphere, ocean, cryosphere, and biosphere have all undergone extensive and swift changes.
- Over many centuries to many thousands of years, the scope of recent changes to the climate system as a whole, as well as the current status of many aspects of the climate system, is unprecedented.
- Between 3.3 and 3.6 billion people reside in environments that are extremely susceptible to climate change.
- The susceptibility of people and ecosystems to climate change varies greatly between and within geographical areas.
- Compared to staying below 1.5°C, many human and ecological systems may face additional serious dangers if global warming temporarily surpasses 1.5°C in the upcoming decades or later.

1.4 AGROFORESTRY CONCEPTS AND AREA

As far as we are aware, agroforestry is a set of very old systems of farming methods that were widely used worldwide. A proper mix of crops, trees, and/or cattle on the same unit of land preserves overall yield productivity in this economically and ecologically sustainable land-use system. Additionally, by increasing carbon sequestration, preserving soil productivity and water tables, and promoting crop diversity to preserve greenery, the agroforestry system lessens the strain on natural forests. There are 1023 million hectares of agroforestry worldwide, which is substantially more than India's 25.32 million hectares [8]. Furthermore, the area covered by agroforestry systems in India is divided into two categories: 20.0 Mha of cultivated land (comprising 7.0 and 13.0 Mha of irrigated and rain-fed areas) and 5.32 Mha of other land, which consists of 2.93 and 2.28 Mha of shifting cultivation and homestead agroforestry, respectively. Similar to this, it is stated that alley cropping, silvopastoral, windbreaks, and riparian buffers accounted for approximately 235.2 Mha of the USA's land area. There are 8.0 million hectares of home gardens throughout South and Southeast Asia [1].

1.5 CLIMATE CHANGE AND AGRICULTURE

Crop production is an economic activity that satisfies the demands of human civilization and is reliant on climate. Because of this, it is evident that agriculture will be significantly impacted by climate fluctuations. Agriculture's contribution to GDP (gross domestic products) is 2.9% globally and 2.0% for high-income countries, which is particularly significant for lower income nations like Ethiopia, Peru, and other Latin American countries [9]. Due to variations in temperature patterns, precipitation patterns, and extreme weather, as well as natural disasters like floods and droughts and soil erosion that promotes the spread of pests and diseases, climate change has an impact on agriculture's production. Climate variability and the innate maturation characteristics of plants both have a major impact on the crop's growth stages [10]. Changes in precipitation patterns encourage water stress situations for agricultural production, which significantly affects farmers' forecasts of their financial returns [11]. Temperature and moisture, two elements that are vital to plant growth, will inhibit the plant's ability to absorb fertilizer and get mineral nutrients, which could ultimately result in a drop in yield and productivity. The occurrence of extreme events and changes in livestock feed are two ways that climate variation affects livestock [10]. The ecosystem of pasturelands is being impacted by climate change, which will worsen the effects on the number of cattle [1].

1.6 WHAT IS CLIMATE ADAPTATION AND THE CHALLENGES

A scientific description of measures to be adopted to lessen the susceptibility of climate change, as proper adaptation to it is the major responsibility toward combatting climate change has been presented by UNEP. Furthermore, the IPCC's fourth suggestion was to simply reduce climate change vulnerability by whatever means or to emphasize ecosystem resilience. It is a truth that adaptation to climate change is happening on schedule and in a variety of ways, including sectorial and spatial. Human civilization has evolved its own mechanisms for adapting to climatic change since the prehistoric era. This includes using a variety of agricultural techniques, choosing more effective irrigation, managing water sustainably, forecasting the climate appropriately, and including the community in disaster management, which includes early prediction. One well-known example of agricultural diversification to address climate change is agroforestry.

There are various ways to improve adaptation to climate change, such as recognizing the problem and taking proactive measures to address it. From a global standpoint, cooperation between rich and developing countries is necessary to address the climate change issue with a positive problem-solving mindset. Developing nations have shared their technology and offered financial support for adapting to climate change [12]. This has aided other nations in entering the climate change battle inside the framework of this cooperative effort [1].

Since climate change is an international problem, global policy was developed in order to advance the adaptation of climate change. Every participating country is required under the United Nations Framework Convention on Climate Change (UNFCCC) Article 4.1b to set up national and regional programs under their control that support climate change adaptation. In the same agenda item for developing nations, Article 4.5 talks about financial support and the exchange of environmentally beneficial technologies. Reducing greenhouse gas emissions and adapting to climate change are two of these methods' main goals. Many other nations are approving their participation in the same events, which will encourage other nations as well [13]. Climate change has been prioritized in the national context through the establishment of strategies, policies, and plans for a variety of developmental initiatives. This is achieved by determining the impact of climate change under the National Adaptation Programme of Action (NAPA) and ensuring that it is implemented properly. NAPA is striving to recognize the need and significance of climate change adaptation as well as to create a system for poor nations [13]. The process of designing a plan frequently dictates whether our goal is mitigation or adaptation.

Every problem has some limitations and difficulties. According to viewpoints on climate change, humans' reliance on nature for resources frequently serves as berries in an appropriate adaptation plan. Additionally, agroforestry's innate capacity to adjust to climate change is essential to the successful execution of climate change adaptation [14]. Among the issues that are most sensitive to climate change are gender inequities, which are also important considerations. For example, poverty, food insecurity, and natural disasters pose serious dangers to women farmers in South Africa [1].

The most important industry where climate change is having a severe impact is agriculture. Diversion from agriculture is crucial to counteracting such an effect. In this context, one kind of paradigm that shows a lot of promise is agroforestry. However, there are drawbacks to using agroforestry models to explore the effects of climate change. One of the main drawbacks is the reduced use in different environments. They are therefore not appropriate examples of the variability of climate change across all systems. As a result, some changes to those models are advised in order to obtain an accurate outcome. The following are some of the primary issues with agroforestry models:

- Because of the intricate structure of agroforestry systems, evaluating the effectiveness of models used to forecast the effects of climate change is challenging.
- The agroforestry system is challenging to employ in all environmental setups due to its considerable diversity.
- Since the effects of climate change are difficult to predict, screening models and evaluating their suitability under different climate scenarios are necessary.
- To properly construct area-specific agroforestry models, more research and development work needs to be done.
- One must be aware of the various structural complexities within the agroforestry model, including plantation crop associations, accurate knowledge of the model's various components, time-oriented analysis for predicting climate change, knowledge of the sources of errors caused by the model's complexity, accuracy, and

precision in data collection, sampling biases, and appropriate data set reliability.
- It is impossible to anticipate future climate conditions, the impact of biotic stress, or increasing CO_2 levels with any degree of accuracy.

1.7 ADAPTATION STRATEGIES

Adjustments in ecological, social, or economic techniques in feedbacks for real or anticipated climate-related prompts and consequences or impacts are referred to as adaptation in the IPCC lexicon. The term pertains to modifications made to procedures, customs, and frameworks in order to mitigate possible harm or capitalize on prospects linked with climate change. The term "adaptations" refers to the measures used by ecosystems and communities to help them adapt to, deal with, or benefit from real or anticipated changes in the climate. The role that agroforestry plays in lessening the susceptibility of agricultural systems (and the rural communities that depend on them for their livelihood) to climate change or variability needs to be addressed as adaptation becomes a science [15]. Agroforestry's adaptive strategies include lowering temperature and raising moisture content to improve crop growth in hot weather; creating a shading effect to reduce crop evapotranspiration and support high biodiversity; reducing the incidence of pests and diseases; and protecting crops from drought. Many recommended techniques for addressing climate change center on adaptation, and agroforestry systems offer opportunities to deal with a variety of adverse effects, including high temperatures, limited rainfall, wind velocity, hail, frost, and drought. This is made possible by the integration of woody perennials into agroforestry, which has a significant capacity to withstand the harsh effects of climate change and, via their coping mechanism, make the entire system sustainable. Below is a brief explanation of a few of the agroforestry systems' adaptive techniques [16].

Agroforestry is one of the best ways to address land degradation, desertification, mitigation of climate change, and food security all at once, according to the most current IPCC special report on climate change and land [17]. Agroforestry may be able to help with both adaptation and mitigation in ways that have previously been suggested but rarely explored [18–21]. One excellent example is how trees affect meso- and microclimates [22].

It is recommended that instead of focusing on co-benefits, agroforestry initiatives could be designed to pursue auxiliary benefits since they may be easier to attain in practice [23, 24]. According to a number of researchs [25–29], agroforestry systems have a major role in climate change adaptation. Trees, for instance, help people coping with food insecurity by supplying commodities or managing soil, hydrological cycles, and microclimate to promote more climate-resilient agricultural output [30, 31]. National Adaptation Plans for poor nations now include the contribution of trees for enduring climate change [32], while there are still few studies that specifically address this topic.

This Special Issue's several articles discussed how, in contrast to treeless systems, agroforestry systems alter microclimates. For instance, it is investigated how, by 2050, climate change in southeast Brazil will affect coffee output [33]. They predicted that the area of coffee plantations in this area would decrease by 66% due to the anticipated rise in the mean annual air temperature. By contrast, by 2050, agroforestry systems with 50% shade coverage will contribute to the survival of 75% of the world's coffee crops. Future research should take into account how agroforestry systems will be impacted by variations in precipitation brought on by climate change and how this would alter the hydrology of the area [33].

The regional variation of fodder yield linked to shade trees—which can also be utilized to shield grazing animals from the anticipated rise in heat stress brought on by climate change was assessed [34]. Large forage alleys were found to be successfully shaded during the hottest part of the day by three dense rows of trees, which had no discernible influence on the productivity or nutritional value of the forage. The anticipated decrease in heat stress for cattle is significantly higher because of the reduction in direct incoming solar radiation, even though the recorded drop in air temperature was less than 1°C [35].

1.8 SCOPING AGROFORESTRY FOR CLIMATE CHANGE

The main industry in low-income nations for rural development and livelihoods is agriculture. However, land pressure and climate change have a negative impact on agricultural systems in emerging nations, endangering food production. Even while thorough techniques of agriculture are supported throughout the whole world, their relative success is exacerbated by reduced productivity brought on by land degradation. There are detrimental reactions to climate change, surety of food supply, and income generated through farm at the regional level as a result of the different environmental effects of agricultural intensification and food production, which have detrimental effects on soil and biodiversity [3]. Moreover, attempts to apply the "green revolution" paradigm within Africa through allotment and addition with the help of fertilizers proved to be expensive and unsustainable, since technology will never be able to completely replace the functions that trees would typically do [36]. For improving management of crop and land through incorporation of trees into process of land usage, diversification is crucial, as the current discussion on sustainable intensification of agriculture emphasizes [36–38]. While combining higher yields with ecosystem services can lead to many sustainable agricultural goals, there are not many choices that simultaneously improve farm production and agro-ecosystem diversity. Certain types of agroforestry are strong candidates for accomplishing both sustainable livelihood and climate change objectives because they have high recycling rates, minimal external input requirements (pro-poor), and good tree, crop, and animal integration [39].

Reforestation and forest conservation are the main components of climate change mitigation across the majority of Africa. However, these initiatives to lessen tropical deforestation (often under the Reducing Emissions from Deforestation and Forest Degradation, plus umbrella) run counter to the necessity of increasing agricultural output in Africa in order to feed the continent's expanding population [20]. Because agroforestry enhances carbon storage and may also increase agricultural output, it may offer mutually beneficial explanations to the apparently tough options between land usage through agriculture and reforestation [40]. According to certain research, smallholder farmers in developing nations can mitigate the effects of climate change with the help of switching back to naturally high-yield procedures that serve better social and ecological purposes [41], all the while addressing the need for adaptation and constructing robust agro-ecological systems that actively sequester carbon. Agroforestry systems are currently gaining popularity as an investment due to their many advantages, as well as their novel methods, which improve the system's resilience to the negative effects of climate change. Evidence that high production levels and economic values of agroforestry products can create financial capital over and above livelihood levels alone, supporting accretion of capital and reinvestment at the farmland magnitude, is a common point of agreement. This is true even though the viability and advantages of mitigation through agroforestry to small agriculturist is still under discussion.

Agroforestry has been shown to be able to increase carbon stocks while also providing benefits for livelihoods; however, the assessment of advantages from different ecosystem facilities going over the typical profit of carbon sequestration presents an increasing challenge that needs to be better understood by the research community.

Smallholder farmers who place a high premium on food security predominate in African agriculture, which is one of its distinguishing characteristics. In these circumstances, climate mitigation strategies will have to show that they enhance food production while also providing benefits for climate adaptation. The state-of-the-art about agroforestry's contribution to adaptation and mitigation of climate within Africa's largely food-centralized production arrangement is presented in this synthesis [20].

1.9 REVIEW OF CURRENT KNOWLEDGE ABOUT THE EFFECTS OF CLIMATE CHANGE ON FOOD SECURITY

The primary conclusions of earlier IPCC reports on the evolution of the climate and its primary physical effects—such as implications for changes in land and ocean temperatures, sea level rise, and ocean acidification—are supported by the most recent report. Additionally, it improves our knowledge of possible regional variations in the amount, intensity, and seasonal distribution of precipitation. Furthermore, we are able to enhance the medium-term and localized estimates of the implications of climate change due to advancements in modeling, data gathering, and utilization. For the purpose of better understanding and projecting possible effects on agricultural systems, these advancements are vital.

The supply of water in many locations is uncertain in the future due to climate change. It will have an impact on precipitation, runoff, and snow/ice melt, which will have an impact on groundwater recharge, hydrological systems, and water quality and temperature. Increased water scarcity brought on by climate change will be a significant obstacle to climate adaptation in many parts of the world. In coastal places, rising sea levels will have an impact on groundwater and surface water salinity.

Extreme event frequency and intensity are expected to change as a result of climate change. Extreme weather events have a significant impact on agriculture already. According to a recent analysis by the Food and Agriculture Organization of 78 post-disaster needs assessments conducted in 48 developing countries between 2003 and 2013, the agriculture sector accounts for 25% of all economic losses and damages caused by medium- and large-scale climate hazards, such as droughts, floods, and storms, in developing nations [42].

After two decades of research, the effects of climate change are already well-documented. Positive effects are less common than negative ones worldwide. Wheat and maize yields have already been adversely impacted by climate change in numerous locations worldwide, according to observations of the effects of climate trends on crop production. Livestock production is impacted by climate change in a number of ways, both directly and indirectly. The greatest effects are shown in the health and productivity of animals as well as in the yields of feed crops and forages. The growth of aquaculture in freshwater and marine habitats, as well as capture fisheries, is impacted by climate change. The physical (sea and inland water surface temperature, ocean circulation, waves, and storm systems) and chemical (salinity content, oxygen concentration, and acidity) changes that are brought about by the slow warming of the atmosphere also have an impact. A rise in the frequency of coral reef bleaching has been noted, endangering the habitats of around 40% of marine organisms.

1.10 LIVELIHOOD UNDER CHANGING CLIMATE

From an economic standpoint, forests are essential to human society. About 300 million rural people depend on NTFPs for fruit and medicinal plant products in addition to tangible benefits like fuel and fodder, which help them maintain their daily livelihoods [43]. However, the effects of climate change and global warming are now readily apparent. High temperatures, erratic rainfall patterns, and an increase in infectious diseases lead to a loss in total production and a crisis affecting farmers' ability to secure a living. Even though the forest has the ability to provide the poor with both material and intangible advantages, unlawful cutting and unsustainable harvesting lead to deforestation, which alters local weather patterns and the global climate. Thus, massive plantations on field bunds and other wastelands/degraded areas, along with other forest products, could potentially mitigate these issues and meet farmers' daily needs for food, fodder, NTFPs, and other forest products while also improving the socioeconomic condition and livelihood security [1].

1.11 ENABLE ADAPTATION THROUGH POLICIES AND INSTITUTIONS

In order to effectively integrate climate change concerns into food and agricultural policies, the aforementioned technical and economic options must be enabled, supported, and complemented by appropriate policies and institutions, as well as a better understanding of the distinctive qualities of agriculture and its crucial role for food security and nutrition in climate policies.

To facilitate food producers' adaptation to climate change, various policies [42] and institutions can be identified, both at the national and international levels:

- Institutions and policies that aid food producers in their attempts to adapt, particularly small-scale food producers.
- Institutions and policies that assist in the group development and execution of adaptation measures within a sector or space (such as a forest or watershed).
- Institutions and policies devoted to the management and mitigation of particular hazards and vulnerabilities, such as invasive species, wildfires, animal illnesses, and plant pests, which may be impacted by climate change.

1.12 CLIMATE CHANGE AND HUMAN INFECTIOUS DISEASES

Variations in temperature, precipitation, wind, sunshine, and other climate variables are examples of climate variations. The availability and modes of transmission of disease pathogens and hosts, as well as their survival, reproduction, and dispersal, may all be impacted by these changes. Variations in the frequency and severity of outbreaks as well as variations in the seasonal and geographic patterns of human infectious illnesses are often indicators of the health repercussions of such impacts.

The multifaceted effects of climate change on several infectious disease categories, such as food-, water-, and vector-borne illnesses, have been extensively documented in the literature.

The term "pathogen" encompasses a broad spectrum of pathogenic entities, such as bacteria, viruses, parasite germs, and fungi. Pathogens may be directly impacted by climate change in ways that affect their survival, reproduction, and life cycle, or they may be indirectly impacted in ways that affect their habitat, surrounding environment, or rivals. As a result, changes may occur in the amount as well as the seasonal and geographic distributions of diseases.

Through its effects on the life cycle of pathogens, temperature may have an impact on disease. A pathogen must first thrive in a specific temperature range in order to develop. For instance, the Japanese encephalitis virus (JEV) ecology is significantly influenced by the two thresholds—the maximum temperature for mosquito development that is 22–23°C and the minimum temperature for JEV transmission that is 25–26°C [44]. Certain diseases have higher mortality rates in extreme temperatures. When the temperature rises over 33–39°C, the expansion of the malarial parasites (*Plasmodium vivax* and *Plasmodium falciparum*) stops. Second, as temperatures rise, infections' ability to reproduce and survive their extrinsic incubation period (EIP) may be affected.

For instance, the EIP for *P. falciparum* is 13 days at 25°C and 26 days at 20°C. Conversely, a lower ambient temperature is probably going to prolong EIP, and as fewer mosquitoes can survive long enough, this could reduce the spread of diseases like dengue. Third, prolonged hot weather can increase the average temperature of food environments and bodies of water, which could create a favorable habitat for algal blooms and the cycles of reproduction of microorganisms. For instance, during the warm summers of 2006, bacteria called *Vibrio* spp., which are indigenous in the North Sea and Baltic, demonstrated an accelerated scale of development. Salmonella is a food-borne illness; when the temperature increases from of 7°C to 37°C, the bacteria reproduce more readily. Finally, a pathogen's ability to proliferate may be inhibited by higher temperatures if their competitors benefit.

For instance, it was discovered that the bacteria that cause the food-borne illness *Campylobacter* spp. are more concentrated in surface water during the winter and at low temperatures. It is thought that this is because UV light prevents *Campylobacter* spp. from surviving and warmer temperatures allow other bacteria to outcompete *Campylobacter* spp.

Changes in precipitation may result from climate change, and this could have an impact on the spread of water-borne infections. Rainfall has a significant impact on the growth of bacteria that cause water-borne illnesses. The growth of fecal pathogens is associated with the rainy season because heavy precipitation can agitate sediments in water, which can result in the buildup of fecal germs [45].

However, unusual rainfall following prolonged droughts can increase pathogens and lead to disease outbreaks. Drought/Lack of precipitation leads to decreased water flows, leading to concentrations of water-borne pathogens [46].

Changes in humidity also affect pathogens of infectious diseases. Airborne diseases such as the flu usually respond more in humid weather. For instance, influenza virus transmission and survival is found to be influenced by absolute humidity and temperature [47]. It is suggested that cold temperatures and low relative humidity promote the influenza virus outspread. Change in moistness in air also affects viruses in water-borne diseases. Typically, the survival of viruses that are water-borne near the water bodies depends on the surface water drying effect. Finally, viruses in vector-borne diseases can be affected by changes in humidity. Humidity has been shown to affect malaria parasite development in anopheline mosquitoes. It was found that humidity and temperature throughout the monsoon season in Yangon and Singapore promoted the spread of dengue among mosquitoes, contributing to dengue outbreaks in these areas.

Sunlight is another important climate variable that can affect infectious disease pathogens. For example, during the cholera season, hours of sunlight and temperature act synergistically to form suitable conditions to spread the *Vibrio cholerae* in oceanic environments.

Wind is a major factor affecting airborne disease pathogens. The literature suggests a definite association between contact of dust particles/adhesion and viability of virus/transmission. The existence of atmospheric dust from the desert in the course of Asian dust storms (ADS) has been reported to be associated with increased concentrations of culturable bacteria, culturable molds, and fungal spores, it was found that influenza A virus concentrations were significantly higher during ADS than usual. Additional studies have shown that pandemic viruses are transported across oceans by dust particles, which may facilitate viral spread between distant hosts.

1.13 ADAPTATION-MITIGATION IN AGROFORESTRY

The combination of adaptation and mitigation is recognized in developing countries to be fundamental, especially in the AFOLU area. Meanwhile, the reality is, no difference is found between production of crops and several other ecosystem techniques or usage of land. Agroforestry as a whole can increase farm profitability by improving and diversifying the production of the tree/plant/shrub sectors, protecting against damage caused by the flow of water or wind, and adding novel results for diversity and flexibility of finance. It can also contribute significantly to mitigating climate change.

Using multifunctional integrated approaches and trees could increase the profits in agroforestry. For instance, for forage, trees can be a source that can sequentially become an essential nutrient for plants. Farm trees furnish edible native fruits [48] and compounds that are non-wood that can be used for food substitutes in times scarcity of food and are a major source of income for many rural communities. Therefore, there is a growing scientific problem regarding methods and tools for assessing the trees utility in different human-to-ecological conditions. In many situations, the agroforestry profits help in capturing GHGs, significantly improving the economic and resource sustainability of agriculture [2].

1.14 AGROFORESTRY IN INDIA

1.14.1 WORLD AND INDIAN CIRCUMSTANCES

It has been reported that the current global agroforestry area amounts to 1.023 million hectares. The current agroforestry area in India is 25.32 Mha, which is 8.2% of the total area of India [8]. This includes 20.0 Mg from cultivated land (7.0 Mg from irrigated areas and 13.0 Mg from rain-fed areas) and 5.32 Mg from other areas such as shifting cultivation (2.28 Mg), vegetable gardens and restoration of problem areas (2.93 Mg). It is estimated the garden area in South and Southeast Asia to be 8.0 Mha [49].

Agroforestry systems in India have a long history and traditional practices. The agroforestry system in India consists of farm tree planting, various indigenous management strategies, indigenous ethno-forestry practices and community forestry on communal lands. Here, the practice of planting trees on agricultural land is relatively ancient, and the trees used for this purpose are as diverse as fruits and vegetables for shade, fodder, fuel, food, and medicinal purposes. Eucalyptus and Populus plantations have been reported at the border of the agricultural regions of Punjab and Haryana. Crop rotation is reported to be dominant in the northeastern region. In contrast, the Taungya system is found in Uttar Pradesh, Kerala, and West Bengal, along with parts of southern India. Home gardens, large areas under cardamom cultivation, and cultivation-based farming systems are found in the eastern Himalayas and northeastern India, respectively (Table 1.1) [1]. The main agroforestry practices in India are found in four major regions [1].

1.14.2 NORTHEAST REGION IN INDIA

Agroforestry as a shifting crop has been proposed. Shifting cultivation is also known as Jhum cultivation in the northeastern region. Farmers are a resource for maintaining biodiversity and socio-economic conditions in rural India. In the current scenario, short fallow periods cause forest degradation and agricultural ecosystem instability. Through agroforestry, it is possible to restore and maintain soil fertility and increase agricultural productivity. Predominant agroforestry systems in the region include agro-horticulture, agriculture, vegetable gardens, forest horticulture, and animal husbandry.

1.14.3 NORTHWESTERN REGION IN INDIA

In the northwestern Indian states of Gujarat and Rajasthan, very low or irregular rainfall and resulting crop losses are common. The potential to reduce the impact of climate extremes and reduce the availability of alternative sources of income for the population can be achieved through agroforestry activities. Together with *Prosopis cineraria*, the plant species constitute the most important agroforestry practice in the region.

1.14.4 WESTERN GHATS REGION OF INDIA

Agroforestry practices in the Western Ghats are an integrated way of growing crops along with tree planting and animal husbandry.

TABLE 1.1 Scenarios of Agroforestry Systems in India [1]

Agroforestry Models/ Agri-silvi practice	Components (Crops with Woody Species)	Nature of Components (a—livestock, w—woody species, h–herb, f—forage)	Agroecological Need
Advance fallow system	Trees are grown and left in fallow land	w—Legume-based short rotation practice, h—Indigenous crops	In areas of shifting cultivation
Taungya method	During initial phase trees grown with agricultural crops	w—Economic tree species, h—Common indigenous crops	Diverse ecological zone
Alley cropping/Hedgerow intercropping system	Timber species planted in hedges, alley and strip plantation of agricultural crops within hedges	w—Short duration legume with coppicing potential, h—Traditional crops	Humid tropics
Multistrata trees-based gardens	Multilayered and diverse plant species	w—Tree species with variable growth and form, h—Occasional crop	Fertile soil region
MPTS (Multipurpose tree species) based agroforestry	Tree plantation in extension areas of farmlands	w—MPTS and other orchard crops, h—Traditional crops	Entire ecological zone (subsistence farming + animal husbandry)
Plantation crop-based system	Multistory crops with plantation crops beneath shade plants + agricultural crops	w—Plantation crops (beverages, orchards, firewood, forage crops), h—Shade bearers' crop	Humid, sub-humid tropics
Kitchen garden	Diversified crops (herbs, shrubs, climbers, trees)	w—Orchard tree species with other woody vegetation	Entire ecological zones
Restorative plants- based practice	Extension plantation with mulching and nitrogen fixers	w—MPTs and orchard plantation, h—Indigenous crops	Undulating regions and difficult site recovery
Shelter-belts, wind-breaks, hedgerows practice	Timber species near agricultural plots	w—Woody plants, h—Traditional crops	Windy region
Energy blocks	Fuelwood species with agriculture	w—Fuelwood timber species, h—Indigenous crop	All ecological zones

Some agricultural systems were dominated by plantation agriculture, including crops such as spices and other beverage crops (coffee and tea), as well as tree seeds and rubber, rice cultivation, home gardens/vegetable gardens, and coconuts.

1.14.5 SOUTH INDIA REGION

South India presents various agroforestry patterns. For example, Kerala represents multistate home gardens and agroforestry, plantation crops, coastal timber from Karnataka and spice-based agroforestry practices and oilseed crops from Tamil Nadu and Andhra Pradesh. In Tamil Nadu, agriculture is combined with intercropping of fruit trees, silk cotton trees, and so on (Table 1.1) [1].

1.15 MITIGATION STRATEGIES

Mitigation refers to technological changes and substitutions that reduce inputs and emissions per unit of life. Mitigation is nothing more than a corrective action to eliminate negative consequences that occur to a limited extent, such as increased CO_2 concentrations. Mitigation through agroforestry involves the storage of carbon dioxide by trees in the atmosphere, also known as carbon sequestration. Carbon sequestration through land use, land-use change, and forestry is an important topic in recent negotiations as measures to mitigate the impact of climate change, and these measures are linked to the clean development mechanism. Simply put, adaptation can happen at the local level, but mitigation happens at the global level. Carbon sequestration by trees and soil is considered one of the most important measures to limit climate change. The main aim of these measures is to reduce the concentrations of GHGs and increase the accumulation of carbon in the soil. Mitigation measures are a win-win option by increasing soil organic carbon content and ultimately improving soil health and quality, thereby increasing crop productivity in agroforestry systems and increasing soil adaptive capacity [50]. Therefore, mitigation options will somehow affect C and/or N cycling in agroecosystems [16].

In addition to delivering numerous co-benefits, particularly for air quality and health, deep, quick, and persistent mitigation and expedited adaptation action execution this decade would lower estimated losses and damages for humans and ecosystems. Besides increasing the danger of stranded assets and cost escalation, delaying mitigation and adaptation measures would lock in high-emission infrastructure, diminish feasibility, and increase losses and damages. The significant upfront costs and perhaps disruptive changes associated with near-term initiatives can be mitigated by a variety of enabling policies.

There are selected mitigation and adaptation options across different systems, including climate responses and adaptation options assessed for their multidimensional feasibility at global scale, in the near term and up to 1.5°C global warming. About 1.5°C is limited, feasibility at higher levels of warming may change, which is currently not possible to assess robustly (Figure 1.1).

1.16 CARBON SEQUESTRATION

The UNFCCC defines carbon sequestration to be the process in which carbon is removed from the atmosphere and storing it. This includes transporting CO_2 from the atmosphere and storing it safely in long-term storage. Carbon captured by trees and stored in aboveground biomass and soil helps reduce greenhouse gas concentrations in the atmosphere. Potential of carbon sequestration systems through agroforestry has been estimated to range from less than 100 million tons of CO_2 per year in 2030 to over 2 billion tons of CO_2 per year 30 years later. Regardless of the exact amount, agroforestry systems tend to sequester significantly more carbon than treeless agricultural systems. Different agroforestry systems sequester different amounts of carbon depending on system type, species composition, soil, and climate. Home gardens contain higher biomass compared to other systems, and dryland agroforestry systems contain more root biomass. Aboveground carbon stocks range from 17 to 36 mg C/ha in tropical vegetable gardens in Kerala [51] and from 21 to 65.6 mg C/ha in folk systems in northern India [52]. Agroforestry area has always been a difficult question for scientists and policy makers in India, but NRCAF and Jhansi estimated it to be about 25.32 million ha or 8.2% of the area of agroforestry systems [8]. Therefore, the total carbon storage potential of Indian agroforests is about 2.4 billion tons. Some of the early studies on the carbon storage potential of agroforestry and alternative land-use systems in India are estimated carbon sequestration at 68–228 mg C/ha.

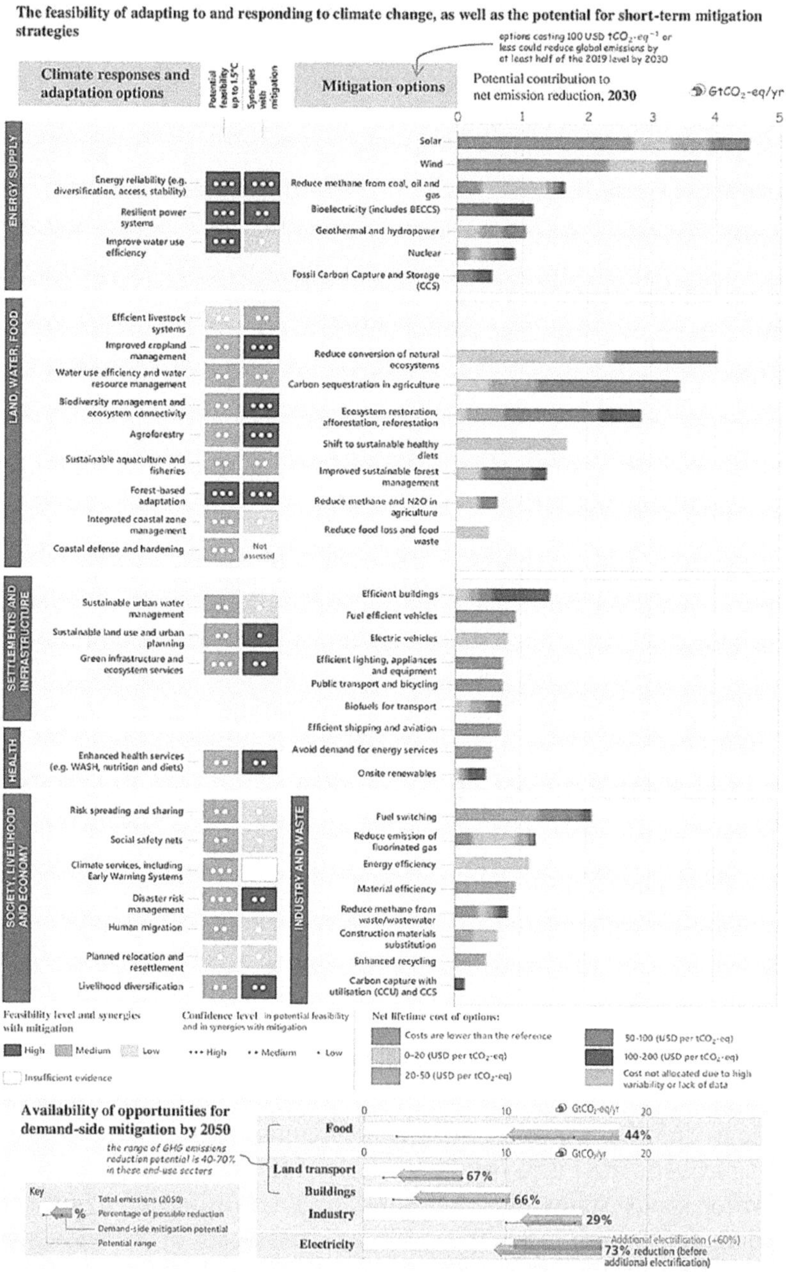

FIGURE 1.1 The feasibility of adapting to and responding to climate change, as well as the potential for short-term mitigation strategies and availability of opportunities for demand-side mitigation by 2050.

This agroforestry can store almost 83.6 mg C/ha. Average carbon stocks from agroforestry practices with fertilizer as a component are estimated at 9, 21, 50, and 63 million ha in semiarid, arid, humid, and temperate regions, respectively. The carbon sequestration potential of biofuel crops was 10.08 Mg C/ha for Jatropha and 34.85 Mg C/ha for Pongamia after 10 years of growth. Not only do these biofuel crops help mitigate climate change by sequestering and storing carbon from agriculture, biomass can also be used to produce renewable energy by replacing fossil fuels, as seeds can be used to produce biofuels. Due to the myriad benefits and potential of agroforestry, it is noted that agroforestry is a "low-hanging fruit" due to its low potential for mitigating climate change. Hence, agroforestry offers the chance to obtain carbon credits under the Kyoto Protocol agreements; it is regarded as a safety net for small-scale farmers [16].

1.17 FOREST, AGROFORESTRY, AND CARBON SEQUESTRATION

It is common practice to use forestry to increase carbon sinks and reduce CO_2 emissions. Forests are a major carbon sink and are widely acknowledged to have a part in the carbon cycle. Agroforestry, or the combination of forestry and agriculture, has several goals, including lowering CO_2 emissions and controlling CO_2 concentrations as well as mitigating and adapting to climate change. It could make the system more flexible in order to address the negative effects of climate change. The agroforestry system's goal is to provide substantial opportunities for creating synergies between adaptation and mitigation techniques. In other words, it makes critical mitigation decisions, such as proper management that affect the amount of carbon stored. As a varied land use, agroforestry encourages carbon sinks. Many plantation programs, including agroforestry, have been advocated in an effort to boost the carbon storage potential of tree species in the terrestrial environment. The most rigorous strategy for preserving human livelihood and overall ecological sustainability is agroforestry, which integrates tree plantations with agricultural systems. The pace of carbon accumulation is accelerated by higher carbon content achieved through such alternative land-use systems compared to native vegetation. Converting from lower biomass-producing land-use activities to higher biomass-producing land-use practices can boost carbon storage.

Although agroforestry practices are not intended to aid in the process of sequestering carbon, research indicates that they can encourage an increase in the amount of carbon in soil, above-ground biomass, and subsurface soil ecosystems. An increase in the terrestrial carbon sink and a decrease in the amount of GHGs in the surrounding environment are the two advantageous functions of agroforestry systems [1].

1.18 BENEFITS OF AGROFORESTRY SYSTEM

Systems of agroforestry have the potential to yield both financial and environmental benefits [1]. A few of the primary benefits of sustaining agroforestry systems are covered in the subsections that follow.

1.18.1 TO INCREASE CARBON STOCK

The immense promise of agroforestry systems lies in their ability to mitigate the effects of a changing climate and store carbon through the use of many plant species and soil. Agroforestry with precise design can function as a powerful carbon sink. Carbon sequestration is the process by which green plants actively absorb carbon dioxide from the atmosphere through photosynthesis and subsequently store it as biomass.

For the same items that are brutally extracted from forest plantations in an unsustainable way, agroforestry practices yield additional wood. By storing carbon in the soil and vegetation, agroforestry can be used to mitigate climate change while also providing a reliable source of income for the farming community. In an agroforestry system, pruning is done on the tree component to lessen species interaction, a type of ecological competition. When the trimmed materials are returned to the soil, they increase the soil's carbon pool and are employed as NTFPs.

1.18.2 TO IMPROVE SOIL FERTILITY

Increasing and sustaining soil fertility is highly dependent on a number of factors, including food security, poverty reduction, environmental preservation, and sustainability. Numerous agroforestry land-use methods, such as agrohorticulture, agropastoral, agro-silvopastoral, and others, are effective means of replenishing the organic matter in the soil. Increased crop rotation yields a higher yield than continuous cultivation. Leaf litter from agroforestry species contributes to the humification process through microbial breakdown. The practice of agroforestry helps to prevent runoff loss and soil erosion. As a result, this has a major impact on reducing water loss and improving soil texture by affecting soil materials, organic matter content, and soil nutrients. Additionally, it aids in the fight against soil acidity and salinization, which promotes soil renovation.

1.18.3 TO GENERATE MORE INCOME

Agroforestry's many constituents provide multiple harvests at different times of the year. An agroforestry system increases soil fertility, generates firewood, and provides feed for livestock. It lowers the possibility of crop failure in this way and gives farmers a backup source of income.

1.18.4 TO INCREASE PRODUCTIVITY

Forest-covered soils possess the capacity to provide higher yields when compared to regular soil types. When fodder grasses and fodder trees are interplanted, the fodder output is higher than when fodder grass is grown alone. Intercropping systems for food crops are more common in South India and in certain states, such as Gujarat, Haryana, Punjab, and Uttar Pradesh.

1.18.5 TO IMPROVE THE AESTHETIC VALUE

Aesthetic value may create a lively setting. Agroforestry techniques can be used to improve agriculture and environmental conditions. A small plot of land used for agroforestry systems encourages biodiversity recovery by enhancing the environment and supporting beneficial bird and insect species that control the number of pests. Using tree biodiversity to improve beauty is one way to add diversity to the landscape. Better nitrogen inputs from nitrogen-fixing plants and trees, less insect pests and related illnesses, more efficient use of solar energy as compared to monoculture systems, and the ability to regulate microclimates are all benefits. Table 1.2 provides a detailed description of the benefits of agroforestry.

TABLE 1.2 Benefits of Agroforestry

Sl. No.	Benefit	Description
1	Increased farm productivity	Agroforestry enhances overall farm productivity by integrating trees and crops, which can lead to higher yields and diverse income sources
2	Soil erosion reduction	The deep roots of trees bind the soil, significantly reducing erosion caused by wind and water
3	Biodiversity enhancement	Agroforestry creates diverse habitats that support various flora and fauna, improving ecological balance
4	Water conservation	Trees help maintain groundwater levels and reduce surface runoff, contributing to better water management
5	Carbon sequestration	Trees absorb CO_2 from the atmosphere, storing it in their biomass and soil, thus mitigating climate change
6	Improved soil health	Organic matter from trees enriches the soil, enhancing its structure and fertility
7	Economic diversification	Farmers can generate additional income through the sale of timber, fruits, and other tree products
8	Pest management	Diverse agroforestry systems can naturally control pests by providing habitats for beneficial organisms
9	Resilience to climate change	Agroforestry systems are more resilient to weather extremes, reducing vulnerability to droughts and floods
10	Aesthetic value	The integration of trees improves the visual appeal of agricultural landscapes, contributing to community well-being

1.19 AGROFORESTRY AND ECOSYSTEM RESILIENCE

Crop diversity, conservation of soil through delayed rotation techniques, home gardens, plantation on boundary, hedgerow intercropping, perennial crops, refined fallows, live fences, or mixed layer agroforestry are just a few of the many land management techniques that make up agroforestry systems. Through microclimate buffering and water flow regulation, agroforestry can significantly increase resistance to unpredictable climates if it is managed correctly (success mostly depends on appropriate implementation) [25].

Agroforestry management strategies comprise pruning of trees and ways in which rivalry below ground can be lessened, especially to get water, so trees can draw from deep groundwater instead of the top-soil moisture that annual crops need. The influence of agroforestry on the regulation of microclimates and other beneficial ecosystem functions is receiving more and more attention. By reducing source pollution that is nonpoint (such as dust), preventing erosion of soil, and establishing habitat for fauna, agroforestry contributes to the preservation and protection of natural resources. It preserves or replenishes soil and water resources while enabling adaptable reactions to quick changes in ecological settings.

Because trees can operate as a buffer against climatic extremes that affect crop development, agroforestry's microclimatic enhancement has a significant impact on agricultural performance. Specifically, the shade effects of agroforestry trees can buffer air saturation deficit and temperature, minimizing supraoptimal temperature exposure, which can put yield and physiological and developmental processes at risk [41]. In agroforestry farms, scattered trees can improve efficiency of water usage, water status, and gas exchange, while promoting underbrush growth by lowering solar radiation incidence, temperature of air, and temperature of soil. Future studies should provide more evidence to support these scientific claims, which are based on a small number of examples.

Agroforestry, as opposed to annual cropping systems, maximizes the use of rainfall, which supports ecosystem services such as water recycling. According to [41], trees use about 25% of the water they transpire during the dry season, meaning they can make use of offseason rainfall, which makes up to 15–20% of total annual rainfall, and soil water that is residual, left over in cropping. Remaining water around 40% is lost to evaporation and 33–40% in deep drainage. Because of the complementarity between trees and annual crops, there are more opportunities for soil moisture uptake, which results in more efficient use of soil resources than in pure monoculture.

With the help of positive responses between vegetation and precipitation via higher albedo and decreased evapotranspiration, trials have been done to show that a drop in plant cover magnifies the decline in rainfall. The significance of controlling tree cover is also discussed in water cycle analysis, as it is one of the direct effects of trees on regional and local rainfall patterns. This demonstrates how agroforestry might help ease Africa's drought [2].

1.20 CONCLUSION AND KEY REMARKS

This study demonstrates how agroforestry systems can easily combine adaptation and mitigation techniques, offering numerous avenues for ensuring food security for low-income farmers, while also assisting in the reduction of climate change. Agroforestry's beneficial effects on the environment and society should make it a priority on international mitigation agendas. To ensure quick acceptance, however, learning sophisticated cultivation techniques and receiving some support are necessary when incorporating trees into cropping systems and/or animal production. The potential to expand agroforestry technologies for better land-use systems is restricted by the shortcomings of extension services in developing African nations. The low funding in the department relative to farming systems that are intensified, have received significant financial support after the colonial period, primarily in export cash crops (monocultures of cotton, groundnut, cocoa, etc.), represents a further structural barrier in scaling the adoption of agroforestry.

In a situation when the main objective is promoting economic and social advantages with the help of agriculture, combining adaptation in agroforestry and mitigation to improve the strength of social and systems of land usage shall be carefully considered at the farm level. Priority activity screening necessitates a comprehensive study that prioritizes meeting the most fundamental local requirements [53–72]. Therefore, if mitigation is considered a win-win strategy under ideal land management practices, adaptation is to be given equal weight, and mitigation techniques must show a distinct advantage of adaptation. Carbon sequestration in Africa should typically be viewed as an added benefit of climate change adaptation and sustainable livelihood methods, not the other way around. Asset accumulation can be used to gauge the progress made toward adapted and sustainable livelihoods, and mitigation strategies ought to be matched with these assets [2].

However, there are obstacles to the expansion of agroforestry methods, including unknowns about future climates, cover and land use, fertility of soil in arid regions, diseases, and pests. Despite extensive efforts to model climatic analogs and future climate impacts, the consequences of climate change on agroforestry systems remain poorly understood. This calls into question how to effectively reduce the detrimental effects of climate change on farming systems as well as what type of management practices and trees will be appropriate in the future [25]. Therefore, in order to evaluate the long- and short-term effects of shifting temperature and rainfall on ecosystem compatibility for existing methods of agroforestry, it is necessary to get better at predicting the range of variability in climate [41]. Very little is known about the quantitative impacts of trees on the regional and local climate, and more research is required to understand the relationships between water recycling and evapotranspiration. Furthermore, it is not evident how much deforestation can be stopped by using agroforestry landscapes to provide ecosystem services like wood energy.

KEYWORDS

- **agroforestry**
- **carbon sequestration**
- **climate change**
- **mitigation**

REFERENCES

1. Jhariya, M. K., Yadav, D. K., Banerjee, A. (Eds.) *Agroforestry and Climate Change: Issues and Challenges.*
2. Mbow, C., Smith, P., Skole, D., Duguma, L., Bustamante, M. (2014). Achieving mitigation and adaptation to climate change through sustainable agroforestry practices in Africa. *Current Opinion in Environmental Sustainability*, 6, 8–14.
3. Krausmann, F., Erb, K. H., Gingrich, S., Haberl, H., Bondeau, A., Gaube, V., Lauk, C., Plutzar, C., Searchinger, T. D. (2013). Global human appropriation of net primary production doubled in the 20th century. *Proceedings of the National Academy of Sciences*, 110(25), 10324–10329.
4. NASA. (2015). http://climate.nasa.gov/solutions/adaptation-mitigation/. (Accessed December 15, 2015).
5. US EPA. (2015). Climate change: Basic information. http://www3.epa.gov/climatechange/basics/. (Accessed December 15, 2015).
6. UN Climate Change. (2015). https://www.un.org/en/global-issues/climate-change.
7. IPCC, (2023). Climate change 2023: Synthesis report. In: Core Writing Team, Lee, H., Romero, J. (Eds.), *Contribution of Working Groups I, II and III to the Sixth Assessment Report of the Intergovernmental Panel on Climate Change* (p. 184). IPCC: Geneva, Switzerland, doi: 10.59327/IPCC/AR6-9789291691647.
8. Dhyani, S. K., Handa, A. K. (2013). Area under agroforestry in India: An assessment for present status and future perspective. *Indian Journal of Agroforestry*, 15(1), 1–1.
9. Ackerman, F., Stanton, E. A. Climate impacts on agriculture: A challenge to complacency.
10. Hoffmann, U. (2013). Agriculture: A key driver and a major victim of global warming. In: *Lead Article in Section B at Hoffmann* (pp. 3–8).

11. OECD (2014). *Climate Change, Water and Agriculture: Towards Resilient Agricultural and Water Systems*. OECD Studies on Water, OECD Publishing.
12. Omambia, C. S., Gu, Y. (2010). The cost of climate change in Tanzania: Impacts and adaptations. *Journal of American Science*, 6(3), 182–196.
13. Mutimba, S., Hepworth, N. Climate change vulnerability and adaptation preparedness in Kenya.
14. Raj, A. B., Jhariya, M. K. (2017). Sustainable agriculture with agroforestry: Adoption to climate change. In: *Climate Change and Sustainable Agriculture* (pp. 287–293).
15. UNFCCC Â. United Nations framework convention on climate change. oct-2013. [On line]. 1992. Available in: http://unfccc. int/2860. php. (Accessed November 01, 2014).
16. Newaj, R., Chavan, S., Prasad, R. (2013). Agroforestry as a strategy for climate change adaptation and mitigation. *Indian Journal of Agroforestry*, 15(2), 41–48.
17. Shukla, P. R., Skea, J., Calvo Buendia, E., Masson-Delmotte, V., Pörtner, H. O., Roberts, D. C., Zhai, P., Slade, R., Connors, S., Van Diemen, R., Ferrat, M. IPCC. (2019). Climate Change and Land: An IPCC special report on climate change, desertification, land degradation, sustainable land management, food security, and greenhouse gas fluxes in terrestrial ecosystems.
18. Schoeneberger, M., Bentrup, G., De Gooijer, H., Soolanayakanahally, R., Sauer, T., Brandle, J., Zhou, X., Current, D. (2012). Branching out: Agroforestry as a climate change mitigation and adaptation tool for agriculture. *Journal of Soil and Water Conservation*, 67(5), 128A–136A.
19. Duguma, L. A., Minang, P. A., van Noordwijk, M. (2014). Climate change mitigation and adaptation in the land use sector: From complementarity to synergy. *Environmental Management*, 54, 420–432.
20. Mbow, C., Van Noordwijk, M., Luedeling, E., Neufeldt, H., Minang, P. A., Kowero, G. (2014). Agroforestry solutions to address food security and climate change challenges in Africa. *Current Opinion in Environmental Sustainability*, 6, 61–67.
21. Rahn, E., Läderach, P., Baca, M., Cressy, C., Schroth, G., Malin, D., Van Rikxoort, H., Shriver, J. (2014). Climate change adaptation, mitigation and livelihood benefits in coffee production: Where are the synergies. *Mitigation and Adaptation Strategies for Global Change*, 19, 1119–1137.
22. Noordwijk, M. V., Bayala, J., Hairiah, K., Lusiana, B., Muthuri, C., Khasanah, N. M., Mulia, R. (2014). Agroforestry solutions for buffering climate variability and adapting to change. In: *Climate Change Impact and Adaptation in Agricultural Systems* (pp. 216–232). CABI: Wallingford UK.
23. Anderson, E. K., Zerriffi, H. (2012). Seeing the trees for the carbon: Agroforestry for development and carbon mitigation. *Climatic Change*, 115, 741–757.
24. Schroth, G., Jeusset, A., Gomes, A. D., Florence, C. T., Coelho, N. A., Faria, D., Läderach, P. (2016). Climate friendliness of cocoa agroforests is compatible with productivity increase. *Mitigation and Adaptation Strategies for Global Change*, 21, 67–80.
25. Nguyen, Q., Hoang, M. H., Öborn, I., van Noordwijk, M. (2013). Multipurpose agroforestry as a climate change resiliency option for farmers: An example of local adaptation in Vietnam. *Climatic Change*, 117(1–2), 241–257.

26. Pandey, R., Meena, D., Aretano, R., Satpathy, S., Semeraro, T., Kumar Gupta, A., Rawat, S., Zurlini, G. (2015). Socio-ecological vulnerability of smallholders due to climate change in mountains: Agroforestry as an adaptation measure. *Change and Adaptation in Socio-Ecological Systems*, 2(1), 26–41.
27. Lasco, R. D., Espaldon, M. L., Habito, C. M. (2016). Smallholder farmers' perceptions of climate change and the roles of trees and agroforestry in climate risk adaptation: Evidence from Bohol, Philippines. *Agroforestry Systems*, 90(3), 521–540.
28. Arenas-Corraliza, M. G., López-Díaz, M. L., Moreno, G. (2018). Winter cereal production in a Mediterranean silvoarable walnut system in the face of climate change. *Agriculture, Ecosystems & Environment*, 264, 111–118.
29. van Noordwijk, M., Coe, R., Sinclair, F. L., Luedeling, E., Bayala, J., Muthuri, C. W., Cooper, P., Kindt, R., Duguma, L., Lamanna, C., Minang, P. A. (2021). Climate change adaptation in and through agroforestry: Four decades of research initiated by Peter Huxley. *Mitigation and Adaptation Strategies for Global Change*, 26(5), 18.
30. Pramova, E., Locatelli, B., Djoudi, H., Somorin, O. A. (2012). Forests and trees for social adaptation to climate variability and change. *Wiley Interdisciplinary Reviews: Climate Change*, 3(6), 581–596.
31. Lasco, R. D., Delfino, R. J., Catacutan, D. C., Simelton, E. S., Wilson, D. M. (2014). Climate risk adaptation by smallholder farmers: The roles of trees and agroforestry. *Current Opinion in Environmental Sustainability*, 6, 83–88.
32. Meybeck, A., Gitz, V., Wolf, J., Wong, T. (2021). *Addressing forestry and agroforestry in National Adaptation Plans: Supplementary Guidelines*. Food & Agriculture Org.
33. Gomes, L. C., Bianchi, F. J., Cardoso, I. M., Fernandes, R. B., Fernandes Filho, E. I., Schulte, R. P. (2020). Agroforestry systems can mitigate the impacts of climate change on coffee production: A spatially explicit assessment in Brazil. *Agriculture, Ecosystems & Environment*, 294, 106858.
34. Castillo, M. S., Tiezzi, F., Franzluebbers, A. J. (2020). Tree species effects on understory forage productivity and microclimate in a silvopasture of the Southeastern USA. *Agriculture, Ecosystems & Environment*, 295, 106917.
35. Cardinael, R., Cadisch, G., Gosme, M., Oelbermann, M., Van Noordwijk, M. (2021). Climate change mitigation and adaptation in agriculture: Why agroforestry should be part of the solution. *Agriculture, Ecosystems & Environment*, 319, 107555.
36. Mueller, N. D., Gerber, J. S., Johnston, M., Ray, D. K., Ramankutty, N., Foley, J. A. (2012). Closing yield gaps through nutrient and water management. *Nature*, 490(7419), 254–257.
37. Garnett, T., Appleby, M. C., Balmford, A., Bateman, I. J., Benton, T. G., Bloomer, P., Burlingame, B., Dawkins, M., Dolan, L., Fraser, D., Herrero, M. (2013). Sustainable intensification in agriculture: Premises and policies. *Science*, 341(6141), 33–34.
38. Gockowski, J., van Asten, P. J. (2012). Agricultural intensification as a climate change and food security strategy for sub-Saharan Africa. In: Wollenberg, E., Nihart, A., Tapio-Bostro, M.L., Grieg-Gran, M. *Climate Change Mitigation and Agriculture* (pp. 382–390). International Center for Research in Agroforestry-International Center for Tropical Agriculture: London, New York.
39. Koohafkan, P., Altieri, M. A., Gimenez, E. H. (2012). Green agriculture: Foundations for biodiverse, resilient and productive agricultural systems. *International Journal of Agricultural Sustainability*, 10(1), 61–75.

40. Mohan Kumar, B., Nair, P. K. (2011). *Carbon Sequestration Potential of Agroforestry Systems: Opportunities and Challenges* (p. 310), Springer Dordrecht.
41. Lott, J. E., Ong, C. K., Black, C. R. (2009). Understorey microclimate and crop performance in a Grevillea robusta-based agroforestry system in semi-arid Kenya. *Agricultural and Forest Meteorology*, 149(6–7), 1140–1151.
42. Gitz, V., Meybeck, A., Lipper, L., Young, C. D., Braatz, S. (2016). Climate change and food security: Risks and responses. *Food and Agriculture Organization of the United Nations (FAO) Report*, 110(2).
43. Basu, J. P. (2014). Agroforestry, climate change mitigation and livelihood security in India. In: *New Zealand Journal of Forestry Science* (Vol. 44, pp. 1–10). Springer International Publishing.
44. Tian, H., Zhou, S., Dong, L., Van Boeckel, T. P., Cui, Y., Newman, S. H., Takekawa, J. Y., Prosser, D. J., Xiao, X., Wu, Y., Cazelles, B. (2015). Avian influenza H5N1 viral and bird migration networks in Asia. *Proceedings of the National Academy of Sciences*, 112(1), 172–177.
45. Jofre, J., Blanch, A. R., Lucena, F. (2010). Water-borne infectious disease outbreaks associated with water scarcity and rainfall events. In: *Water Scarcity in the Mediterranean: Perspectives Under Global Change* (pp. 147–159).
46. Hofstra, N. (2011). Quantifying the impact of climate change on enteric waterborne pathogen concentrations in surface water. *Current Opinion in Environmental Sustainability*, 3(6), 471–479.
47. Xu, B., Jin, Z. Y., Jiang, Z. B., Guo, J. P., Timberlake, M., Ma, X. L., Weng, Q. (2014). Climatological and geographical impacts on global pandemic of influenza A (H1N1). In: *Global Monitoring and Assessment Through Earth Observation* (pp. 233–248).
48. Assogbadjo, A. E., Kakaï, R. G., Vodouhê, F. G., Djagoun, C. A., Codjia, J. T., Sinsin, B. (2012). Biodiversity and socioeconomic factors supporting farmers' choice of wild edible trees in the agroforestry systems of Benin (West Africa). *Forest Policy and Economics*, 14(1), 41–49.
49. Raj, A. (2018). Role of agroforestry in climate change mitigation. *Journal of Pharmacognosy and Phytochemistry*, 7(2), 241–243.
50. Smith, P., Olesen, J. E. (2010). Synergies between the mitigation of, and adaptation to, climate change in agriculture. *The Journal of Agricultural Science*, 148(5), 543–552.
51. Kumar, B. M., Nair, P. R. (Eds.) *Carbon Sequestration Potential of Agroforestry Systems: Opportunities and Challenges*.
52. Rizvi, R. H., Dhyani, S. K., Yadav, R. S., Singh, R. (2011). Biomass production and carbon stock of poplar agroforestry systems in Yamunanagar and Saharanpur districts of northwestern India. *Current Science*, 10, 736–742.
53. Matocha, J., Schroth, G., Hills, T., Hole, D. (2012). Integrating climate change adaptation and mitigation through agroforestry and ecosystem conservation. In: *Agroforestry-the Future of Global Land Use* (pp. 105–126).
54. The WGI Interactive Atlas https://interactive-atlas.ipcc.ch/
55. Bhoi, T. K., Samal, I., Majhi, P. K., Komal, J., Mahanta, D. K., Pradhan, A. K., Saini, V., Nikhil Raj, M., Ahmad, M. A., Behera, P. P., Ashwini, M. (2022). Insight into aphid mediated Potato Virus Y transmission: A molecular to bioinformatics prospective. *Frontiers in Microbiology*, 13, 1001454.

56. Mahanta, D. K., Jangra, S., Ghosh, A., Sharma, P. K., Iquebal, M. A., Jaiswal, S., Baranwal, V. K., Kalia, V. K., Chander, S. (2022). Groundnut bud necrosis virus modulates the expression of innate immune, endocytosis, and cuticle development-associated genes to circulate and propagate in its vector, Thrips palmi. *Frontiers in Microbiology*, 13, 773238.
57. Bhatnagar, S., Mahanta, D. K., Vyas, V., Samal, I., Komal, J., Bhoi, T. K. (2023). Storage pest management with nanopesticides incorporating silicon nanoparticles: A novel approach for sustainable crop preservation and food security. *Silicon*, 16, 471–483.
58. Mahanta, D. K., Komal, J., Samal, I., Bhoi, T. K., Dubey, V. K., Pradhan, K., Nekkanti, A., Gouda, M. N. R., Saini, V., Negi, N., Bhateja, S., Jat, H. K., Jeengar, D. (2023). Nutritional aspects and dietary benefits of "Silkworms": Current scenario and future outlook. *Frontiers in Nutrition*, 10, 1121508.
59. Samal, I., Bhoi, T. K., Vyas, V., Majhi, P. K., Mahanta, D. K., Komal, J., Majhi, P. K., Acharya, L. K. (2023). Resistance to fungicides in entomopathogenic fungi: Underlying mechanisms, consequences, and opportunities for progress. *Tropical Plant Pathology*, 49(2), 1–13.
60. Mahanta, D. K., Bhoi, T. K., Komal, J., Samal, I., Nikhil, R. M., Paschapur, A. U., Singh, G., Kumar, P. V. D., Desai, H. R., Ahmad, M. A., Singh, P. P., Majhi, P. K., Mukherjee, U., Singh, P., Saini, V., Shahanaz, Srinivasa, N., Yele, Y. (2023). Insect-pathogen crosstalk and the cellular-molecular mechanisms of insect immunity: Uncovering the underlying signaling pathways and immune regulatory function of non-coding RNAs. *Frontiers in Immunology*, 14, 1169152.
61. Samal, I., Bhoi, T. K., Majhi, P. K., Murmu, S., Pradhan, A. K., Kumar, D., Saini, V., Paschapur, A. U., Raj, M. N., Ankur, Manik, S., Behera, P. P., Mahanta, D. K., Komal, J., Alam, P., Balawi, T. A. (2023). Combatting insects mediated biotic stress through plant associated endophytic entomopathogenic fungi in horticultural crops. *Frontiers in Plant Science*, 13, 1098673.
62. Komal, J., Desai, H. R., Samal, I., Mastinu, A., Patel, R. D., Kumar, P. D., Majhi, P. K., Mahanta, D. K., Bhoi, T. K. (2023). Unveiling the genetic symphony: Harnessing CRISPR-Cas Genome editing for effective insect pest management. *Plants*, 12(23), 3961.
63. Samal, I., Bhoi, T. K., Mahanta, D. K., Komal, J. (2023). Establishing the role of silicon (Si) in plant resistance to insects: A bibliometric approach. *Silicon*, 16, 2119–2128.
64. Majhi, P. K., Samal, I., Bhoi, T. K., Pattnaik, P., Pradhan, C., Gupta, A., Mahanta, D. K., Senapati, S. K. (2023). Climate-smart agriculture: An integrated approach to address climate change and food security. In: *Climate Change and Insect Biodiversity* (pp. 221–245). CRC Press.
65. Mahanta, D. K., Samal, I., Komal, J., Bhoi, T. K., Majhi, P. K., Ahmad, M. A. (2023). Understanding anthropogenic climate change, its consequences on insect pests, and techniques in forecasting and monitoring pest dynamics: A contemporary scenario. In: *Climate Change and Insect Biodiversity* (pp. 44–67). CRC Press.
66. Bhoi, T. K., Samal, I., Mahanta, D. K., Komal, J., Jinger, D., Sahoo, M. R., Achary, G. C., Singh, S. (2023). Understanding how silicon fertilization impacts

chemical ecology and multitrophic interactions among plants, insects and beneficial arthropods. *Silicon*, 15(6), 2529–2549.

67. Bhoi, T. K., Samal, I., Mahanta, D. K., Komal, J., Majhi, P. K., Ankur. (2024). Exploring Melatonin's potential as an alternative strategy for protecting plants from biotic stresses. In: *Melatonin in Plants: A Pleiotropic Molecule for Abiotic Stresses and Pathogen Infection* (pp. 223–242). Springer Nature Singapore: Singapore.
68. Samal, I., Bhoi, T. K., Pradhan, A. K., Mahanta, D. K. (2023). Plantix app: A success story of artificial intelligence in plant protection. *Van Sangyan*, 10, 24–26.
69. Mahanta, D. K., Bhoi, T. K., Komal, J., Samal, I., Mastinu, A. (2024). Spatial, spectral and temporal insights: Harnessing high-resolution satellite remote sensing and artificial intelligence for early monitoring of wood boring pests in forests. *Plant Stress*, 11, 100381.
70. Samal, I., Mahanta, D. K., Bhoi, T. K., Komal, J., Jatav, H. S., Jatav, S. S., Sathyanarayana, E. (2024). *Approaches of Biochar in Ecosystem Management: Current Scenario and Future Perspectives*. IntechOpen
71. Vyas, V., Bhoi, T. K., Samal, I., Singh, S., Mahanta, D. K. (2024). Revitalizing degraded soils with agroforestry interventions: Opportunities, challenges, and future direction. In: *Agroforestry to Combat Global Challenges: Current Prospects and Future Challenges* (pp. 529–549), Springer Nature.
72. Samal, I., Bhoi, T. K., Mahanta, D. K., Komal, J., Singh, S. (2024). Biorational pest management: Potentials, unintended consequences, and future concerns. In: *Biorationals and Biopesticides: Pest Management* (pp. 47–75).

CHAPTER 2

Beneficial Insects as Agroforestry Ecosystem Guardians

PRASUN KARMAKAR,[1*] AKSHAY MISHRA,[1] and SOUMITA PAL[2]

[1]*Forest Protection Division, Rain Forest Research Institute, Jorhat, Assam, India*

[2]*Department of Agriculture, Government of West Bengal, Santipur, West Bengal, India*

*Corresponding author

ABSTRACT

Agroforestry is a system of land use that contributes to a working landscape that serves as a source of income and provides ecosystem services. Agroforestry ecosystems provide habitats and resources for various animal, plant, and microbial species that aid in sustaining biodiversity conservation in fragmented and deforested landscapes. The guardians of the agroforestry ecosystem are the organisms that, through their structural and functional roles, solely or in association with other species, perform ecological functions that enhance the stability of the ecosystem. The insect' fauna provides numerous such functions in an agroforestry ecosystem. They provide regulating services such as pollination, act as scavengers, and are biological control agents. Insects provide a constant food source for middle- and top-level carnivores, providing a crucial link in the terrestrial food web. Under the ground, ants play a critical role in soil aeration and enrich the soil by recycling nutrients. They also improve water infiltration and circulate air in the soil, which is essential to maintain soil health and for the proper growth of the roots of the plants. In this chapter, the

Agroforestry and Climate Change Mitigation: Harnessing Multitrophic Interactions for Sustainable Plant Protection and Ecosystem Resilience. Hanuman Singh Jatav, Vishnu D. Rajput, Ipsita Samal, Tanmaya Kumar Bhoi (Eds.)

DOI: 10.1201/9781779641397-2

authors will summarize the protective and beneficial ecological functions of insects in agroforestry systems. This chapter also draws attention to how farm management techniques impact the beneficial role of insects in agroforestry systems.

2.1 INTRODUCTION

Agroforestry is a type of mixed cropping system that increases the biodiversity of a region in a planned manner. The agroforestry system is a human-planned artificial biodiversity compared to forest ecosystems that are unplanned and natural. Broadly understood as the presence of tree species in an agricultural landscape, agroforestry is also a strategy of land utilization that serves as a source of income and provides ecosystem services. The total land area under agroforestry is 46% globally, with more than 10% tree cover [1]. The presence of tree species in agroforestry systems is very critical, as they provide a range of services in a terrestrial ecosystem. Trees have a strong above- and below-ground structural presence that provides a modified microclimate for both above- and below-ground organisms. They also modify the temperature, humidity, and soil moisture due to their tree canopy [2, 3]. The soil organic carbon and nutrient profile in an agroforestry system is also increased due to the deposition of litter from the trees and associated grasses [4–6]. These modifications result in specific habitats that have a strong impact on the heterogeneity of the plot microclimate [7]. Interestingly, these microclimates depend on the tree phenology and morphology, which in turn can vary according to different seasons of the year. This results in habitat diversification, modifications to faunal activity, and changes in associated biodiversity [8].

At the same time, the natural equilibrium of the agroforestry system is self-sustained by the complex interactions of various flora and fauna with their environment. Insects are widely dispersed and play a crucial role in all ecosystems [9]. Insects have a structural and functional role, solely or in association with other species, and perform ecological functions that enhance the stability of the ecosystem. They recycle nutrients, pollinate flowers, disperse seeds, provide biocontrol services, and improve soil fertility [10]. These ecological attributes of insects in the ecosystem classify them as ecosystem engineers [11]. The average estimate of ecosystem services provided by insects in the United States alone is US$57 billion [12].

Insects offer a plethora of useful ecosystem services in agroforestry systems, even though their roles are frequently underestimated and often seen by farmers as nuisance causing crop losses. The positive features of insects are rarely discussed in favor of research that concentrates more on insect pest damage and their management. The advantages insects offer through different ecosystem services, however, are probably more than the negative impacts they may have, considering their diversity and the important ecological functions the beneficial insects perform in agroforestry systems [63]. Therefore, recognizing the ecological services that insects offer to agroforestry systems can help to encourage future studies on the conservation of beneficial insects in agroforestry systems and fringe areas by boosting funding and public involvement.

This chapter highlights the beneficial ecological functions of insect fauna in different agroforestry systems. The function and significance of these insects, and, hence, the ecological services they offer, are impacted by farm management methods. This chapter also draws attention to how farm management techniques impact the beneficial role of insects in agroforestry systems.

2.2 ROLE OF BENEFICIAL INSECTS AS AGROFORESTRY ECOSYSTEM GUARDIANS

Insects perform different types of services (1) as agroforestry ecosystem guardians (Figure 2.1), (2) as pollinators, (3) as biological control agents of pests, (4) as seed dispersal agents, and (5) as soil engineers.

A brief description of their services is mentioned below.

2.2.1 AS POLLINATORS

The movement of pollen from the anther of a male flower to the stigma of a female flower is called pollination. Since plants cannot move on their own, the act of pollination requires the involvement of agents. The different pollinating agents are wind, water, animals, and insects. Of all the agents, insects carry out 85% of pollination in flowering plants [13]. Furthermore, 35% of the global production of crops is dependent on the pollination services provided by the insects [14, 15]. The mutualistic

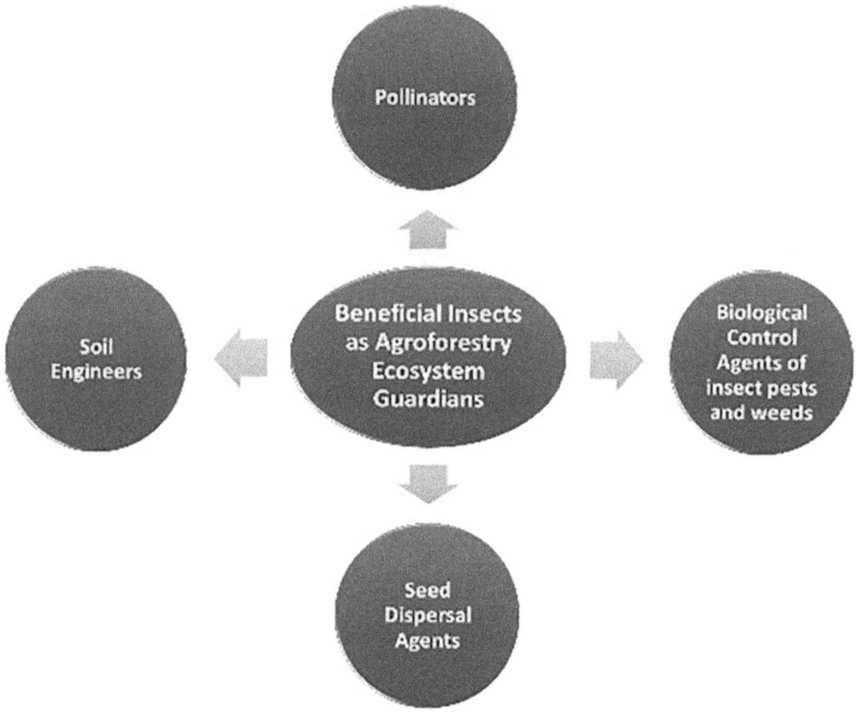

FIGURE 2.1 Role of beneficial insects as agroforestry ecosystem guardians.

interaction between flowering plants and insects developed 140 million years ago [16], with both parties benefiting from each other. The insects benefit from the flowering plants in the form of nectar and pollen as food, and the flowering plants benefit from the insects in the form of pollination.

Insects are a major group of pollinators in all forms of terrestrial ecosystems, including agroforestry ecosystems [17]. Insects were the primary pollinators mentioned in analyses that solely looked at the same crops over time. They accounted for 84% of all the species assessed as efficient pollinators, with birds (12%) coming in second and bats (4%) coming in third. Within 43% of all agricultural pollinator species, bees constituted a significant portion of the population. Beetles accounted for 15% of all crop pollinator species [18].

The nonnegotiable role of insect pollinators in providing maximum productivity of seeds and fruits has been accepted all over the world.

Scientists and researchers have undergone numerous researches all over the globe to establish the facts. In an agroforestry ecosystem, bees, butterflies, syphids, and hawk moths are important agents of pollination. In the bee category, non-*Apis* pollinators are far more efficient than the *Apis* pollinators [19, 20].

In tropical brinjal and ridge gourd-based agroforestry systems, in West Bengal, India, non-*Apis* bees dominated in the pollination service and pollinated 64% of the plants compared to *Apis* bees, which pollinated 36%. The most common pollinator in these systems was *Tetragonula iridipennis*. In temperate and arid agroforestry ecosystems, a type of ground-nesting, the solitary bee is very efficient in pollinating crops such as pulses, fodder, fruits, and other economically important plants. The morphology of the bee in having a hairy body, short and spike-like tongue, scopae, dusky wings, and chewing- and lapping-type mouth parts makes it an efficient pollinator in arid and temperate agroforestry ecosystems. These bees nest under the ground inside the mines and burrows and, hence, are called mining bees. They belong to the genus *Andrena* (Hymenoptera, Andrenidae) and are comprised of 54 species [21].

In Indonesia, researchers compared coffee monoculture systems with coffee-based agroforestry systems and found that the coffee-based agroforestry systems had higher species of bees pollinating them, which in turn resulted in higher pollination and higher yield of coffee beans. The plot with lower bee species (six species) resulted in a 75% seed set, and the plot with more than 20 bee species resulted in a 95% seed set. Here, they also found that the solitary bees were more efficient in pollination compared to the social bees [22].

Insect pollinators in a temperate agroforestry system in the United Kingdom resulted in 2.5 times more seed set compared to a monoculture system; the reason being that the temperate agroforestry system had two times more hoverflies and solitary bees as well as 2.4 times more bumblebee individuals compared to the monoculture system [23]. Solitary bees such as bumblebees do not collect a large amount of nectar like the social bees; instead, they spend more time collecting pollen from the flowers to feed their brood. This could be the reason why solitary bees are considered more efficient pollinators than social bees [24].

Insect pollinators vary according to the type of agroforestry ecosystem and the region where the agroforests are located. Some of the examples are mentioned in Table 2.1.

TABLE 2.1 List of Insect Pollinators in Different Agroforestry Systems

Pollinator Species	Agroforestry Ecosystem	Study Area	References
Tetragonula iridipennis *Nomia elliotii*	Ridge gourd- and brinjal-based agroforestry ecosystem	West Bengal, India	[25]
Xylocopa fenestrata	Brinjal-based agroforestry ecosystem		
Apis dorsata *Apis florea*	Ridge gourd-based agroforestry ecosystem		
Allobaccha brevis, *Eristalinus quinquelineatus*	Coffee agroforestry ecosystem	South West Ethiopia	[26]
Meliponula beccarii			
Andrena barbilabris, *Andrena flavipes*, *Andrena vaga*	Cherry agroforestry ecosystem	Switzerland	[27]
Chelostoma florisomne, *Chelostoma rapunculi*	Cherry agroforestry ecosystem	Switzerland	
Apis mellifera adansonii, *Hypotrigona* sp.	*Parkia biglobosa*-based agroforestry ecosystem	Gambia	[28]
Apis mellifera jemenitica, *Compsomelissa borneri*, *Hypotrigona* spp., *Liotrigona* cf. *bottegoi*	Shea agroforestry ecosystem	Burkina Faso	[29]
Apis cerana	Coffee agroforestry ecosystem	Indonesia	[30]
Lasioglossum sp.	Coffee agroforestry ecosystem	Indonesia	
Xylocopa pubescens	Coffee agroforestry ecosystem	Indonesia	
Apis mellifera	Coffee agroforestry ecosystem	Indonesia, South West Ethiopia	[26, 30]
Xylocopa sp.	Large cardamom-based agroforestry ecosystem	Sikkim and Darjeeling, India	[31]
Forcipomyia sp.	Cocoa agroforestry ecosystem	America, West Africa, and Southeast Asia	[32]

Table 2.1 depicts the dominant pollinators reported by various researchers from different agroforestry ecosystems in different parts of the world. Apidae, Andrenidae, Megachilidae, and Halictidae from the order Hymenoptera and true flies (order: Diptera) are the most common, dominant, and efficient pollinators of the different crop-based agroforestry ecosystems. The Apidae family comprises of dominant pollinators such

as honey bees, carpenter bees, and stingless bees. The sweat bees (Halictidae), scissor bees (*Chelostoma*), and the mining bees (Andrenidae) from the order Hymenoptera are also among the very effective pollinators of the agroforestry ecosystem.

The absence of pollinators from an agroforestry ecosystem leads to a decrease in productivity and overall health of the ecosystem. Such incidents are reported by the large cardamom-growing farmers from Sikkim in Eastern India and Darjeeling and the Eastern Himalayas, where the farmers faced decreased productivity and described the reason to be a decrease in the abundance of pollinators along with other factors such as pest and disease incidences [31]. The factors such as pest and disease incidence could be mitigated using resistance varieties and implementing Integrated Disease and Pest Management practices in the following season, but the absence of pollinators from an ecosystem is not solvable as such as may need rethinking and redesigning of the whole ecosystem. Failure to do so may result in the crop-based agroforestry ecosystem being nonremunerative and may discourage the farmers from further pursuing the same crop-based agroforestry system.

2.2.2 AS BIOLOGICAL CONTROL AGENTS OF INSECT PESTS

Native populations of arthropod biological control agents such as insect predators (ants, carabids, tiger beetles, ladybird beetles, and other Coleopteran beetles), parasitoids (parasitic wasps, flies), and predatory spiders regulate the insect pest abundance in agroforestry ecosystems and keep their population in check. In stable forest ecosystems, natural enemies are likely to be most effective since they can eventually manage most forest pests that do reach epidemic levels and keep the majority of the insect pest population in check below injury levels [33]. In situations when pest levels are low to moderate, insect predators and parasitoids can prevent epidemics, even though they are unlikely to quickly contain extensive pest outbreaks [34]. For instance, low population density of spruce budworm [*Choristoneura fumiferana* (Clern.)] is subject to adequate feeding pressure from native insect predators to mitigate the severity of infestations. Given how challenging it is to forecast the effects of tritrophic interrelationships in agroforestry systems, it is imperative to have a detailed comprehension of the interactions that exist between insect pests, their native biological control agents, and their habitats [35, 36].

The utilization of agroforestry systems, especially those that are intricate, holds significant promise in managing pest populations by enhancing the effectiveness of biological control agents. In many agroforestry systems, biological control by parasites and predators has been shown to be an excellent substitute for chemical pesticides [37]. The minimal degree of infestation in the agroforestry systems is an indicator that the predator–prey complex is in charge naturally and is active both within the field and outside the affected area. Agroforestry-based diverse cropping systems with cover crops are assumed to be more stable and resource efficient [38]. Plant diversity favors the abundance and effectiveness of related natural enemies, which in turn controls the populations of insect herbivores [39].

Many different types of natural enemy complexes make up a sizable portion of the overall fauna in the agroforestry systems (Table 2.2). Since agroforestry (alley cropping) is perennial, it, in conjunction with the fringe area, helps to maintain the natural enemy complex both during and after the main crop's offseason by providing different sources of food, pollen, and nectar, as well as appropriate microhabitats for diverse biological control agents [40–42]. Previous studies showed how perennial habitats are superior to annual crops in terms of enhancing biocontrol [43]. The use of grassy strips in crops can improve aphid biocontrol by elevating the activity–density of predatory arthropods [44].

2.2.3 AS SEED DISPERSAL AGENTS

Seed dispersal by insects is another form of mutualism where the insects benefit from the plants in receiving seeds as a source of food and plants receive the benefit in dispersing their seeds from one location to another. This phenomenon is called seed dispersal. Among the insect group, this phenomenon is the characteristic of ants (myrmecochory) and dung beetles (diplochory). The role of these insects in the seed dispersal phenomenon is briefly presented in the agroforestry systems of Central America and Mexico. In Mexico, myrmecochory is reported in *Cnidoscolus aconitifolius* [54]. Dung beetles (family: Scarabaeidae) are responsible for seed dispersal in *Poulsenia armata* in the Lacandon Maya agroforestry system in Mexico [55], *Bursera simaruba*, a shade tree utilized in Central American agroforestry systems [56], and in cacao and rubber agroforestry systems [57] in the same region of Mexico.

TABLE 2.2 List of Natural Enemies Recorded in Different Agroforestry Systems

Natural Enemy Species	Agroforestry Ecosystem (AFE)	Study Area	References
Ground beetles (*Pterostichus madidus*) and Opiliones	Temperate organic alley cropping system	Southeastern France	[8]
Cotesia sesamiae (Cameron), *Cotesia flavipes* (Cameron), *Cotesia ruficrus* (Haliday), *Glyptapanteles africanus* (Cameron), *Goniozus indicus* (Ashmead), *Chelonus curvimaculatus* (Cameron), *Syzeuctus* sp., *Pediobius furious* (Gahan)	Maize–leucaena agroforestry system	Amoyo in coastal and western Kenya	[45]
96 species of carabids and 63 species of cursorial spiders	Alley cropping agroforestry system	Southwestern France	[46]
334 predatory beetles	Cacao agroforestry systems	Brazilian Atlantic forest	[47]
Rhynchium haemorrhoidale (F.)	Cacao agroforestry systems	Central Sulawesi, Indonesia	[48]
Natural enemies belonging to Mantodea, Hymenoptera, Dermaptera, Coleoptera	Tung- and Subabul-based agroforestry system	Aizawl, Mizoram, India	[37]
Condylostylus sp., *Toxomerus lacrymosus*, *Doru luteipes*, *Ornidia obesa*, *Harmonia axyridis*, *Cycloneda sanguinea*, *Scymnus* sp., *Psyllobora gratiosa*	Vegetable-based agroforestry system	Brazil	[49]
Nebria brevicollis (F.) and *Anchomenus dorsalis* (Pont.)	Temperate agroforestry systems	France	[50]
Oecophylla longinoda (Latr.), *Tetramorium aculeatum* (Mayr)	Cocoa agroforestry systems	Cameroon	[51]
O. longinoda, *T. aculeatum*, and *Crematogaster* spp.	Cocoa agroforestry systems	Africa	[52]
Aclista sp., *Paramesius* sp., *Trichoplasta* sp.	Cocoa agroforestry systems	Colombian Amazon basin	[53]

The ants facilitate dispersal by carrying the seeds to their nests for food and in the process, they protect the plant species from density-dependent mortality. Ants facilitate the plants in two ways. First, by their foraging activity, they carry the seeds and deposit them in their mounds, increasing the abundance and distribution of the species. Second, the mound construction by ants facilitates an environment different from the surroundings, helping in the growth of plants different from the surrounding vegetation [58]. *Messor andrei* is an ant species from the serpentine grasslands of California. The ant species was responsible for the growth of an exotic grass species *Bromus mollis* around the mounds; also, the growth of the native vegetation was less around the mounds. In the same grassland, *Lepidium nitidum*, a forb species that can grow both in mounds and serpentine soil, was seen more abundant near the mounds. Researchers have credited this increased vegetation to the increased temperature around the mounds, but the cause of this increased temperature is still uncertain [58].

Dung beetles do not consume seeds but facilitate secondary seed dispersal called diplochory. The seeds present in the dung are rolled or buried away along with the dung, facilitating germination and dispersal [59]. During the process, they also scarify the seeds, which enhances the germination of the seed. The dung beetle carries the seeds underground and it results in the seeds being protected from predation and increases the chances of germination [60]. Some of the species common in the temperate grassland regions are *Geotrupes spiniger*, *Onthophagus similis*, *Geotrupes niger*, and *Aphodius illiger* [59]. *Copris doriae*, *Onthophagus poenicocerus*, and *Onthophagus* sp. were the dung beetles found in cacao agroforestry systems in Indonesia [61]. Interestingly, the nonendozoochorus seeds, such as oak acorns that are large enough not to be consumed by animals, remain exposed in the ground waiting to be predated and desiccated. These seeds are buried in the ground by the digging activity of the dung beetles, thus preventing their desiccation and predation, increasing their chances of germination. Such a type of passive acorn burial phenomenon was seen by researchers in Mediterranean silvipastoral systems [62].

2.2.3 AS SOIL ENGINEERS

Termites, ants, plant roots, and earthworms enhance the physical and biological properties of soil to create a suitable habitat for the growth and development of other organisms. They perform this activity by decomposing plant and animal waste, which releases important nutrients into the soil [63]. The biogenic, organomineral, and organic matter produced by these organisms remain in the soil space and are utilized by smaller microorganisms that are part of special nutrient cycles in the soil. The earthworm group is well studied and is established as the key soil engineer in humid and temperate ecosystems. But in arid and subarid ecosystems, the role of soil engineer is attributed to the termites [64]. Ants are also reported along with termites, enhancing the biological, physical, and chemical properties of soil. They enhance aggregate formation and maintain the soil structure, influence carbon transformation, and recycle nutrients [65].

In between the sown pastures of *Digitaria decumbens* and *Setaria sphacelata* in the Argentinian grassland agroforestry ecosystem, the mound-building ants, *Camponotus punctatus*, acted as soil engineers and facilitated the drainage, increased porosity, and increased availability of nutrients and organic matter available for the growth and development of the grass species [66].

Termites in cocoa agroforestry systems consume fallen litter and wood remains, contributing to carbon and nutrient recycling [67]. Australian arid agroforestry ecosystem inhabited native termites and ants as soil engineers, where they contributed by increasing the nitrogen pool by 2.5 times [68]. In the Manglid agroforestry systems of Java, *Campodea* sp., *Odontoponera* sp., and *Blatta* sp. were found to be dominant insect detritivores aiding in organic matter decomposition [69]. The humus-feeding termites such as *Cubitermes*, *Jugositermes*, and *Procubitermes* convert the humus into minerals that can be utilized by plants in the cocoa agroforestry ecosystem.

The soil engineering activity of termites and ants can be solely attributed to their nest-building habits. The social organization of these insects obligates them to architect the chambers of each cast of the insects in a different manner. The humidity and soil structure inside these chambers inhabit microorganisms that improve the structural stability of the soil and protect it against water influx [11].

2.3 MANAGEMENT OF AGROFORESTS TO CONSERVE INSECT GUARDIANS

Shade trees' canopy provides a suitable macrohabitat for native beneficial insect fauna, which perform important ecological functions as pollinators, seed dispersal agents, soil engineers, and biological control agents [70–72].

In contrast to enhanced unshaded agroforests, a higher population of predators was observed in more diversified traditional agroforests with shade trees [73]. Shaded agroforestry system facilitates the maintenance of a stable balance between insect pests and their native biological control agents [63]. However, in a tightly maintained system lacking shade, the disturbance of natural enemies is more likely to occur than that of insect pests. A shade canopy may also increase the number of predators that would not be able to live in plantations with direct sunlight [63]. It can also offer a variety of resources, such as decomposing organic substrate, which can serve as a refuge and a source of food for natural enemies [74].

In the context of tropical agroforestry, developed trees can alter predator communities [49]. Researchers have also suggested that, in comparison to monocropping agroecosystem, agroforestry systems with more diversification have lower pest incidence and higher diversity of natural enemies [75]. Trees have a direct impact on conserving and enhancing the population of biological control agents by offering habitats for their egg laying and hibernation and by acting as alternate food sources to natural enemies prior to the incidence or availability of hosts [42, 76]. Maintaining a robust predatory complex likely depends on the existence of the peripheral region of naturally occurring undisturbed plants or tree cover. Tree components in agroforestry systems can alter microclimates [2, 77], which can impact predator daily activity and ground arthropod ecosystems [78]. This is an essential factor for biological pest control. Predatory ground predators such as carabids and ground spiders can also live in undisturbed environments due to the untilled region surrounding the trees in an agroforestry system [79].

Insect pest infestations in agroforestry systems are caused by farm operations that severely impact the ecological diversity of native biological control agents, particularly the cutting down of shade trees and injudicious application of nonselective pesticides [63]. Indiscriminate application of pesticides in agroforests negatively impacts the native beneficial insect fauna, including pollinators [80] and biological control agents [81, 67]. Moreover, injudicious tree felling leads to a decrease in the population of natural enemies such as *Pheidole* sp. in agroforests [51]. Therefore, to conserve the diversity of insect guardians and promote their beneficial ecological services in agroforestry systems, a holistic strategy should be adopted by the farmers that considers the multitrophic relationships between beneficial insect fauna and shade trees in agroforestry ecosystems [63].

This will significantly contribute to the conservation of macrohabitats for beneficial insect guardians. It is recommended that an awareness drive should be carried out for the stakeholders on the habitat management in the agroforestry system and how it contributes toward multitrophic relationships to reduce insect pest incidence as well as a decrease in dependence on pesticides. Additionally, we suggest that farmers and other stakeholders be sensitized on the ecological benefits of insects and their role in sustaining the natural equilibrium of the agroforestry system and on ways for modifying the farm management techniques to conserve these insects in a long run.

2.4 NECESSITIES FOR FUTURE RESEARCH AND PROSPECTIVE VIEWS

This chapter highlights a few research gaps that demand detailed study given as follows:

- Agroforestry systems: Multitrophic relationships and diversity of beneficial insects, with special emphasis on pollinators.
- Methods for optimizing the use of currently available selective insecticides and entomopathogens in agroforestry systems to lessen their detrimental effects on beneficial insects.
- Use of native biological control agents that effectively combat key pests in agroforestry ecosystems while improving ecological indices and environmental benefits.
- More emphasis is needed for multipurpose agroforestry systems and suitable shade trees that will maintain the diversity of native beneficial insects, enhance soil health, and provide supplementary revenue.
- There is a dearth of knowledge on the impact of farm management strategies on beneficial insects in agroforestry systems. It is unclear how farm management techniques affect the insect pest diversity and their native biological control agents. The long-term objective of future research should be to create management recommendations for agroforestry systems, which will include methods for promoting beneficial indigenous insects as a substitute for pesticides.

2.5 CONCLUSION

Nature has created a perfect balance in the ecosystem in which the insects are placed as guardians. This chapter unequivocally demonstrates the vital roles that insects play in agroforestry systems. In an agroforestry system, starting with the plantation of seedlings to the death of the crop/plant/tree, insects fulfill their guardian role without failure, provided there is no unbalancing external intervention. The termites, ants, and other coprophilous (decomposers) beetles nourish the soil and make it ready for the seedlings. The seedling grows into a sapling, and the insect biocontrol agents defend it from the harmful defoliating, stem-boring, root-feeding, and sap-sucking insects. When the plant attains the reproductive stage, the pollinators transfer the pollen from the male flowers to the female flowers, ensuring cross-pollination and increasing the genetic variability. After fertilization, seeds are set, and the dispersal of seeds is done by dispersal agents. After the death of the plant/crop/tree, the wood-feeding termites and beetles break down the cellulose and aid in decomposition. This cycle is self-sustaining in nature. Thus, we summarize the role of beneficial insects in maintaining the natural equilibrium of the agroforestry systems. Therefore, these systems must be effectively maintained, so that beneficial insects, which provide important ecosystem services, are conserved and protected at every stage.

KEYWORDS

- **agroforestry**
- **ecosystem**
- **pollinators**
- **bees**
- **dung beetles**
- **biological control agents**
- **diplochory**
- **myrmecochory**
- **soil engineers**
- **beneficial insect guardians**
- **multitrophic relationships**

REFERENCES

1. Zomer, R. J., Trabucco, A., Coe, R., Place, F., van Noordwijk, M., Xu, J. C. (2014). Trees on farms: An update and reanalysis of agroforestry's global extent and socio-ecological characteristics. In: *Working Paper 179*. World agroforestry Centre (ICRAF) Southeast Asia regional program: Bogor, Indonesia.
2. Chen, J., Saunders, S. C., Crow, T. R., Naiman, R. J., Brosofske, K. D., Mroz, G. D., Brookshire, B. L., Franklin, J. F. (1999). Microclimate in forest ecosystem and landscape ecology. *BioScience*, 49(4), 288–297.
3. Vigiak, O., Sterk, G., Warren, A., Hagen, L. J. (2003). Spatial modeling of wind speed around windbreaks. *Catena*, 52(3–4), 273–288.
4. Cardinael, R., Chevallier, T., Cambou, A., Béral, C., Barthès, B. G., Dupraz, C., Durand, C., Kouakoua, E., Chenu, C. (2017). Increased soil organic carbon stocks under agroforestry: A survey of six different sites in France. *Agriculture, Ecosystems & Environment*, 236, 243–255.
5. Palm, C. A. (1995). Contribution of agroforestry trees to nutrient requirements of intercropped plants. *Agroforestry Systems*, 30(1–2), 105–124.
6. Young, A. (1997). *Agroforestry for Soil Management*, 2nd edn., CAB International: New York.
7. Marsden, C., Martin-Chave, A., Cortet, J., Hedde, M., Capowiez, Y. (2019). How agroforestry systems influence soil fauna and their functions—A review. *Plant and Soil*, 453, 29–44.
8. Martin-Chave, A., Béral, C., Capowiez, Y. (2019). Agroforestry has an impact on nocturnal predation by ground beetles and opiliones in a temperate organic alley cropping system. *Biological Control*, 129, 128–135.
9. Jankielsohn, A. (2018). The importance of insects in agricultural ecosystems. *Advances in Entomology*, 6(2), 62–73.
10. Scudder, G. G. E. (2017). The importance of insects. *Insect Biodiversity*, 2, 9–43
11. Jouquet, P., Dauber, J., Lagerlöf, J., Lavelle, P., Lepage, M. (2006). Soil invertebrates as ecosystem engineers: Intended and accidental effects on soil and feedback loops. *Applied Soil Ecology*, 32(2), 153–164.
12. Losey, J. E., Vaughan, M. (2006). The economic value of ecological services provided by insects. *BioScience*, 56(4), 311.
13. Ollerton, J., Winfree, R., Tarrant, S. (2011). How many flowering plants are pollinated by animals? *Oikos*, 120(3), 321–326.
14. Klein, A. M., Vaissière, B. E., Cane, J. H., Steffan-Dewenter, I., Cunningham, S. A., Kremen, C., Tscharntke, T. (2007). Importance of pollinators in changing landscapes for world crops. *Proceedings of the Royal Society B: Biological Sciences*, 274(1608), 303–313.
15. Eilers, E. J., Kremen, C., Smith Greenleaf, S., Garber, A. K., Klein, A.-M. (2011). Contribution of pollinator-mediated crops to nutrients in the human food supply. *PLoS One*, 6(6), e21363.
16. Wojcik, V. (2021). Pollinators: Their evolution, ecology, management, and conservation. In: Ranz, R. E. D. (Ed.), *Arthropods: Are They Beneficial for Mankind* (pp. 1–22). InTechOpen: London.

17. Kearns, C. A., Inouye, D. W. (1997). Pollinators, flowering plants, and conservation biology. *BioScience*, 47(5), 297–307.
18. Sabino, W., Costa, L., Andrade, T., Teixeira, J., Araújo, G., Acosta, A. L., Carvalheiro, L., Giannini, T. C. (2022). Status and trends of pollination services in amazon agroforestry systems. *Agriculture, Ecosystems & Environment*, 335, 108012.
19. Javorek, S. K., Mackenzie, K. E., Kloet, S. P. V. (2002). Comparative pollination effectiveness among bees (Hymenoptera: Apoidea) on lowbush blueberry (Ericaceae: Vaccinium angustifolium). *Annals of the Entomological Society of America*, 95(3), 345–351.
20. Kremen, C., Williams, N. M., Thorp, R. W. (2002). Crop pollination from native bees at risk from agricultural intensification. *Proceedings of the National Academy of Sciences*, 99(26), 16812–16816.
21. Raina, R. H., Pathak, P., Saddam, B., Parrey, A. H. (2023). Role of mining bees (Hymenoptera: Andrenidae) in pollination ecology in India: A review. *Agricultural Reviews*, 44(1), 66–71.
22. Klein, A. M., Steffan-Dewenter, I., Tscharntke, T. (2003). Pollination of coffea canephora in relation to local and regional agroforestry management. *Journal of Applied Ecology*, 40(5), 837–845.
23. Varah, A., Jones, H., Smith, J., Potts, S. G. (2020). Temperate agroforestry systems provide greater pollination service than monoculture. *Agriculture, Ecosystems & Environment*, 301, 107031.
24. Cato, S. Solitary bees could be more efficient than honey bees, researchers say. Solitary bees could be more efficient than honey bees, researchers say; 2018 https://www.uaex.uada.edu/media-resources/news/2018/june2018/06-07-2018-Ark-solitary-bee.aspx#:~:text=%E2%80%9CBecause%20these%20solitary%20bees%20are (Accessed December 14, 2023).
25. Das, P., Uniyal, V. P. (2023). Diversity of bees in two crops in an agroforestry ecosystem in Kangsabati South Forest division, Purulia, West Bengal, India. *Journal of Threatened Taxa* 15(3), 22889–22893.
26. Geeraert, L., Aerts, R., Jordaens, K., Dox, I., Wellens, S., Couri, M., Berecha, G., Honnay, O. (2018). Intensification of ethiopian coffee agroforestry drives impoverishment of the Arabica coffee flower visiting bee and fly communities. *Agroforestry Systems*, 93(5), 1729–1739.
27. Kay, S., Kühn, E., Albrecht, M., Sutter, L., Szerencsits, E., Herzog, F. (2019). Agroforestry can enhance foraging and nesting resources for pollinators with focus on solitary bees at the landscape scale. *Agroforestry Systems*, 94(2), 379–387.
28. Lassen, K. M., Ræbild, A., Hansen, H., Brødsgaard, C. J., Eriksen, E. N. (2012). Bats and bees are pollinating Parkia biglobosa in the gambia. *Agroforestry Systems*, 85(3), 465–475.
29. Lassen, K. M., Nielsen, L. R., Lompo, D., Dupont, Y. L., Kjær, E. D. (2018). Honey bees are essential for pollination of vitellaria paradoxa subsp. paradoxa (Sapotaceae) in burkina faso. *Agroforestry Systems*, 92(1), 23–34.
30. Supriyadi, S., Dzikrillah, F. D., Arniputri, R. B., Wijayanti, R. (2020). The effect of shade trees in the coffee ecosystem to the population and diurnal activity of insect pollinators. *Revista Internacional on Advanced Science, Engineering and Information Technology*, 10(4), 1743–1749.

31. Vineeta.,Tamang, B., Shukla, S., Chakravarty, C. (2023). The urge of conserving tradition from climate change: A case study of Darjeeling Himalayan large cardamom-based traditional agroforestry farming system. *Nature-Based Solutions*, 3, 100064.
32. Toledo-Hernández, M., Wanger, T. C., Tscharntke, T. (2017). Neglected pollinators: Can enhanced pollination services improve cocoa yields? A review. *Agriculture, Ecosystems & Environment*, 247, 137–148.
33. Mason, R. R. (1987). Nonoutbreak species of forest Lepidoptera. In: Barbosa, P., Schultz, J. (Eds.). *Insect Outbreaks* (pp. 31–57).
34. Dahlsten, D. L., Luck, R. F., Schlinger, E. I., Wenz, J. M., Copper, W. A. (1977). Parasitoids and predators of the Douglas-fir tussock moth, Orgyia pseudotsugata (Lepidoptera: Lymantriidae), in low to moderate populations in Central California. *The Canadian Entomologist*, 109(5), 727–746.
35. Rao, M. R., Singh, M. P., Day, R. (2000). Insect pest problems in tropical agroforestry systems: Contributory factors and strategies for management. *Agroforestry Systems*, 50(3), 243–277.
36. Altieri, M., Nicholls, C. (2018). *Biodiversity and Pest Management in Agroecosystems*. CRC Press.
37. Lalnunsangi, R., Paul, D., Jha, L. K. (2014). Natural enemy complex of some agroforestry systems of Aizawl and their implications in insect pest management. *Energy and Environment Research*, 4(2).
38. Vandermeer, J. (1995). The ecological basis of alternative agriculture. *Annual Review of Ecology and Systematics*, 26(1), 201–224.
39. Altieri, M. A., Letourneau, D. K., Risch, S. J. (1984). Vegetation diversity and insect pest outbreaks. *Critical Reviews in Plant Sciences*, 2(2), 131–169.
40. Jervis, M. A., Kidd, N. A. C. (1966). Phytophagy. In: Jervis, M. A., Kidd, N. A. C. (Eds.), *Insect Natural Enemies* (pp. 375–394), Chapman and Hall: London.
41. Landis, D. A., Wratten, S. D., Gurr, G. M. (2000). Habitat management to conserve natural enemies of arthropod pests in agriculture. *Annual Review of Entomology*, 45(1), 175–201.
42. Begum, M., Gurr, G. M., Wratten, S. D., Hedberg, P. R., Nicol, H. I. (2006). Using selective food plants to maximize biological control of vineyard pests. *Journal of Applied Ecology*, 43(3), 547–554.
43. Simon, S., Bouvier, J.-C., Debras, J.-F., Sauphanor, B. (2010). Biodiversity and pest management in orchard systems. A review. *Agronomy for Sustainable Development*, 30(1), 139–152.
44. Collins, K. L., Boatman, N. D., Wilcox, A., Holland, J. M., Chaney, K. (2002). Influence of beetle banks on cereal aphid predation in winter wheat. *Agriculture, Ecosystems & Environment*, 93(1–3), 337–350.
45. Ogol, C. K. P. O., Spence, J. R., Keddie, A. (1998). Natural enemy abundance and activity in a maize-leucaena agroforestry system in Kenya. *Environmental Entomology*, 27(6), 1444–1451.
46. Boinot, S., Mézière, D., Poulmarc'h, J., Saintilan, A., Lauri, P.-E., Sarthou, J.-P. (2020). Promoting generalist predators of crop pests in alley cropping agroforestry fields: Farming system matters. *Ecological Engineering*, 158, 106041.

47. Novais, S. M. A., Macedo-Reis, L. E., Neves, F. S. (2016). Predatory beetles in cacao agroforestry systems in Brazilian Atlantic forest: A test of the natural enemy hypothesis. *Agroforestry Systems*, 91(1), 201–209.
48. Hoehn, P., Steffan-Dewenter, I., Buchori, D., Tscharntke, T. (2008). Spatiotemporal density patterns of the pest predator Rhynchium haemorrhoidale (F.) along a land-use gradient in cacao agroforestry systems. *Agroforestry Systems*, 76(1), 163–171.
49. Harterreiten-Souza, É. S., Togni, P. H. B., Pires, C. S. S., Sujii, E. R. (2014). The role of integrating agroforestry and vegetable planting in structuring communities of herbivorous insects and their natural enemies in the neotropical region. *Agroforestry Systems*, 88(2), 205–219.
50. Richard, R., Cahon, T., Llandres, A. L., Le Levier, L., Proudhom, G., Casas, J. (2020). Alley cropping agroforestry mediates Carabid beetle distribution at a micro-habitat scale. *Agroforestry Systems*, 94, 309–317.
51. Bisseleua, D. H. B., Begoude, D., Tonnang, H., Vidal, S. (2017). Ant-mediated ecosystem services and disservices on marketable yield in cocoa agroforestry systems. *Agriculture, Ecosystems & Environment*, 247, 409–417.
52. Babin, R. (2018). *Pest Management in Organic Cacao*. CAB International: Wallingford.
53. Pérez-Benavides, A. L., Malagon-Aldana, L., Hanson, P., Serna-Cardona, F. J. (2022). New records and a voucher collection of parasitoid wasps (hymenoptera) inhabiting agroforestry systems in the colombian amazon basin. *Papéis Avulsos de Zoologia*, 62, e202262035.
54. Munguía-Rosas, M. A., Álvarez-Espino, R. X. (2022). What are elaiosomes for? Effects of elaiosomes on ant attraction, seed removal and germination in wild Chaya (Cnidoscolus aconitifolius). *Journal of Arid Environments*, 205, 104826.
55. González-Esquivel, C. E., Briones-Guzmán, C., Tovar-López, E., López-Ridaura, S., Arnés, E., Camacho-Villa, T. C. (2023). Sustainability evaluation of contrasting milpa systems in the Yucatán Peninsula, Mexico. *Environment, Development and Sustainability*, 1–23.
56. Urrea-Galeano, L. A., Andresen, E., Coates, R., Mora Ardila, F., Díaz Rojas, A., Ramos-Fernández, G. (2019). Horizontal seed dispersal by dung beetles reduced seed and seedling clumping, but did not increase short-term seedling establishment. *PLoS One*, 14(10), e0224366.
57. Santos-Heredia, C., Andresen, E., Zárate, D. A., Escobar, F. (2018). Dung beetles and their ecological functions in three agroforestry systems in the lacandona rainforest of Mexico. *Biodiversity and Conservation*, 27(9), 2379–2394.
58. Brown, M. J. F., Human, K. G. (1997). Effects of harvester ants on plant species distribution and abundance in a serpentine grassland. *Oecologia*, 112(2), 237–243.
59. D'hondt, B., Bossuyt, B., Hoffmann, M., Bonte, D. (2008). Dung beetles as secondary seed dispersers in a temperate grassland. *Basic and Applied Ecology*, 9(5), 542–549.
60. Shepherd, V. E., Chapman, C. A. (1998). Dung beetles as secondary seed dispersers: Impact on seed predation and germination. *Journal of Tropical Ecology*, 14(2), 199–215.
61. Shahabuddin, Schulze, C. H., Tscharntke, T. (2005). Changes of dung beetle communities from rainforests towards agroforestry systems and annual cultures in Sulawesi (Indonesia). *Biodiversity and Conservation*, 14(4), 863–877.

62. Leiva, M. J., Sobrino-Mengual, G. (2023). Cattle dung and bioturbation by dung beetles improve oak seedling establishment in mediterranean Silvopastoral ecosystems. *New Forests*, 54(2), 289–309.
63. Ambele, C. F., Bisseleua, H. D. B., Djuideu, C. T. L., Akutse, K. S. (2023). Managing insect services and disservices in cocoa agroforestry systems. *Agroforestry Systems*, 97, 965–984.
64. Jouquet, P., Chaudhary, E., Kumar, A. R. V. (2018). Sustainable use of termite activity in agro-ecosystems with reference to earthworms. A review. *Agronomy for Sustainable Development*, 38(1), 3.
65. Fahad, S., Chavan, S. B., Chichaghare, A. R., Uthappa, A. R., Kumar, M., Kakade, V., Pradhan, A., Jinger, D., Rawale, G., Yadav, D. K., Kumar, V., Farooq, T. H., Ali, B., Sawant, A. V., Saud, S., Chen, S., Poczai, P. (2022). Agroforestry systems for soil health improvement and maintenance. *Sustainability*, 14(22), 14877.
66. Folgarait, P. J., Perelman, S., Gorosito, N., Pizzio, R., Fernández, J. (2002). Effects of Camponotus punctulatus ants on plant community composition and soil properties across land-use histories. *Plant Ecology*, 163(1), 1–13.
67. Ambele, F. C., Bisseleua Daghela, H. B., Babalola, O. O., Ekesi, S. (2018). Soil-dwelling insect pests of tree crops in sub-Saharan Africa, problems and management strategies-A review. *Journal of Applied Entomology*, 142(6), 539–552.
68. Ernakovich, J. G., Evans, T. A., Macdonald, B., Farrell, M. (December 2016). The effect of ecosystem engineers on N cycling in an arid agroecosystem. In: *Proceedings of the 2016 International Nitrogen Initiative Conference, Solutions to Improve Nitrogen Use Efficiency for the World* (pp. 4–8). Melbourne, Australia.
69. Hani, A., Suhaendah, E. (2019). Diversity of soil macro fauna and its role on soil fertility in manglid agroforestry. *Indonesian Journal of Forestry Research*, 6(1), 61–68.
70. Sonwa, D. J., Weise, S., Adesina, A., Nkongmeneck, A. B., Tchatat, M., Ndoye, O. (2005). Production constraints on cocoa agroforestry systems in West and Central Africa: The need for integrated pest management and multi-institutional approaches. *The Forestry Chronicle*, 81(3), 345–349.
71. Guenat, S., Kaartinen, R., Jonsson, M. (2017). Shade trees decrease pest abundances on brassica crops in Kenya. *Agroforestry Systems*, 93(2), 641–652.
72. Maleque, M. A., Ferdous, J., Shitel, A. A., Ahmed, J., Islam, A., Mondal, M. F., Hassan, K., Razzak, A., Ishiga, H. (2023). Role of shade trees in conserving beneficial arthropods of biocontrol importance in tea ecosystem. *Agroforestry Systems*, 98, 21–36.
73. Klein, A., Steffan-Dewenter, I., Tscharntke, T. (2002). Predator–prey ratios on cocoa along a land-use gradient in Indonesia. *Biodiversity and Conservation*, 11(4), 683–693.
74. Greenberg, R., Bichier, P., Cruz Angón, A., MacVean, C., Perez, R., Cano, E. B. (2000). The impact of avian insectivory on arthropods and leaf damage in some Guatemalan coffee plantations. *Ecology*, 81(6), 1750–1755.
75. Stamps, W. T., Linit, M. J. (1997). Plant diversity and arthropod communities: Implications for temperate agroforestry. *Agroforestry Systems*, 39(1), 73–89.
76. Ferracini, C., Bertolino, S., Bernardo, U., Bonsignore, C. P., Faccoli, M., Ferrari, E., Lupi, D., Maini, S., Mazzon, L., Nugnes, F., Rocco, A., Santi, F., Tavella, L. (2018).

Do Torymus Sinensis (Hymenoptera: Torymidae) and agroforestry system affect native parasitoids associated with the asian chestnut gall wasp? *Biological Control*, 121, 36–43.

77. Morecroft, M. D., Taylor, M. E., Oliver, H. R. (1998). Air and soil microclimates of deciduous woodland compared to an open site. *Agricultural and Forest Meteorology*, 90(1), 141–156.
78. Tuf, I. H., Dedek, P., Veselý, M. (2012). Does the diurnal activity pattern of carabid beetles depend on season, ground temperature and habitat? *Archives of Biological Sciences*, 64(2), 721–732.
79. Weber, D., Pfannenstiel, R. S., Lundgren, J. G. Diel. (2009). Predation pattern assessment and exploitation of sentinel prey: New interpretations of community & individual behaviors. In: *The Proceedings of the Third ISBCA* (pp. 485–494), Christchurch, New Zealand.
80. Brittain, C. A., Vighi, M., Bommarco, R., Settele, J., Potts, S. G. (2010). Impacts of a pesticide on pollinator species richness at different spatial scales. *Basic and Applied Ecology*, 11(2), 106–115.
81. Prabhuraj, A., Patil, B. V., Bhemanna., Prabhakar, T. V., Singh, N. (2016). MOOC on Integrated Pest Management (IPM): A new online venture in agricultural education system. In: *Proceedings of the Pan-Commonwealth Forum 8 (PCF8), Commonwealth of Learning (COL) and Open University Malaysia (OUM).*

CHAPTER 3

Environmental Stress Mitigation and Adaptability Enhanced by Multitrophic Interaction

SUCHARITA PANDA,[1] RAJESWARI DAS,[2] MITA MEHER,[3]
ADITYA PRATAP SINGH,[4*] and GOLMEI LANGANGMEILU[5]

[1]*Department of Agroforestry, School of Agriculture, GIET University, Gunupur, Rayagada, Odisha, India*

[2]*Department of Soil Science and Agricultural Chemistry, School of Agriculture, GIET University, Gunupur, Rayagada, Odisha, India*

[3]*Department of Agricultural Extension, School of Agriculture, GIET University, Gunupur, Rayagada, Odisha, India*

[4]*Department of Plant Breeding and Genetics, School of Agriculture, GIET University, Gunupur, Rayagada, Odisha, India*

[5]*Department of Agronomy, School of Agriculture, GIET University, Gunupur, Rayagada, Odisha, India*

*Corresponding author

ABSTRACT

Environmental stresses impair biological systems by affecting ecological stability, dynamics, and sustainability. Such stresses may result from natural causes such as floods, cyclone, drought, and volcanic activity or as anthropogenic factors such as pollution, disease outbreak, land-use changes, and overexploitation of natural resources. Research have established that interaction among different trophic levels of the ecosystem plays a significant

Agroforestry and Climate Change Mitigation: Harnessing Multitrophic Interactions for Sustainable Plant Protection and Ecosystem Resilience. Hanuman Singh Jatav, Vishnu D. Rajput, Ipsita Samal, Tanmaya Kumar Bhoi (Eds.)

DOI: 10.1201/9781779641397-3

role in pest regulation, development of biotic resistance, erosion control, climate regulation, adaptive response of species, and overall ecosystem sustainability. High biodiversity in multitrophic system results in diverse communities equipped to adapt to changing conditions as different species respond differently to climate stressors. In a multistoried system, species may exhibit phenotypic plasticity, which allows the species to adapt to fluctuations in nutrient availability, thereby optimizing their fitness under different nutrient condition. Agroforestry systems can be cited as one of the best examples of multitrophic interaction as an adaptation toward nutrient stress. Efficient water-use strategies influenced by interactions, such as mycorrhizal associations, lead to the adaptation of plants for drought conditions. Mutualistic relationships can assist species in migration or expanding their range in response to changing climate conditions. It can also affect seasonal phenology by facilitating interactions such as pollination or predator–prey relationships, which plays a pivotal role in ecosystem functioning. Understanding the intricate relationship between various trophic levels and their way of interaction can be leveraged for attaining sustainable environmental stress mitigation.

3.1 INTRODUCTION

The Earth's ecosystems, intricate and interdependent, are shaped by the intricate tapestry of life woven through the myriad interactions among organisms at different trophic levels [1]. In the contemporary epoch, marked by the imprint of human activities on the planet's ecosystems, the study of these interactions has become imperative. The term "multitrophic interactions" encapsulates the dynamic relationships existing among organisms spanning various trophic levels—ranging from primary producers to consumers and decomposers [2, 3]. These interactions underpin the functionality and resilience of ecosystems, influencing processes such as nutrient cycling, pest control, and overall ecological stability.

As humanity faces the challenges of the Anthropocene, characterized by profound and often detrimental alterations to the environment, the comprehension of multitrophic interactions assumes heightened importance. Unprecedented alterations induced by human activities, including climate change, habitat destruction, and pollution, necessitate a profound understanding of the intricate connections that sustain life on our planet [4, 5]. It is within this context that this chapter seeks to unravel the multifaceted roles played by multitrophic interactions in mitigating environmental stressors, safeguarding biodiversity, and fortifying ecosystems against perturbations. Beyond the conventional notions of predator–prey relationships, multitrophic interactions encompass a rich tapestry of associations—mutualistic symbioses, decomposer activities, and trophic cascades—that collectively determine the health and adaptability of [6].

This exploration extends from the microscopic realms of soil ecosystems, where unseen interactions profoundly impact nutrient cycling, to the sweeping landscapes where large predators exert cascading effects on entire ecosystems. By delving into specific examples and case studies, this chapter endeavors to elucidate the nuanced contributions of multitrophic interactions across diverse environmental stressors. Climate change, nutrient limitations, drought, and wildfire disturbances serve as focal points for understanding how these interactions operate in complex and often unexpected ways. The goal is to unveil the underlying complexity of these relationships and illuminate their fundamental roles in shaping ecological resilience. In doing so, it aspires to contribute to the collective effort to secure a sustainable and harmonious future for the diverse life forms that call Earth home.

3.2 ENVIRONMENTAL STRESSES: TYPES, CAUSES, AND IMPLICATIONS

Plants have experienced evolutionary modifications over several centuries, influenced by escalating abiotic and biotic stressors resulting from heightened human activities spanning the past 200 years. These activities have altered habitats, introducing various harsh stressors such as diminished biodiversity and ecosystem pollution [7, 8]. Plant stress, characterized by external conditions negatively impacting plant growth, development, or productivity, elicits a diverse array of responses, including modified gene expression, alterations in cellular metabolism, changes in growth rates, and fluctuations in crop yields. The manifestation of plant stress typically signifies abrupt shifts in environmental conditions. Nonetheless, stress-tolerant plant species exhibit acclimation to specific stresses over time when exposed to such conditions.

Plant stress can be categorized into two main types: abiotic stress and biotic stress. Abiotic stress, originating from the environment, encompasses both physical and chemical factors. However, biotic stress affecting crop plants involves biological agents such as diseases and insects. Plants regularly contend with multiple unfavorable or adverse environmental conditions. Abiotic stresses, including salt, drought, heat, cold, heavy metals, ozone, UV radiation, and nutrient deficiencies, exert harmful effects on plant growth and productivity. These impacts have gained increasing significance due to the direct or indirect consequences of climate change. Plants exhibit diverse responses to abiotic stresses, spanning gene expression, physiology, plant architecture, and primary and secondary metabolism [9]. The intricate adjustments enable plants to endure and/or adapt to challenging conditions. The intricacy of plant response is further shaped by factors such as the duration and intensity of stress, plant genotype, the combination of different stresses, exposed tissue and cell type, and the developmental stage at which plants perceive the stress. Consequently, a deeper understanding of how plants perceive, respond to, and adapt to stress conditions is crucial, both in natural and anthropogenic environments.

Biotic stress in plants stems from the activities of living organisms, specifically viruses, bacteria, fungi, nematodes, insects, arachnids, and weeds. These biotic stress agents directly exploit their host, depleting nutrients and potentially leading to the death of plants. Biotic stress can escalate into a significant issue, causing pre- and postharvest losses. Despite the absence of an adaptive immune system, plants have evolved sophisticated strategies to counteract biotic stresses. The defense mechanisms against biotic stresses are genetically controlled, governed by the plant's genetic code. The plant genome harbors numerous resistant genes encoded to combat various biotic stresses. This stands in contrast to abiotic stress, which arises from nonliving factors such as salinity, sunlight, temperature, cold, floods, and drought, negatively impacting crop plants. The type of biotic stress imposed on crop plants is determined by the prevailing climate, influencing the crop species' ability to resist specific stressors. Numerous biotic stresses can impact photosynthesis; for instance, chewing insects can reduce leaf area, while virus infections can diminish the rate of photosynthesis per leaf area. Understanding the interplay between plants and their biotic environment is crucial for managing and mitigating the impact of biotic stress on crop plants.

Abiotic stresses, including drought (water stress), excessive watering (waterlogging), extreme temperatures (cold, frost, and heat), salinity, and mineral toxicity, have adverse effects on the growth, development, yield, and seed quality of crops and other plants. Predictions suggest that future freshwater scarcity will escalate, leading to an increased intensity of abiotic stresses. Consequently, there is an urgent need to develop resilient crop varieties that can withstand these abiotic stresses to ensure food security and safety in the coming years. A plant's primary defense against abiotic stress lies in its roots. The likelihood of surviving stressful conditions is higher when the soil supporting the plant is healthy and biologically diverse. One notable response to abiotic stress, such as high salinity, involves the disruption of the Na^+/K^+ ratio in the plant cell's cytoplasm. The phytohormone abscisic acid (ABA) plays a crucial role in the plant's adaptation to environmental stress, including high salinity, drought, low temperature, or mechanical wounding.

Some stresses can cause metabolic dysfunctions in plants, leading to injury. Mild or short-term stress can allow plants to recover, as the effects are temporary. However, severe stresses can result in the death of crop plants by inhibiting flowering, seed formation, and inducing senescence, classifying such plants as stress-susceptible. Human efforts in breeding plants have focused on enhancing their performance and yield attributes to meet the increasing global food demand for growing populations, addressing both the needs of humanity and the imperative of food security [7, 10].

Biotic and abiotic environmental stressors stand as significant factors capable of constraining plant growth and productivity, resulting in substantial losses in yield and biodiversity [11, 12]. Biotic stress originates from insect pests, diseases, and competition with other plants, whereas abiotic stress arises from nonliving factors such as extreme temperatures, drought, heavy metals, nutrient deficiencies, and mineral toxicity, all adversely affecting crop and plant species productivity [13–15]. These stresses impact plant physiology and biochemical processes, inducing adverse changes in growth and developmental processes [16, 17]. Understanding the effects of both biotic and abiotic environmental hazards on plants is crucial for developing sustainable ecosystems with plants that exhibit greater resistance to stress combinations.

Plants have evolved a range of strategies to adapt to biotic and abiotic stressors [18]. These mechanisms encompass physiological, biochemical, and molecular responses, including the production of stress hormones, osmotic adjustment, and the activation of stress-related genes [17, 19]. Plant breeding programs have been instituted to enhance the resilience of plants, enabling them to withstand both biotic and abiotic stresses. Researchers have delved into the molecular and physiological mechanisms governing plant responses to stress, identifying specific genes, proteins, and enzymes activated in response to stress, and elucidating the role of plant hormones such as ABA in mediating plant responses to stress combinations [20].

Paralleling advancements in techniques such as CRISPR/Cas9, the selection and breeding of plants are integral components in the development of stress mitigation and tolerance potential in plants. This intersection between genetic modification and traditional breeding approaches is particularly relevant in understanding and enhancing multitrophic interactions involving plants. However, the interplay of two or more stressful events could have even more adverse effects on plant growth and development, highlighting the complexity of multitrophic interactions [21].

Consider, for instance, the intricate relationship between drought stress and disease infestation in plants. Drought stress not only directly impacts plant physiology but also influences the interaction of various disease infestations, resulting in weakened plant defense mechanisms. This weakened defense, in turn, may lead to plant growth stagnation and/or proliferation of pathogens in the ecosystem. The multifaceted nature of these interactions underscores the need for a comprehensive understanding of how multiple stressors can synergistically affect plant health.

Moreover, emerging research suggests that plant water stress can negatively influence aphid performance, adding another layer of complexity to plant–insect interactions [22]. The nuanced connections between drought stress and the prevention of sucking-type insects in plants necessitate further investigation to unravel the mechanisms involved. The intricate web of multitrophic interactions demands a holistic approach to comprehend the underlying dynamics and develop strategies for sustainable plant health management.

The convergence of climate change-induced extreme environmental events, such as rising temperatures, adverse soil conditions, limited water availability, and the introduction of anthropogenic pollutants, poses a distinctive and formidable challenge for plant species [21, 23]. While our understanding of plant responses to individual stressors is extensive, knowledge regarding the combined effects of multiple stress factors occurring simultaneously, known as multifactorial stress combinations, remains limited [24].

These simultaneous abiotic and biotic stressors intricately influence plant productivity through complex physiological and biochemical processes [25]. The effects extend to growth and survival, mediated by molecular pathways and hormone regulation critical for stress acclimation [26]. Comprehensive studies are imperative to gain a thorough understanding of these interactions and their implications for improved stress tolerance in plants. As we navigate the complexities of multitrophic interactions, bridging the gap between genetic modifications, traditional breeding, and ecological dynamics becomes crucial for developing resilient plant varieties capable of withstanding the multifaceted challenges posed by environmental stressors.

3.3 ROLE OF MULTITROPHIC INTERACTION IN MITIGATION OF ENVIRONMENTAL STRESS

Relationships between creatures from different trophic levels in an environment, such as plants, herbivores, and predators, are examples of multitrophic interactions. These interactions have an impact on energy transfer, species diversity, and the stability of ecosystems [27]. Changes in multitrophic dynamics can have an impact on nutrient cycling, community structure, and biodiversity, eventually determining ecosystem health and resilience. Multitrophic interactions are critical in reducing environmental stress by promoting ecological equilibrium [28]. Predator–prey relationships, for example, regulate herbivore populations, preventing overgrazing and preserving plant diversity. Furthermore, interactions between plants and beneficial microorganisms improve nutrient uptake and stress tolerance. These complicated relationships help to increase ecosystem resilience by minimizing the effects of disturbances such as climate change and pollution [29]. The interaction of trophic levels fosters biodiversity by strengthening the resilience and sustainability of ecosystems confronted with a variety of environmental problems.

Multitrophic interactions, which are essential in ecological systems, are classified by trophic levels in terms of alleviating environmental stress. Predator–prey interactions are those in which predators control herbivore numbers, preventing overgrazing and ensuring ecosystem stability. The reintroduction of gray wolves (*Canis lupus*) to Yellowstone National Park in 1995 provided a once-in-a-lifetime scientific opportunity. It enables a comparison of ecosystem responses to climate change with and without human-altered species composition, highlighting wolf-induced changes in carrion availability and how they affect scavenger dynamics [30]. Natural predators on rodents include birds of prey, particularly owls and hawks, which serve as important ecological regulators. These avian predators serve an important role in rodent population management in agricultural settings, reducing the demand for chemical rodenticides. This ecological strategy not only reduces pesticides' harmful environmental impacts but also promotes sustainable pest control practices, thereby improving overall ecosystem health and balance [31, 32].

Similarly, in herbivore–plant interaction, herbivores promote plant variety by avoiding single-species dominance and increasing plant resilience. Parasitoid–host interaction will be another example in which parasitoids manage host populations, preventing outbreaks and ensuring biodiversity and stability. Synthetic insecticides, like lambda-cyhalothrin, can disrupt biological control of insect pests by affecting natural enemies and fostering insecticide resistance. The study with the aphid parasitoid wasp *Aphidius colemani* revealed sublethal doses of lambda-cyhalothrin disturbed host-searching and oviposition behavior, particularly with resistant aphids [33]. Such insecticide impacts may undermine biological control effectiveness, emphasizing the need for integrated pest management to address resistant pest populations and preserve natural enemy populations. Mutualistic interactions, such as pollination, improve ecosystem health by facilitating plant reproduction and nutrient cycling. Symbiotic connections between microbes and hosts help with nutrition cycle and stress tolerance. Relationship between legumes and rhizobium is one such interaction in nature which have evidence of atmospheric nitrogen fixation in legumes even under various unfavorable environmental conditions such as salt stress, nutrient stress, and draught stress [34, 35]. Fungi as decomposers play an important function in ecosystems by releasing critical nutrients such as nitrogen and phosphorus during detritus breakdown. This improves soil fertility and promotes plant health. Furthermore, certain fungi operate as biocontrol agents by interacting antagonistically with plant diseases.

Evidence supports their significance in plant stress relief, stressing fungi's multiple ecological benefits in nutrient cycling and plant–microbe interactions [36, 37].

Trophic Cascades is a series of indirect interactions in a food web that changes population sizes throughout trophic levels, beginning at the top. It can be "top-down" (driven by predators) or "bottom-up" (driven by resources), influencing biodiversity, ecosystem structure, and environmental factors, which is important for efficient ecological management and conservation [38]. Herbivores benefit from predator-driven cascades that promote balance and resilience. Positive species interactions have seen to reduce environmental stress. Positive species interactions promote stress coping and ecological health. Understanding and preserving these connections are critical for environmental stress mitigation.

Plant growth is hampered by abiotic stressors that are increased by climate change and agricultural malpractice. Excessive use of fertilizer and pesticides harms ecosystems. Eco-friendly solutions, such as multitrophic plant–microbe interactions like arbuscular mycorrhizal fungi (AMF) as biofertilizers, are critical for increased agricultural output. AMF inoculation improves host plant tolerance to heat, salinity, drought, metals, and severe temperatures by activating tolerance mechanisms and avoiding the downregulation of critical metabolic pathways. AMF, as natural root symbionts, supply essential inorganic nutrients, increasing growth and productivity in both unstressed and stressed environments. Using AMF as a biofertilizer may help plants adapt to changing environments [39–41]. Few works demonstrated the potential of Streptomyces rhizobacteria as effective plant growth promoters under salinity stress. Key characteristics of the characterized isolates included indolic compound and siderophore synthesis, ACC deaminase activity, and salt tolerance [42]. The strains showed considerable increase of maize plant development and relief of salinity-induced negative effects. The findings highlighted Streptomyces PGPR's prospective function in boosting crop resilience to salt stress, providing useful insights for sustainable agriculture in tough situations [43]. There are various multitrophic interactions that have their contribution toward alleviation of environmental stress in one or other ways and have been discussed briefly in this section (Table 3.1).

TABLE 3.1 Investigating the Role of Multitrophic Interactions in Environmental Stress Alleviation

S.l. No.	Type of Interaction		Ecosystem	Role in Environmental Stress Mitigation	References
1	Predator–prey interactions	Wolves and elk	Yellowstone National Park	Elk are preyed upon by wolves. Wolf reintroduction in Yellowstone resulted in a decrease in elk populations. This, in turn, helped vegetation to recover, limiting overgrazing and boosting plant species regeneration. Other species in the ecosystem benefited from the regenerated vegetation.	[30]
		Birds of prey and rodents	Agricultural ecosystems	Rodents as natural predators include owls and hawks, among other birds of prey. These birds lessen the demand for chemical rodenticides by aiding in the management of rodent populations in agricultural habitats. This lessens the negative effects of pesticides on the environment while promoting sustainable pest control.	[31, 32]
		Fungi-feeding nematodes and nematodes feeding fungi	Forest ecosystem	In forest settings, several nematodes play a natural predatory role against plant-pathogenic nematodes and vice versa, hence regulating soil health. This promotes the general well-being of plants and trees.	[44, 45]
2	Parasitoid–host interactions	Aphid parasitoid wasps (*Aphidius colemani*) and aphids	Agricultural ecosystems	*A. colemani* wasps use their ovipositor to lay eggs inside aphids. Wasp larvae grow by eating aphid tissues and avoiding essential organs. The pupated wasp crawls out of the mummified aphid, looking for new hosts. *A. colemani* provides natural pest management in agriculture, minimizing the need for toxic chemicals. Aphid control improves crop health by boosting yields and reducing ecological stress. This promotes biodiversity and helps to keep an ecosystem resilient.	[33]
		Chalcid wasps (e.g., *Megastigmus* spp.) and seeds of coniferous trees (e.g., pine seeds)	Forest ecosystem	Chalcid wasps lay their eggs inside the seeds of coniferous trees. Wasp larvae devour the contents of the seed, reducing seed viability. This parasitism aids in the management of tree population development, hence preserving a healthy forest ecosystem and lowering the risk of overpopulation-induced stress. Furthermore, it increases biodiversity by preventing individual tree species from dominating, so contributing to the overall ecosystem health and resilience.	[46, 47]

TABLE 3.1 *(Continued)*

S.l. No.	Type of Interaction		Ecosystem	Role in Environmental Stress Mitigation	References
3	Mutualistic interactions	Leguminous plants (e.g., soybeans, clover) and nitrogen-fixing bacteria (e.g., Rhizobia)	Terrestrial ecosystem	Nitrogen-fixing bacteria, such as Rhizobia, collaborate with leguminous plants to transform atmospheric nitrogen into a nutrient source. This mutualism improves soil fertility, stimulates plant growth, and reduces the need for synthetic fertilizers. Their sustainable nitrogen cycle reduces environmental stress and promotes healthier habitats.	[34, 35]
		Coral-algae symbiosis corals and symbiotic algae (Zooxanthellae)	Marine ecosystem	Corals shelter symbiotic algae, offering a secure habitat and necessary elements for photosynthesis. In reciprocation, algae supply corals with nutrients. This mutualistic bond enhances coral reef health, fortifying resilience against environmental stressors and supporting the delicate balance within marine ecosystems.	[48]
4	Decomposer interactions	Nutrient cycling by fungi	Forest ecosystem	Fungi's enzymatic activities during detritus decomposition leads to the release of vital nutrients such as nitrogen and phosphorus, improving soil fertility and plant health. Furthermore, several fungi act as biocontrol agents, demonstrating antagonistic interactions with plant diseases and thereby alleviating plant stress.	[36, 37]
5	Trophic cascade	Predator-mediated control of herbivores	Forest ecosystem	Increased top predator abundance, as seen by an increase in wolf numbers, has a regulating effect on herbivorous species such as deer. This phenomenon reduces the impact of overgrazing on vegetation. As a result, reduced herbivory promotes greater plant species regeneration, increasing total biodiversity and preventing environmental damage.	[49, 50]
		Reintroduction of top predators such as sharks or large fish	Aquatic ecosystem	Top predators, such as sharks or huge fish, can be reintroduced to limit the number of mid-level predators (mesopredators) such as lesser fish. This keeps mesopredators from overconsuming herbivores, which helps keep seagrass beds and coral reefs healthy. Healthy seagrass beds help to sequester carbon and support a variety of marine species.	[51]

TABLE 3.1 *(Continued)*

S.l. No.	Type of Interaction		Ecosystem	Role in Environmental Stress Mitigation	References
6	Plant–microbe interactions	Mycorrhizal association (*Glomus* spp. with *Zea mays*)	Crop ecosystem	Mycorrhiza mycorrhiza (mycorrhiza) is a symbiotic relationship between fungi and plants. Examples of symbiotic mycorrhiza include Glomus spp., Zea mays, Cereal. Fungal mycorrhiza that explore a larger area of soil, solubilizing phosphorous for transfer to maize via arbuscules, improving nutrient use efficiency. Hyphal extensions increase water absorption, enhancing drought tolerance. This symbiotic relationship promotes sustainable agriculture, reduces synthetic fertilizer requirements, and increases crop yields, highlighting the practical benefits of mycorrhiza in farming.	[39, 40, 41]
		Actinomycetes and antibiosis (*Streptomyces* spp. and higher plants)	Soil ecosystem	Streptomyces produces antibiotics such as streptomycin that inhibit soil-borne pathogens to control plant diseases and induce systemic resistance. Streptomyces root colonization improves the plant's defense mechanisms. This reduces the use of chemical pesticides and promotes sustainable agriculture. Practical applications of commercial biopesticides derived from Streptomyces can be seen. However, challenges such as antibiotics specificity and resistance development need to be addressed, highlighting the need for balanced and judicious application in environmental stress mitigation.	[42, 52]
		PGPR-plant interaction (*Pseudomonas fluorescens* and wheat)	Soil-plant ecosystem	PGPR, a plant growth-promoting rhizobacterium from the genus Pseudomonas, reduces environmental stress in wheat by solubilizing phosphorous, producing growth-promoting hormones and inducing systemic resistance. It improves nutrient uptake, root growth, and stress tolerance. Its biocontrol mechanism reduces reliance on chemical inputs and mitigates soil-borne pathogens. Field trials demonstrate an increase in wheat yields, demonstrating the practical application of PGPR to sustainable agriculture while emphasizing considerations of strain specificity and environmental variability.	[13, 52]

3.4 MULTITROPHIC INTERACTION INFLUENCING ADAPTABILITY OF ENVIRONMENTAL STRESS

3.4.1 ECOSYSTEM FUNCTION AS ENHANCED BY MULTITROPHIC BIODIVERSITY

Multitrophic biodiversity is particularly evident in various ecological processes such as nutrient cycling, pest control, and overall ecosystem stability [53]. Different species within various trophic levels play specialized roles in decomposition and nutrient recycling such as decomposers like bacteria and fungi break down dead organic material, making nutrients available for primary producers. This intricate web of interactions ensures a continuous flow of essential elements, sustaining the productivity of the ecosystem. The natural balance in biodiversity in different trophic levels, especially involving predators and prey, helps regulate populations of potentially pests and in controlling herbivore populations, preventing overgrazing or excessive damage to plants. Different ecosystems have various species of pollinators, including bees, butterflies, and birds. This diversity ensures that a wide range of plant species can reproduce, maintaining floral biodiversity. This, in turn, supports the availability of food resources for other trophic levels and contributes to the resilience of the ecosystem. Different species may respond differently to environmental changes, making the ecosystem as a whole more adaptable to disturbances. Biology and ecosystem function relations in agroecosystems seem to be universal and robust to disturbances [54]. Redundancy in species' functions ensures that even if one species is affected, others can compensate, preventing a collapse in ecosystem functioning. The variety of species in multiple trophic levels contributes to the provisioning of essential ecosystem services such as clean water, air purification, and climate regulation. Land-use effects on biodiversity and functions are often indirectly manifested through trophic cascades [55, 56]. A diverse ecosystem is better equipped to provide these services efficiently, benefiting both the environment and human well-being.

3.4.2 ADAPTATION TOWARD CLIMATE STRESSORS

The intricate web interaction of various species across different trophic levels within ecosystems can significantly influence adaptation toward climate stressors in several ways. Multitrophic interactions create dependencies among species. For example, a change in the abundance or behavior of one species due to climate stressors may have cascading effects on others. This interdependence can influence the adaptive strategies of species as they navigate the challenges posed by climate change (Figure 3.1). The distribution and abundance of species within trophic levels can be altered by climate stressors. This, in turn, affects the dynamics of multitrophic interactions. Species may adapt by adjusting their foraging patterns, migration routes, or reproductive strategies in response to these changes, influencing the overall resilience of the ecosystem. Climate change will influence the relative importance, frequency and composition of functional groups, their trophic interactions, and processes controlling these interactions. Specific climate change factors generally have a beneficial influence on pathogen suppression and stress tolerance [1]. Climate stressors impose selective pressures on species, leading to evolutionary responses.

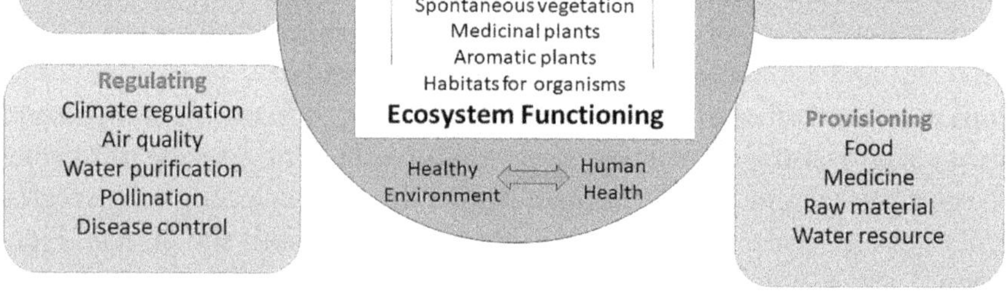

FIGURE 3.1 Biodiversity, ecosystem function, and services.

Multitrophic interactions contribute to the evolution of traits that enhance resilience to climate-related challenges [57]. For instance, plants may evolve to produce compounds that make them less palatable to herbivores under changing climatic conditions, influencing both plant and herbivore populations. Climate stressors can trigger shifts in community composition within trophic levels. When certain species become more or less abundant, the nature of interactions between predators, prey, and competitors evolves. Species may adapt by developing new associations or adjusting their ecological roles within the community to optimize their survival and reproduction [58]. Changes in climatic conditions may affect the balance of symbiotic relationships, such as mutualism or parasitism, influencing the ability of species to cope with stressors. For example, symbiotic partners may coevolve to better tolerate or exploit changing environmental conditions. The adaptation of species within multitrophic interactions contributes to the overall functioning of ecosystems under climate stress. For example, the ability of pollinators to adapt to altered flowering times or the capacity of predators to regulate herbivore populations can influence the resilience and stability of ecosystems facing climate-related disturbances. Conservation strategies that consider the interconnectedness of species and their adaptive responses across trophic levels can contribute to the long-term sustainability of biodiversity in the face of ongoing climate change.

3.4.3 ADAPTATION TOWARD NUTRIENT STRESSORS

The dynamics among different species across trophic levels shape how organisms respond to limited nutrient availability or excess nutrients, influencing nutrient cycling, ecosystem stability, and species interactions. Multitrophic interactions influence nutrient cycling as different species contribute to the breakdown of organic matter and nutrient release. Adaptation to nutrient stressors involves adjustments in the efficiency of nutrient cycling processes. For instance, decomposer organisms may play a key role in breaking down organic material, releasing nutrients that become accessible to primary producers [59]. Interactions between plants and soil organisms, such as mycorrhizal fungi and bacteria, are critical for nutrient acquisition [60]. Plants release exudates that attract beneficial soil microbes, facilitating nutrient uptake. Multitrophic interactions influence the composition of these microbial communities, impacting plant adaptation to nutrient stress. Some plants may establish associations with specific microbes that enhance nutrient availability. Herbivores play a role in nutrient cycling by consuming plant material. Adaptations in herbivore feeding behavior or plant defensive mechanisms can occur in response to nutrient stress [61]. For example, plants may invest more resources in chemical defenses or alter their nutrient content to deter herbivores under nutrient-limited conditions.

Predators may adapt to changes in prey availability by adjusting their foraging strategies or shifting their preferences among available prey species. This adaptation is crucial for maintaining the balance between trophic levels in the face of nutrient limitations. Multitrophic interactions involving microbes, such as bacteria and fungi, impact nutrient transformations. Microbial communities may adapt to nutrient stress by adjusting their metabolic activities, influencing the overall nutrient status of the ecosystem. Some species may be better adapted to thrive under nutrient-limited conditions, while others may decline [62]. These changes in community structure can have cascading effects on multitrophic interactions and ecosystem functioning.

3.4.4 ADAPTATION TOWARD DRAUGHT STRESSORS

The complex relationships among different species across trophic levels play a crucial role in shaping how organisms respond to water scarcity, affecting various aspects of ecological dynamics. Drought stress often leads to changes in plant physiology, affecting factors such as water content and nutrient concentrations [63]. Herbivores may adapt by altering their feeding preferences or behavior in response to changes in plant quality [64]. This adaptation can influence population dynamics within both plant and herbivore communities. Drought stress can alter the abundance and distribution of prey species, influencing predator–prey interactions. Predators may adapt by adjusting their foraging strategies or expanding their range in search of suitable prey. This adaptive response is crucial for maintaining the balance between predator and prey populations under drought conditions. Drought stress can impact mutualistic relationships such as those between plants and pollinators or mycorrhizal fungi. Plants may adapt by altering flowering patterns or releasing specific compounds to attract drought-tolerant pollinators or form associations with fungi that enhance water uptake. These adaptations help sustain crucial ecological processes despite water limitations. Soil microbes play a vital role in nutrient cycling and water retention. Drought stress can affect the composition and activity of microbial communities [65]. Plants may adapt by forming symbiotic relationships with specific microbes that enhance their ability to access water or nutrients under water-scarce conditions. Drought stress can intensify competition for limited water resources among plant species. Some plants may adapt by developing deeper root systems to access water from lower soil layers or by altering their growth strategies to outcompete others. These adaptations influence the composition and structure of plant communities. Multitrophic interactions contribute to the overall resilience of ecosystems facing drought stress [1]. Biodiversity across trophic levels can enhance the ability of ecosystems to withstand and recover from the impacts of drought. Diverse communities are more likely to contain species with varying tolerance levels, ensuring a more robust response to changing environmental conditions. Understanding the interconnectedness of species within multitrophic interactions is crucial for predicting and managing the impacts of drought stress on ecosystems [58]. Conservation strategies that consider these complex relationships can contribute to the resilience and long-term sustainability of ecosystems facing water scarcity.

3.4.5 ADAPTATION TOWARD WILD FIRE DISTURBANCE

The intricate web of relationships among different species across trophic levels shapes how organisms respond to the ecological changes triggered by wildfires. Wildfires influence plant communities, triggering changes in vegetation composition and structure. Some plant species may adapt to fire through mechanisms such as serotiny (releasing seeds in response to heat) or resprouting from underground structures. These adaptations are influenced by interactions with herbivores, pathogens, and mutualistic associates [66]. Wildfires can alter the availability and quality of vegetation for herbivores. Herbivores may adapt by adjusting their foraging behavior, shifting preferences to newly emerging postfire vegetation, or relocating to areas less affected by fire. These adaptations influence population dynamics within herbivore communities and impact interactions with plants. Wildfires can disrupt predator–prey relationships by changing the availability of resources and altering habitat structure. Predators may adapt by adjusting their hunting strategies or shifting their home ranges in response to postfire conditions. Herbivores, in turn, may face altered predation risks in the transformed landscape. Wildfires influence soil microbial communities by altering nutrient availability and organic matter content.

Plants may adapt through interactions with mycorrhizal fungi that play a crucial role in nutrient uptake and ecosystem recovery postfire. Mutualistic relationships with specific microbes can enhance the resilience of plants to fire disturbances. Wildfires can impact the availability of nesting sites and shelters for various species, including birds, mammals, and insects [67]. Species may adapt by modifying their nesting behaviors, choosing locations that are less prone to fire impacts. This adaptation can influence the survival and reproductive success of these organisms. Multitrophic interactions contribute to the overall resilience of ecosystems facing wildfire disturbances. Biodiversity across trophic levels enhances the ability of ecosystems to recover from the impacts of fire [68]. Diverse communities are more likely to contain species with varied responses to fire, ensuring a more robust and adaptable ecosystem.

3.5 FACTORS AFFECTING MULTITROPHIC INTERACTION IN ENVIRONMENT

Multitrophic interactions, the intricate relationships among species across different trophic levels, represent a cornerstone of ecological dynamics. These interactions are shaped by a multitude of factors, with environmental conditions playing a central and influential role. Among these factors, climate change emerges as a primary and overarching force impacting multitrophic dynamics [1]. The rise in global temperatures and the increasing unpredictability of weather patterns associated with climate change have profound effects on species across trophic levels [58]. These shifts influence the behavior, distribution, and abundance of organisms, leading to intricate changes in multitrophic interactions. Temperature alterations can affect the metabolic rates, reproductive behaviors, and developmental patterns of various species, ultimately influencing their roles within the multitrophic network [57].

Changes in precipitation patterns, another hallmark of climate change, contribute to the disruption of synchrony in species interactions. Timing is critical in multitrophic relationships, and alterations in precipitation can affect the availability of resources, breeding conditions, and the overall phenology of species. For instance, a delay in the onset of rainy seasons may impact the flowering patterns of plants, subsequently affecting herbivores and their predators [69]. Habitat loss and fragmentation, often linked to climate change, represent additional challenges to multitrophic interactions. The transformation of landscapes through deforestation, human development, and altered land use directly impacts the availability of suitable habitats and resources for various trophic levels. Species compositions may undergo shifts, leading to mismatches in multitrophic relationships. Fragmented habitats can also hinder the movement of species, disrupting the natural flow of interactions and potentially leading to isolated subpopulations. Human activities, extending beyond habitat alteration, contribute significantly to the challenges faced by multitrophic interactions. Pollution, including air and water pollution, introduces stressors that can affect the health and behavior of species within ecosystems. Land-use changes, such as the conversion of natural habitats to agriculture, urban areas, or industrial zones, further limit the available space and resources for diverse trophic interactions.

Mitigating the impacts of these environmental factors requires a deep understanding of the intricate web of multitrophic interactions. Conservation efforts must not only focus on preserving individual species but also consider the interconnected relationships that sustain ecosystems. Adaptive management strategies are imperative, taking into account the dynamic nature of multitrophic interactions in the face of ongoing environmental changes. Furthermore, scientific research plays a crucial role in unraveling the complexities of multitrophic dynamics under climate change. Advanced monitoring technologies, molecular techniques, and data analytics offer unprecedented insights into the responses of species across trophic levels. Genomic studies enable the exploration of adaptive mechanisms, evolutionary responses, and the genetic basis of multitrophic interactions in changing environments [70].

3.5.1 CLIMATE CHANGE VERSUS MULTITROPHIC INTERACTION

The intricate web of multitrophic interactions faces unprecedented challenges in the era of climate change, as alterations in environmental conditions exert profound impacts on the structure and functioning of ecosystems globally. Climate change, driven by anthropogenic activities, introduces disruptions in temperature, precipitation patterns, and the frequency of extreme weather events, all of which significantly influence species within different trophic levels. One of the primary consequences of climate change is the rise in global temperatures [71, 72]. Warming temperatures have far-reaching effects on the phenology of plants, influencing the timing of crucial events such as flowering. As plants adjust their flowering times in response to temperature changes, herbivores dependent on these plants for sustenance face challenges in synchronizing their life cycles. This temporal mismatch can lead to reduced food availability for herbivores, impacting their reproductive success and overall population dynamics.

Furthermore, the geographical ranges of species are undergoing shifts in response to changing climatic conditions. As temperatures rise, many species are compelled to migrate to higher latitudes or elevations in search of suitable habitats. This range shift has cascading effects on established multitrophic relationships [57]. For instance, herbivores adapted to specific plants may struggle to keep pace with the shifting distributions, disrupting the finely tuned predator–prey interactions within ecosystems. Extreme weather events, intensified by climate change, pose additional threats to multitrophic interactions. Heatwaves, storms, and other climatic disturbances can lead to habitat destruction, alter resource availability, and disrupt the delicate balance among trophic levels. The repercussions of such events can reverberate through ecosystems, affecting the abundance and distribution of both predators and prey. These disturbances highlight the vulnerability of multitrophic interactions to the cascading effects of climate change [55].

The interconnected nature of multitrophic interactions amplifies the challenges posed by climate change. Disruptions at one trophic level can propagate through the entire network, affecting the stability and resilience of ecosystems. The need for adaptive strategies is imperative to preserve biodiversity and maintain ecosystem stability in the face of these environmental transformations. Conservation efforts should not only address individual species but also consider the intricate relationships that define multitrophic interactions.

3.5.2 RANGE SHIFT IN PLANT VERSUS MULTITROPHIC INTERACTION

The rising global temperatures compel plants to embark on migrations to higher latitudes or elevations in pursuit of more suitable habitats. While this adaptive response is a survival strategy for individual plant species, its repercussions are profound and multifaceted, echoing through the intricate tapestry of multitrophic interactions. The spatial and temporal misalignment triggered by these plant range shifts poses a considerable challenge to the finely tuned relationships among plants, herbivores, pollinators, and predators [73]. As plants venture into new territories, the herbivores intricately adapted to specific plant species may find themselves in a race against ecological time (Figure 3.2). The struggle to keep pace with the shifting distributions of their preferred plant hosts can lead to disruptions in herbivore foraging patterns and potentially instigate changes in herbivore abundance [74].

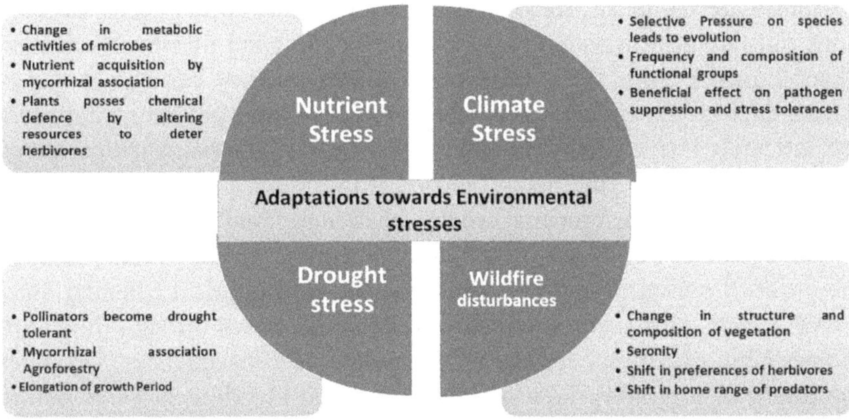

FIGURE 3.2 Adaptations toward different environmental stresses.

The ripple effects extend beyond herbivores to the predators that depend on them for sustenance. Predators, finely tuned to the availability and distribution of particular prey species, face their own set of challenges as the landscape transforms. The shifting dynamics of herbivore populations, dictated by the changing locations of their preferred plant hosts, may force predators to adapt their hunting strategies or face a mismatch in the availability of prey [75]. This ecological domino effect underscores the critical need for a holistic understanding of multitrophic interactions in the context of dynamic plant distributions induced by climate change. The interconnectedness of species within these interactions amplifies the consequences of spatial and temporal mismatches. As we witness the unfolding drama of plant range shifts, there is an urgent call for research that delves into the intricate mechanisms governing these multitrophic relationships. Such research is essential not only for comprehending the ecological implications but also for informing conservation strategies that seek to preserve the delicate balance of diverse ecosystems in the face of ongoing climate transformations. The future of multitrophic interactions hinges on our ability to unravel and respond to the intricate ecological narratives playing out across shifting landscapes [76].

3.5.3 SEASONAL PHENOLOGY OF MULTITROPHIC INTERACTION AS AFFECTED BY MULTITROPHIC INTERACTION

The intricate dance of seasonal phenology within multitrophic interactions unfolds as a dynamic story shaped by the hand of climate change. Phenological events, the synchronized timings of key ecological occurrences such as flowering, herbivore emergence, and predator migration, are delicately interwoven with climatic variations. The rising temperatures and shifting climate patterns of our planet introduce a new chapter in this ecological narrative, one marked by challenges and disruptions to the harmonious cadence of multitrophic relationships. As the mercury climbs, the onset of spring, a pivotal moment in the phenological calendar, may arrive earlier. This seemingly subtle shift reverberates through the entire multitrophic ensemble, setting off a cascade of consequences. Consider the plight of migratory bird species arriving at breeding grounds before their insect prey has reached peak abundance. The temporal misalignment between the predators and their prey creates a dissonance in the once finely tuned rhythm of nature, leading to food shortages and consequential impacts on reproductive success [77].

A parallel drama unfolds in the botanical realm as the timing of flowering plants and the emergence of pollinators fall out of sync under the influence of climatic variations. The intricate dance between plants and their pollinators, a partnership honed over millennia, faces disruption. This temporal mismatch threatens pollination efficiency, a critical process in the reproductive success of flowering plants. Seeds, the currency of plant reproduction, become pawns in the climatic chessboard, with potential implications for the future dynamics of plant populations. These shifts in seasonal phenology underscore the vulnerability of multitrophic interactions to the nuanced impacts of climate change. The interconnectedness of species within these interactions amplifies the consequences of even subtle shifts in timing. The ecological symphony, once orchestrated with precision, is at risk of discordant notes. The recognition of this vulnerability is a clarion call for adaptive management strategies, a plea for interventions that consider and mitigate the potential disruptions to these intricate ecological processes.

In the face of climate-induced changes, the future of multitrophic interactions hinges on our capacity to understand, anticipate, and respond to the altered temporal dynamics of species within ecosystems. The ecological clock is ticking, and as stewards of the planet, our challenge is not only to comprehend the intricacies of these temporal shifts but also to weave a narrative of resilience that safeguards the delicate ballet of life unfolding across seasons.

3.6 FUTURE PROSPECTS

In envisioning the future prospects of multitrophic interactions in the environment, we find ourselves at the nexus of scientific inquiry, technological innovation, and a collective responsibility to navigate the challenges imposed by an ever-evolving world. The intricate web of relationships among organisms across trophic levels beckons exploration, and as we embark on this journey, several key areas present themselves for further investigation and concerted action.

Advancements in technology and monitoring represent a beacon of promise for unraveling the complexities inherent in multitrophic dynamics. The integration of state-of-the-art technologies, ranging from remote sensing to molecular techniques and big data analytics, offers a quantum leap in our ability to comprehend the intricacies of these interactions. Continuous monitoring and real-time data collection empower researchers to scrutinize ecological changes with unprecedented precision, providing nuanced insights into how species respond across trophic levels to environmental stressors. This technological arsenal equips us to be custodians of the environment, arming us with the tools necessary to monitor and understand the subtle shifts within ecosystems. The evolving climate landscape necessitates the development of climate-smart conservation strategies. Future research endeavors should focus on crafting conservation approaches that acknowledge the dynamic nature of multitrophic interactions. This entails considering the adaptive capacity of species within ecosystems, implementing habitat restoration initiatives, and formulating conservation plans that anticipate and address the impacts of climate change on multitrophic relationships. It is an era where foresight becomes an invaluable asset, and strategies must be malleable to adapt to the unpredictable twists that climate change may introduce to ecological interactions.

In the vast expanse of agricultural landscapes, multitrophic interactions emerge as pivotal players in the quest for sustainable pest management. Future research should delve into the intricacies of these interactions, unlocking the potential for eco-friendly pest control methods. Understanding the delicate balance between natural predators, herbivores, and crops can inform the development of agricultural practices that minimize reliance on chemical inputs, fostering resilient agroecosystems that endure and thrive. As we confront environmental challenges that transcend geographic boundaries, the future beckons us to embrace global collaboration and policy initiatives. Multitrophic interactions, by their very nature, disregard borders, necessitating international cooperation to address shared environmental concerns. The prospects on the horizon involve the cultivation of global partnerships, the dissemination of knowledge across borders, and the formulation of policies that champion biodiversity conservation and sustainable resource management on a planetary scale. It is a call for unity in safeguarding the intricate dance of life that unfolds across ecosystems worldwide.

The future prospects of multitrophic interactions in the environment are as vast and diverse as the ecosystems they inhabit. The convergence of technological prowess, climate foresight, agricultural ingenuity, and global cooperation holds the key to unraveling the intricacies of these interactions and charting a course toward a more sustainable and harmonious coexistence with the natural world. As we gaze into the future, it is a landscape shaped by a profound understanding of multitrophic relationships and an unwavering commitment to preserving the delicate balance that sustains life on Earth.

3.7 CONCLUSION

The intricate tapestry of multitrophic interactions emerges as a linchpin in shaping the very fabric of ecosystems, exerting profound influences on a spectrum of ecological processes. From the nuanced dynamics of nutrient cycling and pest control to the adaptive responses of species to environmental stressors, multitrophic interactions weave a narrative that underscores their fundamental role in the intricate dance of life within ecosystems. This manuscript has cast a spotlight on the pivotal significance of multitrophic relationships in mitigating diverse environmental stresses, ranging from the macroscopic scale of climate change to the microscopic intricacies of nutrient limitations, drought, and the disruptive forces of wildfire disturbances. The essence of multitrophic interactions lies in the fine equilibrium they maintain, delicately orchestrating the functioning of ecosystems. This equilibrium, manifested through predator–prey relationships, mutualistic associations, and the myriad connections within trophic levels, forms a resilient and interconnected web that buttresses the health of our planet. From the majestic predators regulating herbivore populations to the microscopic interactions between plants and symbiotic microbes, each thread in this intricate web plays a crucial role in sustaining biodiversity and bolstering ecosystem resilience.

In the face of unprecedented environmental challenges, the imperative to comprehend and conserve multitrophic interactions becomes paramount. Unraveling the complexities of these relationships is but the first step; the true challenge lies in devising and implementing strategies that mitigate the impacts of environmental stressors. It calls for a holistic approach that integrates scientific understanding, innovative solutions, and global collaboration. The future trajectory of multitrophic interactions hinges on our collective ability to adapt, innovate, and foster international cooperation. Only through such concerted efforts can we aspire to ensure the continued health and functioning of our ecosystems, safeguarding the intricate connections that underpin the vitality of our planet for generations yet to come.

KEYWORDS

- **adaptability**
- **climate change**
- **multitrophic interaction**
- **stress mitigation**

REFERENCES

1. Chakraborty, S., Pangga, I. B., Roper, M. M. (2012). Climate change and multitrophic interactions in soil: The primacy of plants and functional domains. *Global Change Biology*, 18(7), 2111–2125.
2. Shikano, I. (2017). Evolutionary ecology of multitrophic interactions between plants, insect herbivores and entomopathogens. *Journal of Chemical Ecology*, 43, 586–598.
3. Briones, M. J. (2018). The serendipitous value of soil fauna in ecosystem functioning: The unexplained explained. *Frontiers in Environmental Science*, 6, 149.
4. Scanes, C. G. (2018). Human activity and habitat loss: Destruction, fragmentation, and degradation. In: *Animals and Human Society* (pp. 451–482). Academic Press.
5. Roy, S. C. (2023). Anthropogenic disturbances on climate change, global warming, ecosystem and COVID 19 pandemic. In: *Economic, Environmental and Health Consequences of Conservation Capital: A Global Perspective* (pp. 155–170). Springer Nature: Singapore.
6. Eisenhauer, N., Schielzeth, H., Andrew, D. *et al*., (2019). A multitrophic perspective on biodiversity–ecosystem functioning research. *Advances in Ecological Research*, 61, 1–54. https://doi.org/10.1016/bs.aecr.2019.06.001
7. Han, G.-Z. (2019). Origin and evolution of the plant immune system. *New Phytologist*, 222, 70–83.
8. Westwood, M., Cavender, N., Meyer, A., Smith, P. (2021). Botanic garden solutions to the plant extinction crisis. *Plants, People, Planet*, 3(1), 22–32.
9. Arbona, V., Manzi, M., Zandalinas, S. I., Vives-Peris, V., Pérez-Clemente, R. M., Gómez-Cadenas, A. (2017). Physiological, metabolic, and molecular responses of plants to abiotic stress. *Stress Signaling in Plants: Genomics and Proteomics Perspective,* 2, 1–35.
10. Swarup, S., Cargill, E. J., Crosby, K., Flagel, L., Kniskern, J., Glenn, K. C. (2021). Genetic diversity is indispensable for plant breeding to improve crops. *Crop Science*, 61(2), 839–852.
11. Rillig, M. C., Lehmann, A., de Souza Machado, A. A., Yang, G. (2019). Microplastic effects on plants. *New Phytologist*, 223(3), 1066–1070.
12. Zohaib, A., Tabassum, T., Jabbar, A., Anjum, S. A., Abbas, T., Mehmood, A., Irshad, S., Kashif, M., Nawaz, M., Farooq, N., Nasir, I. R., Rasool, T., Nadeem, M., Ahmad, R. (2018). Effect of plant density, boron nutrition and growth regulation on seed mass, emergence and offspring growth plasticity in cotton. *Scientific Reports*, 8(1), 7953.
13. Ahmad, H. M., Fiaz, S., Hafeez, S., Zahra, S., Shah, A. N., Gul, B., Aziz, O., Mahmood-Ur-Rahman, Fakhar, A., Rafique, M., Chen, Y., Yang, S. H., Wang, X. (2022). Plant growth-promoting rhizobacteria eliminate the effect of drought stress in plants: A review. *Frontiers of Plant Science,* 13, 875774. https://doi.org/10.3389/fpls.2022.875774.
14. Lin, M., Lin, T., Sun, C., Jones, L., Sui, J., Zhao, Y., Liu, J., Xing, L., Ye, H., Zhang, G., Li, X. (2020). Using the Eco-Erosion Index to assess regional ecological stress due to urbanization—A case study in the Yangtze River Delta urban agglomeration. *Ecological Indicators*, 111, 106028.

15. Fichman, Y., Mittler, R. (2020). Rapid systemic signaling during abiotic and biotic stresses: Is the ROS wave master of all trades. *The Plant Journal*, 102(5), 887–896.
16. Freschet, G. T., Violle, C., Bourget, M. Y., Scherer-Lorenzen, M., Fort, F. (2018). Allocation, morphology, physiology, architecture: The multiple facets of plant above- and below-ground responses to resource stress. *New Phytologist*, 219(4), 1338–1352.
17. Shaar-Moshe, L., Blumwald, E., Peleg, Z. (2017). Unique physiological and transcriptional shifts under combinations of salinity, drought, and heat. *Plant Physiology*, 174(1), 421–434.
18. Gong, Z., Xiong, L., Shi, H., Yang, S., Herrera-Estrella, L. R., Xu, G., Chao, D. Y., Li, J., Wang, P. Y., Qin, F., Li, J., Ding, Y., Shi, Y., Wang, Y., Yang, Y., Guo, Y., Zhu, J. K. (2020). Plant abiotic stress response and nutrient use efficiency. *Science China Life Sciences*, 63, 635–674.
19. Jain, P., Pandey, B., Singh, P., Singh, R., Singh, S. P., Sonkar, S., Gupta, R., Rathore, S. S., Singh, A. K. (2021). Plant performance and defensive role of glycine betaine under environmental stress. In: *Plant Performance Under Environmental Stress: Hormones, Biostimulants and Sustainable Plant Growth Management* (pp. 225–248).
20. Raza, A., Razzaq, A., Mehmood, S. S., Zou, X., Zhang, X., Lv, Y., Xu, J. (2019). Impact of climate change on crops adaptation and strategies to tackle its outcome: A review. *Plants*, 8(2), 34.
21. Choudhury, F. K., Rivero, R. M., Blumwald, E., Mittler, R. (2017). Reactive oxygen species, abiotic stress and stress combination. *The Plant Journal*, 90(5), 856–867.
22. Kansman, J. T., Basu, S., Casteel, C. L., Crowder, D. W., Lee, B. W., Nihranz, C. T., Finke, D. L. (2022). Plant water stress reduces aphid performance: Exploring mechanisms driven by water stress intensity. *Frontiers in Ecology and Evolution*, 10.
23. Balfagón, D., Zandalinas, S. I., dos Reis de Oliveira, T., Santa-Catarina, C., Gómez-Cadenas, A. (2022). Reduction of heat stress pressure and activation of photosystem II repairing system are crucial for citrus tolerance to multiple abiotic stress combination. *Physiologia Plantarum*, 174(6), e13809.
24. Zandalinas, S. I., Mittler, R. (2022). Plant responses to multifactorial stress combination. *New Phytologist*, 234(4), 1161–1167.
25. Lopez-Delacalle, M., Camejo, D. M., García-Martí, M., Nortes, P. A., Nieves-Cordones, M., Martínez, V., Rubio, F., Mittler, R., Rivero, R. M. (2020). Using tomato recombinant lines to improve plant tolerance to stress combination through a more efficient nitrogen metabolism. *Frontiers in Plant Science*, 10, 1702.
26. Dudhate, A., Shinde, H., Tsugama, D., Liu, S., Takano, T. (2018). Transcriptomic analysis reveals the differentially expressed genes and pathways involved in drought tolerance in pearl millet [Pennisetum glaucum (L.) R. Br]. *PLoS One*, 13(4), e0195908.
27. McWilliams, C., Lurgi, M., Montoya, J. M., Sauve, A. Montoya, D. (2019). The stability of multitrophic communities under habitat loss. *Nature Communications*, 10, 2322. https://doi.org/10.1038/s41467-019-10370-2
28. Wim, H. V., Louise, E. M. V., Jeffrey, A. H. Felix, L. (2001). Linking above- and belowground multitrophic interactions of plants, herbivores, pathogens, and their antagonists. *Trends in Ecology & Evolution*, 16(10), 547–554 https://doi.org/10.1016/S0169-5347(01)02265-0.

29. Dillen, M., Smit, C., Buyse, M., Höfte, M., De Clercq, P., Verheyen, K. (2017). Stronger diversity effects with increased environmental stress: A study of multitrophic interactions between oak, powdery mildew and ladybirds. *PLoS One,* 12(4), e0176104. https://doi.org/10.1371/journal.pone.0176104
30. Wilmers, C. C., Getz, W. M. (2005). Gray wolves as climate change buffers in Yellowstone. *PLoS Biology*, 3(4), e92. https://doi.org/10.1371/journal.pbio.0030092
31. Jenner, B. M., Lewis, T. R. (2023). Environmental conditions and vehicle disturbance influence stress behaviors in a working Harris's hawk (*Parabuteo unicinctus*). *Journal of Applied Animal Welfare Science*, 27(2), 373–385. https://doi.org/10.1080/10888705.2023.2253143
32. Malhi, Y., Lander, T., Roux, E., Stevens, N., Macias-Fauria, M., Wedding, L., Girardin, C., Ågård Kristensen, J., Sandom, C. J., Evans, T. D., Svenning, J., Canney, S. (2022). The role of large wild animals in climate change mitigation and adaptation. *Current Biology*, 32(4), R181–R196.
33. Alfaro-Tapia, A., Alvarez-Baca, J. K., Fuentes-Contreras, E., Figueroa, C. C. (2021). Biological control may fail on pests applied with high doses of insecticides: Effects of sub-lethal concentrations of a pyrethroid on the host-searching behavior of the aphid parasitoid *Aphidius colemani* (Hymenoptera, Braconidae) on aphid pests. *Agriculture*, 11, 539. https://doi.org/10.3390/agriculture11060539
34. Das, R., Pradhan, M., Sahoo, R. K., Mohanty, S., Kumar, M. (2021). Isolation, identification and role of novel endosymbiotic bacterium rhizobium pusence in root nodule of green gram cv.OUM-11-15 (Vigna radiata L.) under salinity stress. *Legume Research*, 44(12), 1512–1520. https://doi.org/10.18805/LR-4239.
35. Goyal, R. K., Habtewold, J. Z. (2023). Evaluation of legume–rhizobial symbiotic interactions beyond nitrogen fixation that help the host survival and diversification in hostile environments. *Microorganisms*, 11, 1454. https://doi.org/ 10.3390/microorganisms11061454
36. Liu, Y., Li, X., Kou, Y. (2020). Ectomycorrhizal fungi: Participation in nutrient turnover and community assembly pattern in forest ecosystems. *Forests*, 11, 453. https://doi.org/10.3390/f11040453
37. Zhang, S., Hu, W., Xu, Y., Zhong, H., Kong, Z., Wu, L. (2022). Linking bacterial and fungal assemblages to soil nutrient cycling within different aggregate sizes in agroecosystem. *Frontiers in Microbiol*ogy, 13, 1038536. https://doi.org/10.3389/fmicb.2022.1038536
38. Morris, T., Letnic, M. (2017). Removal of an apex predator initiates a trophic cascade that extends from herbivores to vegetation and the soil nutrient pool. *Proceedings of the Royal Society B: Biological Sciences*, 284, 20170111. https://doi.org/10.1098/rspb.2017.0111.
39. Begum, N., Qin, C., Ahanger, M. A., Raza, S., Khan, M. I., Ashraf, M., Ahmed, N., Zhang, L. (2019). Role of arbuscular mycorrhizal fungi in plant growth regulation: Implications in abiotic stress tolerance. *Frontiers of Plant Science*, 10, 1068. https://doi.org/10.3389/fpls.2019.01068
40. Sun, C., Yang, Y., Zeeshan, M., Qin, S., Ma, J., Liu, L., Yang, J., Zhou, X., Huang, J. (2021). Arbuscular mycorrhizal fungi reverse selenium stress in Zea mays seedlings by improving plant and soil characteristics. *Ecotoxicology and Environmental Safety*, 228, 113000. https://doi.org/10.1016/j.ecoenv.2021.113000.

41. Wahab, A., Muhammad, M., Munir, A., Abdi, G., Zaman, W., Ayaz, A., Khizar, C., Reddy, S. P. P. (2023). Role of Arbuscular Mycorrhizal Fungi in Regulating Growth, Enhancing Productivity, and Potentially Influencing Ecosystems under Abiotic and Biotic Stresses. *Plants (Basel, Switzerland)*, 12(17), 3102. https://doi.org/10.3390/plants12173102.
42. Nozari, R. M., Ortolan, F., Astarita, L. V., Santarém, E. R. (2021). Streptomyces spp. enhance vegetative growth of maize plants under saline stress. *Brazilian Journal of Microbiology*, 52(3), 1371–1383. https://doi.org/10.1007/s42770-021-00480-9
43. Silva, G. C., Kitano, I. T., Ribeiro, I. A. F. Lacava, P. T. (2022). The potential use of actinomycetes as microbial inoculants and biopesticides in agriculture. *Frontiers in Soil Science,* 2, 833181. https://doi.org/10.3389/fsoil.2022.833181
44. Poveda, J., Abril-Urias, P., Escobar, C. (2020). Biological control of plant-parasitic nematodes by filamentous fungi inducers of resistance: *Trichoderma*, mycorrhizal and endophytic. fungi. *Frontiers in Microbiology,* 11, 992. https://doi.org/10.3389/fmicb.2020.00992
45. Topalovic, O., Geisen, S. (2023). Nematodes as suppressors and facilitators of plant performance. *New Phytology,* 238, 2305–2312. https://doi.org/10.1111/nph.18925
46. Paulson, A. R., von Aderkas, P., Perlman, S. J. (2014). Bacterial associates of seed-parasitic wasps (Hymenoptera: *Megastigmus*). *BMC Microbiol*ogy, 14, 224 https://doi.org/10.1186/s12866-014-0224-4
47. Santini, A., Battisti, A. (2019). Complex insect-pathogen interactions in tree pandemics. *Frontiers in Physiology*, 10, 550. https://doi.org/10.3389/fphys.2019.00550
48. Torres, A. F., Valino, D. A. M., Ravago-Gotanco, R. (2021). Zooxanthellae diversity and coral-symbiont associations in the Philippine archipelago: Specificity and adaptability across thermal gradients. *Frontiers in Marine Science,* 8, 731023. https://doi.org/10.3389/fmars.2021.731023
49. Silliman, B. R., Angelini, C. (2012). Trophic cascades across diverse plant ecosystems. *Nature Education Knowledge*, 3(10), 44.
50. Chan, A. A. Q., Banks-Leite, C. (2020). Habitat modification mediates the strength of trophic cascades on oak trees. *Perspectives in Ecology and Conservation*, 18(4), 313–318, https://doi.org/10.1016/j.pecon.2020.09.002.
51. Rasher, D. B., Hoey, A. S., Hay, M. E. (2017). Cascading predator effects in a Fijian coral reef ecosystem. *Scientific Reports*, 7(1), 15684. https://doi.org/10.1038/s41598-017-15679-w
52. Mishra, A. K., Das, R., George Kerry, R., Biswal, B., Sinha, T., Sharma, S., Arora, P., Kumar, M. (2023). Promising management strategies to improve crop sustainability and to amend soil salinity. *Frontiers in Environmental Science,* 10, 962581. https://doi.org/10.3389/fenvs.2022.962581
53. Mulder, C., Boit, A., Mori, S., Vonk, J. A., Dyer, S. D., Faggiano, S., Geisen, S., González, A. L., Kaspari, M., Lavorel, S., Marquet, P. A., Rossberg, A. G., Sterner, R. W., Voigt, W., Wall, D. H. (2012). Distributional (In) congruence of biodiversity–ecosystem functioning. *Advances in Ecological Research,* 46, 1–88. ISSN 0065-2504. http://dx.doi.org/10.1016/B978-0-12-396992-7.00001-0.

54. Craven, D., Isbell, F., Manning, P., *et al.* (2016). Plant diversity effects on grassland productivity are robust to both nutrient enrichment and drought. *Philosophical Transactions of the Royal Society B: Biological Sciences*, 371, 20150277.
55. Barnes, A. D., Allen, K., Kreft, H., (2017). Direct and cascading impacts of tropical land-use change on multi-trophic biodiversity. *Nature Ecology and Evolution*, 1, 1511–1519.
56. Chen, J. T., Wang, M. Q., Li, Y., *et al.* (2023). Functional and phylogenetic relationships link predators to plant diversity via trophic and non-trophic pathways. *Proceedings Biological Sciences*, 290, 20221658.
57. Putten Van der, W. H., Macel, M., Visser, M. E. (2010). Predicting species distribution and abundance responses to climate change: Why it is essential to include biotic interactions across trophic levels. *Philosophical Transactions of the Royal Society B: Biological Sciences*, 365(1549), 2025–2034.
58. Jamieson, M. A., Trowbridge, M. A., Raffa, K. F., Lindroth, R. L. (2012). Consequences of climate warming and altered precipitation patterns for plant-insect and multitrophic interactions. *Plant Physiology*, 160, 1719–1727.
59. Dubey, R. K., Tripathi, V., Dubey, P. K., Singh, H. B., Abhilash, P. C. (2016). Exploring rhizospheric interactions for agricultural sustainability: The need of integrative research on multi-trophic interactions. *Journal of Cleaner Production*, 115, 362–365. https://doi.org/10.1016/j.jclepro.2015.12.077
60. Audet, P. (2012). Arbuscular mycorrhizal symbiosis and other plant- soil interactions in relation to environmental stress. In: *Environmental Adaptations and Stress Tolerance of Plants in the Era of Climate Change* (pp. 233–264). https://doi.org/10.1007/978-1-4614-0815-4_11
61. Scharf, I., Lubin, Y., Ovadia, O. (2011). Foraging decisions and behavioural flexibility in trap-building predators: A review. *International Biology Review*, 86, 626–639. https://doi.org/10.1111/j.1469-185X.2010.00163.x
62. Francis, B., Aravindakumar, C. T., Brewer, P. B., Simon, S. (2023). Plant nutrient stress adaptation: A prospect for fertilizer limited agriculture. *Environmental and Experimental Botany*, 213, 105431. https://doi.org/10.1016/j.envexpbot.2023.105431.
63. Anjum, S. A., 1, Xie, X., Wang, L., Saleem, M. F., Man, C., and Lei, W. (2011). Morphological, physiological and biochemical responses of plants to drought stress. *African Journal of Agricultural Research*, 6(9), 2026–2032. ISSN 1991-637X ©2011 https://doi.org/10.5897/AJAR10.027
64. Hamann, E., Denney, D., Day, S., Lombardi, E., Jameel, M. I., MacTavish, R., Anderson, J. T. (2021). Plant eco-evolutionary responses to climate change: Emerging directions. *Plant Science*, 304, 110737. Western, D. (2001). Human-modified ecosystems and future evolution. *Proceedings of the National Academy of Sciences*, 98(10), 5458–5465.
65. Sanaullah, M., Blagodatskaya, E., Chabbi, A., Rumpel, C., Kuzyakov, Y. (2011). Drought effects on microbial biomass and enzyme activities in the rhizosphere of grasses depend on plant community composition. *Applied Soil Ecology*, 48, 38–44. https://doi.org/10.1016/j.apsoil.2011.02.004
66. Malison, R. L., and Baxter, C. V. (2010). The fire pulse: Wildfire stimulates flux of aquatic prey to terrestrial habitats driving increases in riparian consumers. *Canadian Journal of Fisheries and Aquatic Sciences*, 67, 570–579 doi:10.1139/F10-006

67. Moghli, A., Santana, V. M., Baeza, M. J., Pastor, E., Soliveres, S. (2022). Fire recurrence and time since last fire interact to determine the supply of multiple ecosystem services by Mediterranean forests. *Ecosystems*, 25, 1358–1370. https://doi.org/10.1007/s10021-021-00720-x.
68. Bowman, D. M. J. S., Perry, G. L. W., Higgins, S. I., Johnson, C. N., Fuhlendorf, S. D., Murphy, B. P. (2016). Pyrodiversity is the coupling of biodiversity and fire regimes in food webs. *Philosophical Transactions of the Royal Society*, B371, 20150169. http://dx.doi.org/10.1098/rstb.2015.0169
69. Elzinga, J. A., Atlan, A., Biere, A., Gigord, L., Weis, A. E., Bernasconi, G. (2007). Time after time: Flowering phenology and biotic interactions. *Trends in Ecology & Evolution*, 22(8), 432–439.
70. Hamann, E., Blevins, C., Franks, S. J., Jameel, M. I., Anderson, J. T. (2021). Climate change alters plant–herbivore interactions. *New Phytologist*, 229, 1894–1910. https://doi.org/10.1111/nph.17036
71. Hitz, S., Smith, J. (2004). Estimating global impacts from climate change. *Global Environmental Change*, 14(3), 201–218.
72. Atta, K., Mondal, S., Gorai, S., Singh, A. P., Kumari, A., Ghosh, T., Roy, A., Hembram, S., Gaikwad, D. J., Mondal, S., Bhattacharya, S., Jha, U. C., Jespersen, D. (2023). Impacts of salinity stress on crop plants: Improving salt tolerance through genetic and molecular dissection. *Frontiers in Plant Science*, 14, 1241736.
73. Harvey, J. A., Malcicka, M. (2015). Climate change, range shifts and multitrophic interactions. In: *Biodiversity in Ecosystems-Linking Structure and Function* (pp. 85–109).
74. Subedi, B., Poudel, A., Aryal, S. (2023). The impact of climate change on insect pest biology and ecology: Implications for pest management strategies, crop production, and food security. *Journal of Agriculture and Food Research*, 14, 100733.
75. Jablonski, D. (2008). Biotic interactions and macroevolution: Extensions and mismatches across scales and levels. *Evolution*, 62(4), 715–739.
76. Perović, D. J., Gámez-Virués, S., Landis, D. A., Wäckers, F., Gurr, G. M., Wratten, S. D., You, M. S., Desneux, N. (2018). Managing biological control services through multi-trophic trait interactions: Review and guidelines for implementation at local and landscape scales. *Biological Reviews*, 93(1), 306–321.
77. Durant, J. M., Hjermann, D. Ø., Ottersen, G., Stenseth, N. C. (2007). Climate and the match or mismatch between predator requirements and resource availability. *Climate Research*, 33(3), 271–283.

CHAPTER 4

Investigating Detritivores Roles in Nutrient Cycling and Soil Health in Agroforestry

VISHWA PRAJAPATI* and S. N. MURTHY MIDDE*

ICFRE-Arid Forest Research Institute (ICFRE-AFRI), Jodhpur, Rajasthan, India

*Corresponding author

ABSTRACT

Sustaining soil fertility through biological processes necessitates careful management of plant residues regarding their quantity, quality, and timing. An effective method in pursuit of this objective is agroforestry, offering the potential to regulate plant residue decomposition and promote robust soil ecosystems. Soil organisms, integral to nutrient cycling and overall soil and plant health, are influenced by human practices, shaping nutrient cycling patterns. Trees are emphasized as pivotal for revitalizing soil fertility in agroforestry systems. They augment the availability of organic materials and nutrients, mitigate nutrient losses, and control input timing and quality. This chapter emphasizes a preliminary representation of a plant/soil organic matter cycle in agroforestry, highlighting the promise of agroforestry in integrating annual crop production with long-term soil fertility preservation. The primary aim of this chapter is to comprehend how detritivores such as ants, earthworms, and nematodes influence the soil food web and the cycling of carbon, nitrogen, and phosphorus in soils under different management conditions. By shedding light on these intricate relationships, the chapters contribute to more sustainable and ecologically sound agricultural practices, underscoring the vital role of agroforestry in achieving these objectives.

Agroforestry and Climate Change Mitigation: Harnessing Multitrophic Interactions for Sustainable Plant Protection and Ecosystem Resilience. Hanuman Singh Jatav, Vishnu D. Rajput, Ipsita Samal, Tanmaya Kumar Bhoi (Eds.)

DOI: 10.1201/9781779641397-4

4.1 INTRODUCTION

4.1.1 DETRITIVORES AND THEIR ECOLOGICAL ROLE

Detritivores are animals that get their food from eating dead plants and animal remains as they break down. They are a vital part of land-based ecosystems [1]. These creatures help to break down and recycle nature's waste. This helps nutrients go in circles, which makes the environment healthy and long lasting. So, they are very important for how nature works. The breaking down of dead plants and animals shows the role that insects play in nature. Detritivores start the process of changing big organic stuff into simpler parts and this gives important nutrients back to nature [2]. Types of creatures are grouped as detritivores. These include earthworms, bugs (such as beetles and ants), shellfish, and tiny living organisms like bacteria or fungi [3]. Within this working team, each garbage eater group has a lot of special skills in nature. They all help break down waste but do it differently and uniquely. Organic matter is turned into humus, which helps with soil structure and nutrient holding. This happens when small organisms break down trash from plants or animals [4]. By eating and breaking down germs found in decaying living things, animals also help control the number of harmful bacteria [5]. This part of their nature work underlines how detritivores help control organic pests. This keeps the balance and healthiness in ecosystems generally.

4.1.2 IMPORTANCE OF DETRITIVORES IN NUTRIENT RECYCLING

Nutrient loop in nature systems is a moving process needed for keeping balance and sustainability. Detritivores, which help with recycling nutrients, are very important in this complicated chain of actions. They start breaking down complex organic matter into simpler forms during this process, releasing important nutrients such as nitrogen, phosphorus, and carbon from dead plants and animals [1, 6]. Recycling these nutrients well is key for keeping life cycles going in natural systems. Detritivores change bad nutrients into good ones so that plants can use them. They do this by eating and interacting with bacteria inside their stomachs [7]. This process called mineralization is important for keeping nutrients in the soil and helping plants grow well. Beyond getting minerals, creatures that eat dead matter affect the movement of nutrients. Detritivores help plants get nutrients faster by breaking down complex organic things into simpler forms.

This important process affects how plants grow and the overall health of nature. It is also key to managing nutrients in ecosystems [8].

Detritivores also help make humus, a stable part of soil that comes from organic material. Detritivores break down plant stuff, making soil better for holding water and nutrients. This builds up humic substances in the ground [9]. Having humus in the soil makes it better at holding and giving out nutrients. This helps keep the soil fertile for a long time. Moreover, creatures that eat dead stuff help in moving around carbon. Carbon is important for how dirt's "living" part moves and changes over time. When organisms break down organic stuff, this releases carbon dioxide. This changes how much carbon is in ecosystems [10]. This part about detritivore actions is very important in the world's carbon moving and fighting climate change. Detritivores work by adding minerals, changing nutrients, and making humus is significant for keeping the right balance of food within nature. Knowing and understanding what detritivores do are very important for making land use plans that healthily help nutrient movement. This also helps make ecosystems strong against changes or problems they may face.

4.1.3 OVERVIEW OF AGROFORESTRY SYSTEMS AND THEIR RELEVANCE TO DETRITIVORE ACTIVITY

Agroforestry systems, where trees or bush-like plants are mixed with crops and animals on purpose, make good use of land. They offer many benefits all at once. In agroforestry places, different kinds and types make many little environments that strongly affect what eats dead things. Knowing the connection between forest farming systems and detritivores is very important for making nutrient flow better, and breaking down natural things like plants or animals easier. This keeps our surroundings going well forever. Agroforestry systems include some ideas such as alley cropping, silvopasture, and windbreaks. These are just a few of the many options available [11]. The way trees and plants are arranged, along with the different types of plant life, create a varied space that gives many homes and tools for animals. This change keeps a wide insect system going such as earthworms, beetles, and ants. Tiny creatures help too [12].

The value of using plants with trees for farming is that there are many pieces of dead stuff to eat. Having plants that keep coming back in farm forest spaces makes sure we always have leaves, twigs, and other plant parts to give dirt [11]. These things are great food for trash eaters, making them even more active and plentiful. The mixed group of trash eaters, in turn, helps break down this living material. The way things are arranged in farming communities with trees can impact the movement and spread of creatures that eat dead plants. Trees with their big leaves and roots change small weather patterns, which affects how bugs that eat trash are active or varied [13]. For example, the cover given by trees in farming with forest systems may change where scavengers hang out. Some types like to stay close when it is dark, while others do well away from shadows.

Moreover, agroforestry systems help make detritivore habitats more complex. The different layers of plants, like understory ones and cover crops, provide food sources for animals that eat dead things. These creatures have many places to feed, make babies, and stay safe [12]. This complicated home makes more kinds of trash-eating creatures, which keeps their survival strong and helps them work better with each other. The need for decomposers in farming systems is directly related to breaking down organic waste. The connections between trash eaters and little bugs in the breakdown process decide how quickly nutrients are made for plants to take [6]. So having trash eaters in farm-forest places helps to make the system work better and last longer (Figure 4.1).

FIGURE 4.1 Detritivores that power the nutrient cycle in agroforestry, returning essential elements to the soil for healthy plant growth.

4.2 DIVERSITY AND FUNCTIONS OF DETRITIVORES

4.2.1 TAXONOMIC DIVERSITY OF DETRITIVORES IN AGROFORESTRY

The soil ecosystem is key to helping land use such as making things grow, turning waste from plants into minerals, recycling, and providing needed nutrients. It also forms big holes in the ground for water entry and helps break down bad stuff while forming a solid base below our feet. Detritivores are animals that focus on breaking down dead plant stuff and other organic trash. These creatures are significant because they help recycle nutrients back into nature and break things down. Detritivores are found in all forms, representing many kinds of animals. It is divided into two groups according to their size: macrofauna and microfauna. Big insects that are more than 2 mm in size fall under the category of macrofauna. This includes earthworms, springtails, beetles, ants, among others [14]. Earthworms are those slippery helpers that work hard to make soil better by turning it over. These worms are part of the animal class called Oligochaeta. Pill bugs or woodlice belong to the order Isopoda. These small armored bugs move around, happily eating decayed leaves and wood. Millipedes are those animals with many legs that come from the Diplopoda class. They are great at recycling, and breaking down organic stuff with their strong jaws. Microfauna consists of small creatures that have a body width between 1 and 100 μm [15]. These organisms belong to different groups such as bacteria, fungi, and worms like nematodes (Table 4.1).

TABLE 4.1 Roles **of Different Detritivore Groups in Agroforestry Systems**

Sr. No.	Detritivore Type	Role in Agroforestry	Examples	References
1	Macrofauna	Break down large organic matter (e.g., leaves, branches)	Earthworms, millipedes, snails, termites	[16]
2	Mesofauna	Further fragment organic matter, making it accessible to microbes	Springtails, isopods, beetles, ants	[17]
3	Microfauna	Decompose organic matter into nutrients readily available for plants	Nematodes, mites, protozoa, fungi	[18]
4	Microbial decomposers	Break down complex molecules into simple forms usable by plants	Bacteria, fungi, actinomycetes	[19]

4.2.2 ROLES AND FUNCTIONS OF DETRITIVORES IN ORGANIC MATTER DECOMPOSITION

Detritivores make up a wide selection of living things. Different types of animals help break down food bits because each one has a special body and natural ways that make it easier to decompose things. Earthworms are known for eating dead stuff and helping break it down. They are very important in this process. Agroforestry greatly increases the number of worms on a farm [20, 21]. They help grow more types too. They help break down organic waste, promote tiny bug activity, and make nutrients available [22, 23]. Arthropods include beetles and springtails. Arthropod community dynamics contribute to improved soil health in the context of agroforestry [24, 25]. They also help speed up breaking down of plant matter to make nature-made stuff by cutting it into pieces or tearing it apart—all these are important for ground-organized matter decomposition. Microorganisms like bacteria and fungi also split up tough things into simpler forms by using little helpers called [8, 26].

Detritivores help make the soil better, keep water from going away, and control pests. This shows how important they are for many things in nature like what we call ecosystem services. The important job of cleaner animals is breaking down garbage from nature, focusing on their key parts for nutrient sharing, soil quality, and overall health of life systems. Understanding these processes is very important. We need them to develop ways of managing that use the good things detritivores do for keeping strong and useful nature areas at their best health.

4.2.3 IMPACT OF DETRITIVORE DIVERSITY ON NUTRIENT RELEASE AND SOIL STRUCTURE

The complicated balance between the variety of litter-eating animals, nutrients coming out, and how soil is built up plays a key role in land ecosystem activity. Diversity among decomposers means differences in how things work and what jobs they play across different groups of animals or plants. This is very important for shaping nature's way of breaking down old stuff like leaves and dead organisms better and with more successful results. Detritivores are groups of animals that break down dead matter. Their numbers and kinds affect how quickly needed nutrients become available for plants to take in [8, 27]. Bohlen and team's research in 2004, along with Wardle in the same year, shows how waste-eating insects like nonnative invasive earthworms change forest ecosystems in a heating climate [8, 27]. They focus on the complex relations of life under the ground as well as above it for processes connected to recycling nutrients such as minerals at the same time, creatures that eat decayed matter help to form soil in many ways. Earthworms, which are famous for helping make the environment better, help improve soil by making holes that let water go in. They also support more air to get into and create stable clumps of earth stuff [28, 29].

Understanding how different types of detritivores affect soil nutrients and structure is very important. It helps us make better practices for caring for land in a way that can last over time. Knowing how the different types of animals that eat dead things help with these processes can aid in making ways to improve services from nature, increase food production, and stop bad changes. This review not only helps in understanding ecological interactions better but also shows the real-world uses for managing crops and ecosystems sustainably. In the end, relations between animal diversity that eat plants and nutrients show how all parts of an environment work together. It also points out that we need a whole view when studying nature or managing land properly.

4.3 DETRITIVORES AND NUTRIENT CYCLING IN AGROFORESTRY

4.3.1 DETRITIVORE-MEDIATED DECOMPOSITION PROCESSES

Waste eaters like worms, bugs, and little living things break down what is natural by biting or grinding it up [2]. This machine stops working, which helps microbes to break down the remains of plants and animals more easily. Collaboration between tiny living things is vital for how decomposers break down waste. After the first breakup, detritivores continue to handle dead material by eating and pooping it out. When the detritivore eats organic matter, it goes through its digestive system and bacteria help break down complex polymers into simpler compounds [4]. The poop, also called fecal pellets, has partly broken down plant stuff mixed with tiny living creatures. Waste from detritivores is a place where microbes are very active. These groups of tiny organisms help break down complex materials. They turn them into simple substances that plants and soil need as food [9]. The made-up dirt improves soil stuff that comes from plants and changes how water sticks around, the shape of the ground, and nutrients are available. The work of detritivores also affects what is in the decomposed organic matter. For instance, earthworms pick and choose different food parts. This leads to changes in the amounts of carbon and nitrogen left behind [10]. This choice affects how organic material gets added into the soil, which in turn impacts nutrient cycles.

The speed of breakdown caused by detritivores depends on many things like the type of creatures, environment, and quality of stuff made from living organisms. Different kinds of detritivores have different choices for their meals and how fast they break down things [3]. Temperature, moisture, and soil features also have very important roles in controlling how quickly and well detritivores help break down waste [6]. Knowing these complicated processes is very important for understanding how nutrients move and change in ecosystems. Detritivore-assisted decay not only helps to release nutrients but also forms soil structure and productivity.

4.3.2 INFLUENCE OF DETRITIVORES ON NUTRIENT MINERALIZATION AND AVAILABILITY

Detritivores, by interacting with organic material in complex ways, have a very strong impact on the release and accessibility of nutrients within ecosystems. Detritivores start processes that change organic nutrients into inorganic forms. Then, plants and other living things can use them easily. This shows how important they are for nature when it comes to the chemistry mostly found on land areas. When detritivores break down and handle dead things, they make tiny sites filled with leftover bits from plants. These spots are easier for small microbes to rot away [7]. This starts a chain of tiny bacteria stuff, which lets out important nutrients such as nitrogen, phosphorus, and sulfur from the ground. Microscopic bugs in the detritivore's tummy and nearby ground help break apart this stuff more, freeing up nutrients that become minerals they can use [4]. Detritivores affect how much nutrients are available, but not just near where they work. For example, earthworms are known to spread organic material and nutrients in the soil by digging holes and eating food. They also help move these things through their tunnels called biopores [7]. This change helps make nutrients more available for plant roots and it also adds different levels of nutrient availability at various locations.

Also, detritivores help keep the balance of nutrients during breakdown. How worms and other waste eaters choose what they eat can change the carbon/nitrogen balance in any leftover organic materials [10]. This changed chemical makeup may lead to a long-term impact on nutrient rotation and the way plants take in food. The effect that scavengers have on nutrients is dependent on different things, like the type of worms or bugs and what they eat. It also depends upon where it happens and how good or fresh their food is. Understanding how nutrient mineralization works with insects that eat dead things is very important for controlling nature systems and farming in a way we can all support long term. By making more food available, detritivores help the soil get richer and make plants grow better. This makes all living things in nature healthier too.

4.3.3 RELATIONSHIP BETWEEN DETRITIVORE ACTIVITY AND SOIL FERTILITY ENHANCEMENT

The actions of detritivores directly influence how fertile the soil is by breaking down organic materials. As creatures that eat and break down dead plants or animals, they help free up useful substances such as nitrogen and phosphorus in the earth [2]. The turning of organics into minerals by scavengers helps plants get the nutrients they need. This is the main way for that to happen. The breaking down of organic matter done by tiny living things helps to create humus, a solid kind of dirt stuff. This comes from what plants and animals give off as waste. Humus makes the soil better by improving how it is built, keeping water in, and making plants take up more good stuff from the ground [9]. The mixing of soil particles caused by humus helps plants grow their roots and soak up nutrients better. Waste-eating creatures, especially earthworms, help a lot in making holes and tunnels underground. These small holes in the soil help air to get through, let the water soak in, and carry nutrients around [7]. The actions of trash eaters make soils well-made and good for tiny living things and roots. This also helps to boost soil richness.

Adding soil-enriching feces made by decomposers is very important for making the soil better. These poop particles, filled with partly rotten living stuff and tiny microbe groups, help make small nutritious spots [4]. The bacteria groups with worm poop help break down more stuff, giving the soil what it needs to grow plants well. The link between detritivore action and soil health is also affected by the variety of community members who eat dead materials. Different types of trash-eating creatures, including different kinds and working groups, help with more tasks in breaking down waste stuff and sharing nutrients [3]. Having different garbage recyclers helps keep the soil healthy. This makes sure that processes that help improve fertility in the soil can bounce back easily after being disrupted or damaged. Soil fertility can be affected by various environmental factors such as heat, water, and plant life in the soil. Detritivores (animals that eat dead plants of animals) play a part too but how big their influence is could change with these conditions around them. For example, the activity of earthworms in improving soil health might change with temperature and how wet or dry it is [6]. Knowing how these environmental changes work is important to guess and control the ways that soil workers help make fertile dirt.

4.4 DETRITIVORES AND SOIL HEALTH IN AGROFORESTRY SYSTEMS

4.4.1 DETRITIVORES AS INDICATORS OF SOIL HEALTH IN AGROFORESTRY

In farm-forest systems, detritivores do more than break down organic material. They also help show how healthy the dirt is by giving important signs of soil condition. Agroforestry combines trees and plants to make a good space for animals that eat dead things. When you put trees and plants together, it makes a lot of small homes for little creatures [30]. This also changes the number of large dead-eating bugs in that area. It is important to know the different kinds of creatures that eat dead matter in farm-forest systems. Detritivores help break down dirt and also tell us about the health of the ground they are on. Learning from how plants act to farming with trees can be them as signs of soil health. This helps us manage land more sustainably over time. Agroforestry is getting more important in farming that is good for the earth. Using insect-eating types of markers can help watch and manage these systems better, making them strong with lots of output too.

Detritivores are quick to show if the soil is doing well. They mirror any changes in how good the soil conditions are, or its supply of nutrients and quality of organic matter [31]. Their numbers, variety, and what they do help us see how healthy the soil environment is. For example, a large number of worms living in the soil can show that there is good air and rich nutrients. Checking bugs that eat dead plants in farming forests can give a complete look at how soil dirt breaks down, the flow of nutrients, and its strength. Besides being markers, creatures in farming with woodland systems help make good agriculture go on in different ways. Their work makes nutrients move around better, helps to form strong dirt groups, and improves soil quality [28]. Also, creatures that eat waste help control bugs and provide services for nature. This matches the aims of farming that lasts long term (Figure 4.2 and Table 4.2).

(a) (b)

FIGURE 4.2 Contrast between detritivore-depleted soil (a) showing dryness, compaction, and reduced vegetation, and detritivore-rich soil, (b) exhibiting higher organic matter, improved soil health, and thriving vegetation.

TABLE 4.2 Comparison of Soil Characteristics Between Detritivore-Depleted and Detritivore-Rich Soils

Detritivore-Depleted Soil (Left side)	Detritivore-Rich Soil (Right side)
This soil has a compact structure and devoid of visible organic matter such as leaf litter or animal residues.	This soil exhibits a crumbly structure with abundant organic matter, indicating active decomposition by detritivores.
The absence of detritivores leads to limited decomposition and nutrient cycling, resulting in a nutrient-poor environment.	The presence of healthy root structures signifies efficient nutrient uptake and plant growth.
The lack of organic matter also inhibits water retention and soil aeration, further hindering plant growth.	The diverse array of detritivores, represented in the microscopic view, suggests a balanced and functional soil ecosystem.
Visible signs of erosion suggest increased susceptibility to wind and water runoff, potentially leading to land degradation.	This soil likely exhibits improved water retention, aeration, and resistance to erosion due to the presence of organic matter and healthy microbial activity.

4.4.2 CONTRIBUTION OF DETRITIVORES TO SOIL STRUCTURE AND MOISTURE RETENTION

Detritivores, including various creatures like worms and bugs as well as tiny life forms, cause major changes to soil water flow and make it easier for plants. They also help improve the structure of dirt holding more moisture. Earthworms are a type of creature that helps make dirt healthier. They do this by digging and eating in the ground, making them key parts for keeping ecosystems balanced. Earthworm holes make a system of paths in the dirt, helping water go inside faster. They stop surface rainwater from rushing away and also play a part in controlling how wet or dry the earth is overall. It also makes it easier for roots to get deep into the soil and keeps all kinds of airflow through better. The earthworm complex building makes strong soil clumps, making the ground better shaped and filled with holes [28]. These better soil conditions are important for spreading nutrition, helping tiny organisms work and letting plants explore the dirt. All of these things make up strong soils that can handle pressure well.

The function of detritivores in keeping moisture is also very important. Organic matter decomposition helped by detritivores lets out organic water-holding compounds. When natural stuff rots, it helps to build soil carbon from living things. This acts like a sponge and makes the land hold more water [29]. Detritivores, organic waste, and soil dampness work together in a detailed dance. This complex system shows how these creatures can change the water properties of soils. Furthermore, creatures that eat garbage are very important for moving nutrients around. Detritivores, nutrient cycling, and moisture control work together. These organisms are important for keeping the overall health of natural areas in balance.

4.4.3 MITIGATION OF SOIL DEGRADATION THROUGH DETRITIVORE ACTIVITIES

Controlling soil quality damage by letting creatures eat dead stuff is very important in taking care of land the right way. This shows how necessary these animals are for keeping dirt good and making sure all living things can stand strong together. Detritivores, like worms and bugs plus tiny living things, help a lot to stop soil from getting worse. These creatures have many good roles for nature's well-being. Detritivores help to break down dead plants and other living things by eating rotting materials. This important action not only recycles necessary food parts but also improves soil shape by helping to create strong groups. The breakdown of living stuff lets out important foods that plants can take in. It also makes humus, which enhances how much water the soil holds and raises its all-round goodness [27].

Earthworms and other diggers help improve the air in the soil. They also make bigger or smaller groups of dirt clumps called aggregates. Detritivore actions make the soil better and stop it from getting too squeezed. It helps roots go into the ground easily and lets more water soak in quickly. This stops soil from wearing away and makes the ground even stronger [28]. Detritivores play a major role in nutrient recycling by breaking down organic material and releasing it back to the plants. This work not only keeps the soil fertile but also stops nutrient problems that can cause bad quality. The work of trash eaters helps plants grow, keeps tiny life forms going, and makes the dirt world healthier. They do this by moving nutrients around [8]. Besides cycling nutrients, creatures that eat dead leaves help keep soil from washing away by making the dirt solid and increasing ground cover. They help plants to grow better, and the waste they add to the dirt helps hold soil pieces together. This lowers water flow on surfaces and stops ground land from moving away easily. This also helps to keep the soil safe [29]. Detritivores also help keep water in the soil better. They do this by creating stable clusters of dirt pieces and adding more organic material to them. This ability is really important to deal with dry weather conditions. It makes sure plants have enough water and helps environments better handle changes in the climate [22].

4.5 MANAGEMENT STRATEGIES TO ENHANCE DETRITIVORE ACTIVITY

4.5.1 AGROFORESTRY PRACTICES THAT PROMOTE DETRITIVORE POPULATIONS

Agroforestry systems with many layers of plants make lots of minihomes for bugs. These detritivores live in these texts and help clean up dead stuff from roots to soil face structure (litter layer). The different kinds of plants help creatures that eat decaying matter find their place. This helps make species more diverse and they work in many ways [32]. Adding legume trees to farming and forest systems helps bugs in dead plant habits because there are better quality leaves or what we call litter around. Soil from trees, which is rich in nitrogen, helps break down materials faster. This makes more kinds of creatures that eat dead plants have a better chance to move or grow [33]. Farming with trees helps the little germs in the soil, making it a friendly place for bug eaters. Better breakdown of living stuff by tiny bugs gives animal scavengers a food-filled base, helping their numbers [34].

4.5.2 ROLE OF ORGANIC AMENDMENTS AND MULCHING IN SUPPORTING DETRITIVORE COMMUNITIES

Adding natural stuff to the ground and putting down a layer of material on top help living things that feed on dead plants or animals by affecting how small life forms in dirt work together, which affects what nutrients are available. The ones that clean up leftovers help make farming last longer by making soils healthier and keeping nature strong.

4.5.2.1 ORGANIC AMENDMENTS AND DETRITIVORE COMMUNITIES

4.5.2.1.1 Role of Compost

Compost, a common natural help for plants, brings in different tiny bug friends and food into the ground. This helps to improve detritivore activity or the use of dead things by animals [35].

4.5.2.1.2 Manure and Detritivores

Using manure makes the soil rich in natural stuff, which helps increase pest numbers and kinds because there is more food for them [36].

4.5.2.2 MULCHING AND DETRITIVORE COMMUNITIES

4.5.2.2.1 Impact of Straw Mulching

Using straw on the ground helps hold water and keeps it cool. This makes a good area for insects that live off dead things [37].

4.5.2.2.2 Cover Crop Residues

Adding cover crop waste as mulch improves pest-eating insect groups by giving them a good nutrient place and keeping the soil together [38].

4.5.3 CONSERVATION APPROACHES FOR ENHANCING DETRITIVORE HABITATS IN AGROFORESTRY SETTINGS

Taking care of nature in farming with trees can help make homes for trash eaters better. Having different types of plants, keeping dead wood and tree stumps, and adding ground cover plants are very important for helping the population of creatures that eat waste material.

4.5.3.1 ROLE OF AGROFORESTRY DIVERSIFICATION

Putting different tree kinds and layers of plants in farming fields makes the land more mixed, which helps creatures that eat dead things. The different types of plants give feast-eating animals many places to live, which helps bring more kinds and variety in species [11].

4.5.3.2 INCORPORATION OF DEADWOOD AND SNAG RETENTION

Keeping dead trees and logs in farm forestry systems helps bugs that eat decaying material. These places help keep organic stuff available and make hiding spots for creatures that get energy from dead plants or animals. This happens at different times in their life pattern [39].

4.5.3.3 INTEGRATING COVER CROPS

When farmers add cover crops to their farming practices with trees, it also makes homes for small creatures that eat dead things. This adds more stuff into the soil and helps improve how soil is put together along with giving nutrients back around [40].

4.6 CHALLENGES AND FUTURE PERSPECTIVES

4.6.1 IDENTIFYING CHALLENGES IN INTEGRATING DETRITIVORES EFFECTIVELY IN AGROFORESTRY

Adding detritivores to farm forests is very important for making farming areas more sustainable and toughened. This will also increase their total output. Detritivores, which include creatures like worms and bugs along with tiny living things, are very important because they help in breaking down dead stuff. They also aid the movement of nutrients around nature's cycle while making soil even better than before. But some problems stop the easy mixing of junk eaters into farm-forest systems.

One big problem comes from making farming stronger and using lots of pesticides, both of which can hurt the creatures that eat dead things (detritivores). Intense farming usually messes up detritivore homes and the things they normally do. Pesticides, even friendly ones for nature, can hurt how many or different kinds there are of these creatures [41]. It is important to find a middle ground between farming more and protecting scavengers, so they keep helping nature. Another issue is the big type of farming in forest systems. This may not give enough variety to garbage-eating creatures and change their communities. Monoculture places usually have a more even area. This might help some rubbish-eating creatures while not allowing others to be in the same space. We need to fix this problem by creating agroforestry plans that first think about how they will keep different types of plant life. This helps make a better place for insects and other small creatures called detritivores [11].

The best time for organic materials, like crop leftovers and ground covers, can affect detritivore work. If organic materials are used too much or wrongly, it can make things bad for those tiny living creatures that help improve dirt. This might affect the way they return nutrients to the soil and fix it up [42]. So, knowing what trash eaters like best from stuff that is made of organic materials is very important. It helps to have these things when they want it most so some animals who eat trash can do their work better. Climate change and the breaking up of habitats create extra problems for connecting detritivores, or animals that eat rotting matter, in farming with trees. When the weather changes, it might make suitable homes hard to find for trash-eating animals. Also, when places with a lot of green grow apart from each other, this can separate groups and stop them from moving between areas [31]. We need plans to include bits of garbage catchers. We must look at how climate change might affect them and fix areas for these bug communities so they keep going strong in the environment.

Education and telling people become very important parts of dealing with these problems. Farmers and landkeepers may not know enough about detritivores' jobs or the good things they do. This could make it hard to put them properly into farmer-tree systems [43]. Teaching and training that explain the jobs decomposers do for nature while giving simple ways to protect them can help people see how valuable they are. This could lead us to use more of these helpful creatures in farming practices which is good for everyone involved with growing food.

4.6.2 TECHNOLOGICAL ADVANCEMENTS AND RESEARCH GAPS IN UNDERSTANDING DETRITIVORE ROLES

We have made a big change in understanding garbage eaters or detritivores. This is because of new tools that have helped us learn more about these hidden animals that play an important role in keeping nature balanced and healthy. DNA barcoding and metagenomics, crucial molecular methods, have changed how we recognize and study detritivore groups. For example, DNA metabarcoding lets us quickly read genetic codes from nature. This gives a detailed view of different bugs that eat garbage and how they could work in their environment [44]. Also, with new technologies like satellite and drone pictures, we can see more space and time better. This helps us watch where detritivores live in a detailed way. These tools show information about how detritivores act, where they like to live, and react when land use changes are big over many places [45]. But as we study the many parts of the detritivore world, some important things still need more look at them. Even though we have made good progress in studying communities at the local level, a big problem is still figuring out how different dead material-using creatures work together and overlap. It is important to study the roles of different bug types and find out if some bugs might take over jobs in the community. This helps us predict how these bugs may react to changes around them. Also, the complex ways detritivores and tiny living things relate to each other are realized. However, we need more studies for a better understanding of how they work together [46].

Another big problem in research is the effect that world changes, like climate change and how people use land differently might have on communities of animals that eat dead things. Understanding how creatures that eat waste change and react to these changes is very important for guessing if they can last. It is also needed so we know what might happen when humans interact with ecosystems because of it. Plus, while detritivores are well known for their roles in recycling nutrients, there is a need to study how they also contribute to other important services related to ecosystems. By looking at how detritivores affect soil structure, water holding, and bug control, we will understand more about their many different impacts on the health of nature. This helps improve that knowledge. Using new technology has boosted the study of garbage eaters. It provides us with smart ways to look at a whole community and what it does in detail. However, the study's shortcomings show us that we need to look more at how different detritivore groups work and their connections with tiny life forms. We also have to understand that they are responding to worldwide changes in our world while delivering services throughout all ecosystems. Fixing these holes will not just make our knowledge about detritivore ecology better but also help in creating smarter plans for preserving and managing ecosystems that last.

4.6.3 FUTURE DIRECTIONS FOR HARNESSING DETRITIVORE POTENTIAL IN SUSTAINABLE AGROFORESTRY

Adding detritivores to good farming practices is a forward-thinking method for making soil better, helping nutrients circulate naturally, and strengthening the overall well-being of ecosystems. As we think about what the future has in store for tree farms, some important paths come up. These use new technology tools and wise-natured ideas to get more from trash-eating bugs within these systems. A key part of planning new farmland practices in the future is to include nature-focused methods. By putting together many types of trees and plants in the undergrowth, we can make areas like nature that support a lot of bug-eating animals. This creates homes for them to live happily with good biodiversity. This way is good for a variety of trash eaters and makes all nature's help better like controlling bugs, helping flowers pollinate, and improving diet quality [47]. Different functions, an important part of detritivore groups, should be a key point for future farming with trees. It is necessary to know the unique parts that different detritivore species play in recycling nutrients, breaking down organic material, and improving soil structure. By choosing and taking care of trash-eating insects based on what they do, farm woods can encourage a team effort among the bug eaters that will most help nature's services [48].

The use of new technology is helping us know and control detritivores in agriculture better. Technologies that are far away, DNA decoding, and networks of sensors give live information about detritivores. These species understand their world better in real time when the environment changes around them. These tools let farmers and people who look after land use smart ways to manage agroforestry systems. They do this by watching how detritivores (the animals that eat dead plants) work with the system over time [49]. As the weather gets worse due to global warming, farmers need more strong farming methods that can survive changing conditions. This means choosing tree types that are good for changing weather conditions, putting in farming with trees methods to protect against bad storms, and knowing how insect eaters help plan ways to adapt plants when things get hot or rainy [50]. Getting support and teaching people are vital parts of making trash eaters work well with farming forest mixes. Working with farmers, people who make laws, and local groups is very important for getting practices that help bugs eat dead stuff to use. Teaching programs that show the good effects of detritivores, which help soil health and crop growth sustainability, can make people like them more. This will likely lead to their wide use over a long time in farming practices with lots of trees [51].

The best way to use insects and small animals in eco-friendly farming practices is by taking a big-picture approach. This includes plans that protect lots of animal kinds, thinking about how different types of flora and fauna work together or apart, using cool new technology tools like tractors improved robot-like toys for scientists' research studies also engaging badly paid people who grow food crops satisfactory by combining these new ways, agroforestry systems can make detritivores more useful. This would give farms better results and help the environment in a good way without using too many resources.

4.7 CASE STUDIES AND PRACTICAL APPLICATIONS

4.7.1 REAL-WORLD EXAMPLES SHOWCASING SUCCESSFUL DETRITIVORE-BASED PRACTICES IN AGROFORESTRY

Detritivores in farming worldwide have made big changes. In Ghana, a country known for its cocoa making, ways to plant trees and farm at the same time have been put in place so they do not harm nature while still staying productive. Adding special nitrogen-fixing trees, like Inga types, with cocoa farms makes a place helpful for trash eaters like worms and bugs. These trash-eating critters are very important for breaking down dead stuff, recycling nutrients, and making soil better. The success of putting together farming and forestry helps make more cocoa and healthier soil. It also reduces the need for outside inputs in a positive way [52]. This shows how adding bugs or decay-loving creatures to agroforestry improves both farm output and environmental strength.

Colombia is using silvopastoral systems, a new farming method that puts trees and livestock grazing together. Trees like *Leucaena leucocephala* and *Gliricidia sepium* have helped to make soil healthier by breaking down dead plants and animals. These beings in nature help with nutrients moving around, making a friendly connection that helps both the dirt and makes food better for animals like cows or sheep. This all-around method not only supports fair animal farming practices but also helps to bring back nature and make the environment harder. At the same time, it proves that involving decomposing creatures in agricultural work has lots of good effects [53].

In Malawi, a way of farming that mixes trees with soil-boosting abilities like *Gliricidia sepium* has been found. This helps crop plants like maize grow healthily without harm to the environment. These counterparts help break down tree stuff that was cut off in these systems. This adds to recycling nutrients and improving soil quality. The outcome has made soil richer, given higher maize harvests, and needed fewer outside fertilizers. This all-round approach does not just deal with food safety problems; it also shows that eating stuff like insects and plants can help protect our environment. It makes sure we have strong systems to produce good things for us in the future [54].

These real-world examples show strong proof of the big changes that happen with practices involving garbage-eating creatures in farming and forestry. Detritivores or waste eaters like insects and small animals work well with nature. They help make farms in West Africa to South America better for crops while also taking care of the land and helping it stay lively with different kinds of plants and wildlife. These programs are like guides that help the whole world try to make farm forestry more enduring, eco-friendly, and lasting.

4.7.2 PRACTICAL APPLICATIONS AND LESSONS LEARNED FROM IMPLEMENTING DETRITIVORE-FOCUSED STRATEGIES

Using detritivores in farm ways has shown useful lessons on how to make the land last longer. It also shows that these methods can be used for different things. One clear example is from California's vineyards, where cover crops have been used as a plan that helps bugs eat dead plants. Adding cover plants helps protect against soil loss; it also adds natural stuff to the ground in a way that is good for insect-eating creatures. Earthworms help a lot to break apart leftovers from plants called cover crops, give important food for soil, and make the dirt better. The use of this plan has shown farmers who grow grape plants how important it is to pick cover crops that match what animals called detritivores need, understand the season changes these creatures have, and manage carefully between breaking down cover crop parts and possible competition with vine plants [55].

In Southeast Asia, especially in Indonesia, trees that fix nitrogen into the soil for other plants to use are used. This is an important example study from there too. In this case, trees like *Acacia* and *Gliricidia* help fix nitrogen in the soil. They then provide their food matter to keep bug populations alive. The things learned from these systems show how important it is to carefully choose trees for planting, understand exactly what happens with nutrients in nature, and add practices that benefit decomposers in regular farm fields. Good results from this method are larger crop yields, better soil health, and not needing as many outside fertilizers. This shows how well these strategies that focus on detritivores work in farming areas with warm weather [56].

Also, in a part of Africa called sub-Saharan and specifically Zambia, using farming methods that help slow down soil decay shows one way to make the dirt better by focusing on bugs that eat dead stuff. We call all these methods at once, which makes soil better for small animals and our cycles of nutrients. This helps us keep dirt together while making food easily available to plants again. This project's main points show that we need to change plans for different places and how much it matters when farmers learn about the good things about bug eaters. Also, it helps see how important community getting together can be in these situations. Good results include better ground construction, more water holding, and greater plant growth. These give us useful information about how well strategies focused on organisms like scavengers work in different farming areas [57].

These on-the-ground uses highlight the value of choosing and changing garbage eater tactics to meet particular farming world scenarios. The lessons from this show how important it is to think about local nature rules, encourage people in the community to get involved, and give farmers education all the time. As farming keeps changing to be more eco-friendly, using tactics that help insects and animals in the dirt can lead to better soil health. It also supports a wide range of living things found on Earth while keeping our world's food systems strong for many years ahead.

4.7.3 POTENTIAL ECONOMIC AND ENVIRONMENTAL BENEFITS OF DETRITIVORE-CENTRIC AGROFORESTRY APPROACHES

Using plentiful natural creatures in farming with forests can help. It is a kind act that last many years, which makes good parts for us now and helps fix problems we have today while growing our food. These methods focus on using creatures such as earthworms, bugs, and tiny bacteria in farm areas. Their goal is to make the soil healthier by improving ways food gets reused and making nature tougher again. From what is written in science books, these methods have many possible good effects on money and the environment. One of the main money benefits is getting better soil health and a cycle of nutrients. Detritivores have an important part in breaking down plant and animal waste, which helps to make nutrients available for soil use. This natural thing cuts down the use of fake foodstuff, maybe saving a lot for farmers [58]. Additionally, the better soil structure caused by things that eat dead plants helps to hold more water. This might lower how often we need to give water and improve using it well in farming areas [59]. A type of farming called agroforestry that focuses on eating trash can also show possible money advantages by controlling insects and stopping illnesses. Having a group of different bug eaters in an area makes the environment balanced. They help control insects that can become pests and reduce our need for strong chemical poisons [60–78]. This helps farmers lower their costs for getting stuff, but it also makes the environment safer from risks linked to using harmful chemicals.

In addition to financial benefits, farming with a focus on trash eaters can give big environmental rewards. Keeping different types of plants and animals alive is an important result. When you combine trees with many kinds of green stuff, it makes a home that lots of plant life forms can live in. This helps nature bounce back, adding to the protection of insects that pollen flowers and other creatures good for us [60]. Another important natural advantage is the capture of carbon and stopping big changes in weather conditions. Trees in farm-with-forest systems capture carbon from the air, holding it down. They take in a lot of unwanted air stuff named CO_2 from our skies. While working, those creatures that eat junk help break down living stuff and turn it into a steady kind of carbon in the earth. This two-way system helps fight climate change by lowering gas emissions that lead to global warming and adding more carbon atoms to farms [58].

4.8 CONCLUSION

In the complex pattern of farming and forestry systems, insects that eat dead plants become crucial players. They help move nutrients around in nature while also affecting the health of the soil. It becomes clear that animals that eat garbage are not just part of the ecosystem but also help make it last.

1. Diverse detritivore assemblages: At farm systems with trees, all sorts of groups called decomposers live there. These include worms that dig in the ground or creepy bugs and tiny living beings. Having many trash-eating species helps make sure nutrient cycles work well. This is because the different types of decomposers can take over when one type is not able to do it alone.
2. Impact on nutrient cycling: In tree-focused agroforestry, the mix of trees and different plants helps to create a big group of creatures called detritivores. These are animals like insects or bugs that live among leaves or dirt. Detritivores have a strong effect on how nutrients get moved around in the soil of farm forests. Detritivores, or decomposers of dead plants and animals, break down things making soil more nutritious by giving out important minerals like nitrogen. This makes the ground better for growing new life when organic matter breaks apart into pieces in nature's way.
3. Soil health indicators: Detritivores show how well the soil is doing in places where farm and forest areas overlap. Their being around and what they do shows how good the soil's living world is, making them useful for taking care of soil in a lasting way.
4. Influence on soil structure: Earthworms and other detritivores help improve soil structure by tunnel digging. They do this while

including natural stuff into the mix. There is a different way to say all of that, but using only 2000 common words in English is not possible for everyone. This makes water go deep into the ground and stay there, helping roots grow better in soil that can be full of air. It cuts down on digging up land areas to till it less often or lay across plants as protection against soil loss from wind erosion. It helps make soils filled with nutrients for crops, and this also stops drought disaster effects by making fields safer to use This can mean using less water to grow plants and better use of existing water resources, especially in areas where there is not enough drinking or clean water.

5. Organic matter decomposition dynamics: Eaters of waste break down plants and animals into smaller parts by tearing them apart, then help tiny living things interact with the broken pieces. Soil structure, water holding, and the reachability of organic stuff become better after this step.
6. Nutrient cycling and soil fertility: Detritivores help to break down organic things, freeing important nutrients such as nitrogen and phosphorus back into the dirt. Dead animals break down living things in soil and this helps to pass nutrients. This is very important for keeping the dirt rich.
7. Humus formation and carbon sequestration: We get farms with trees that also have creatures that eat trash. This helps to create a good type of soil matter for many plants. Humus, poor in carbon compounds, controls climate change by trapping or holding back a gas called "carbon" found above the ground and also trapped inside things like plants. Trees are very important because they get carbon dioxide from the air. Also, animals that eat decaying material help break down plant matter and turn it into a stable form of ground carbon when mixed with soil. This two-part process helps lock up carbon, making garbage-eating farm forestry a helpful friend in fighting against the getting hotter planet.
8. Microbial interactions and biodiversity: Trash-eating animals work with tiny bugs in the soil when they break down garbage. This connection helps tiny living creatures in the dirt, makes many types of life healthier, and kept balanced. This leads to long-lasting good soil conditions overall.

9. Optimizing agroforestry productivity: It is really important to understand the part detritivores play. This helps us make agroforestry more effective, a kind of farming where crops and trees both grow together on one land area. By creating a good environment for detritivores, people who work on this can make sure the food cycle works better. This will also improve soil structure and lead to growing more crops.
10. Pest regulation and disease suppression: Trash eaters help control pests without directly doing so by affecting the quantity and actions of natural enemies. In farming with tree systems, many different kinds of bug-eating animals lead to a well-balanced environment where bugs are naturally managed. This lessens the need for chemical pest control, reducing expenses and risks of polluting nature. Also, the breaking down of living things by creatures that eat them can help stop diseases in the soil. This gives a natural and long-lasting way to deal with sicknesses without making them worse or using harmful chemicals.
11. Resilience to environmental changes: Systems blending farming and tree planting that count the part of detritivores help them to handle environmental changes better. Detritivores help make soil strong, stop it from moving away, and keep the environment steady by supporting good ways of looking after soil that lasts for a long time.
12. Economic and environmental benefits: Adding detritivore functions to how we manage farms with trees helps the money and nature. Better soil health and more nutrients help farming easier. Saving our environment from bad changes is best with good care of the dirt long-term. Better food recycling, better soil quality, and long-lasting land care lead to more farming output. This also helps the environment adapt well. In agroforestry, besides regular harvests from crops, there can be extra items like wood for building or cooking firewood. Plus, you might find edible stuff such as fruit and nuts in these systems too but they are not made of trees alone—more things are involved here! This mix of different sources makes money grow and also helps farms become stronger by making them less dependent on only one type of crop.
13. Conservation strategies: Agroforestry helps detritivores or creatures like bugs and worms make spaces where different plants and animals can live safely. Trees and smaller plants offer spaces for trash eaters and good creatures like insects. A higher number of

wild animals helps to keep nature steady, strong, and safe. It also protects important insects that help plants grow. Saving different species also goes with world goals, helping to make living together healthy farmlands and nature spaces.

Agroforestry, the blending of trees and bushes with farming systems, is becoming known for its many high-value effects. These include more varied nature life, healthier soil quality, and a stronger ability to resist climate change impacts. However, among the factors that make agroforestry successful, how detritivores play a role is still not fully examined. Knowing that detritivores are important in farming with trees, we need to push for more studies and use of this helpful practice.

1. Acknowledging knowledge gaps: Even though we have made great strides in learning about detritivore roles, there are still things that we do not know. We need to study more about how detritivores work together, the different things they do, and their effect on what goes on around them.
2. Integrating molecular approaches: Molecular methods can separate the small parts of detritivore action linked to microbes and genes. Using fancy ways with tiny parts will make us know more about bug-eating creatures and their part in how dirt bacteria act.
3. Exploring detritivore–plant interactions: Looking at how plants and detritivores, the organisms that eat dead or decaying matter, interact is very important. It helps us understand how these creatures affect a plant's health and also their well-being in return. This type of study will help us understand how detritivores, or organisms that eat dead plant and animal material, are involved in feedback loops within agroforestry ecosystems.
4. Long-term monitoring and experiments: Watching and running tests are very important to understand the changes in detritivore populations over time because they affect how healthy soil is. These efforts will give useful knowledge about how strong and steady these processes are that use scavengers.
5. Scaling-up from micro to landscape level: It is important to connect small-scale research with larger landscape movements. Increasing research work will help us know more ways that detritivores contribute to different farm and forest setups. This understanding can be applied across various types of land areas.

6. Knowledge transfer and implementation: Turning research results into helpful instructions is very important for successful use. Workshops, programs, and working with worker people can make sharing of science knowledge easier. This helps farmers use practices that are focused on breaking down dead things in their forest farming setup.

As we begin to study detritivores in farming systems, all different types of information from nature studies, experiments using tiny organisms, and managing things must be combined. This will help make our work better for the future when looking closer at how insects break down waste into gold resources throughout farms under forests too.

KEYWORDS

- **nutrient cycling**
- **soil ecology**
- **soil quality**
- **sustainable agroforestry**

REFERENCES

1. Swift, M. J., Heal, O. W., Anderson, J. M. (1993). *Decomposition in Terrestrial Ecosystems*. Oxford University Press.
2. Moore, J. C., et al. (2014). Detritus, trophic dynamics, and biodiversity. *Ecology Letters*, 7(7), 584–600.
3. Decaëns, T., et al. (2003). Earthworms as indicators of soil biodiversity and land use impacts in the Neotropical Savannas of Colombia: taxonomic and functional diversity. *Pedobiologia*, 47(5–6), 508–514.
4. Anderson, J. M., Domsch, K. H. (1978). A physiological method for the quantitative measurement of microbial biomass in soils. *Soil Biology and Biochemistry*, 10(3), 215–221.
5. Lavelle, P., et al. (2006). Soil function in a changing world: The role of invertebrate ecosystem engineers. *European Journal of Soil Science*, 57(2), 159–193.
6. García-Palacios, P., et al. (2013). Microbial responses to simulated global change fluctuate between soil layers and are sensitive to season. *Soil Biology and Biochemistry*, 62, 115–118.

7. Bardgett, R. D., et al. (2005). Plant species and nitrogen effects on soil biological properties of temperate upland grasslands. *Functional Ecology*, 19(2), 369–377.
8. Wardle, D. A., Bardgett, R. D., Klironomos, J. N., Setälä, H., van der Putten, W. H., Wall, D. H. (2004). Ecological linkages between aboveground and belowground biota. *Science*, 304(5677), 1629–1633.
9. Coleman, D. C., et al. (2004). The microbial loop in soils. *Soil Biology and Biochemistry*, 36(3), 325–335.
10. Swift, M. J., et al. (1996). Decomposition and organic matter turnover. In: *Microbial Ecology in Sustainable Agroecosystems*. CRC Press.
11. Jose, S. (2009). Agroforestry for ecosystem services and environmental benefits: An overview. *Agroforestry Systems*, 76(1), 1–10.
12. Altieri, M. A. (2004). Linking ecologists and traditional farmers in the search for sustainable agriculture. *Frontiers in Ecology and the Environment*, 2(1), 35–42.
13. Lal, R. (2018). *Agroforestry and Environmental Sustainability*. Springer.
14. Bedano, J. C., Domínguez, J. (2016). Terrestrial isopods (Oniscidea) as biological indicators of ecological disturbance in terrestrial ecosystems. *Environmental Monitoring and Assessment*, 188(8), 426.
15. Neher, D. A. (1999). Microarthropods in soil ecology: Progress and prospects. *Biology and Fertility of Soils*, 29(4), 163–171.
16. Lavelle, P., Spain, A. V. (2001). *Soil Ecology and Ecosystem Services*. Elsevier.
17. Wall, D. H., Wardle, D. A., Callaway, R. M., Valentine, L. E. (2012). Decomposers are sensitive to the interactive effects of grazing and fertilizer in a long-term grassland soil experiment. *Oecologia*, 169(4), 803–817.
18. Gessner, A., Swaney, M. P., Sheppard, L. A. (2010). Diversity meets decomposition. *Trends in Ecology & Evolution*, 25(6), 329–336.
19. Bardgett, R. D., Wardle, D. A. (2010). Above-ground and below-ground biodiversity effects on soil function. *Nature Education*, 1(1), 15.
20. Edwards, C. A., Bohlen, P. J. (1996). *Biology and Ecology of Earthworms*. Chapman and Hall.
21. Brown, G. G., Edwards, C. A. (2001). Influence of trees on earthworm populations in soils under alley cropping. *Agroforestry Systems*, 53(1), 197–202.
22. Edwards, C. A. (2004). The importance of earthworms as key representatives of the soil fauna. In: *Earthworm Ecology* (pp. 3–11). CRC Press.
23. Hendrix, P. F., Callaham Jr, M. A., Drake, J. M. (1986). Decomposition of tree leaf litter on the soil surface. *Pedobiologia*, 29(6), 367–377.
24. Barrios, E. (2007). Soil biota, ecosystem services, and land productivity. *Ecological Economics*, 64(2), 269–285.
25. Altieri, M. A. (1999). The ecological role of biodiversity in agroecosystems. *Agriculture, Ecosystems & Environment*, 74(1–3), 19–31.
26. Swift, M. J., Heal, O. W., Anderson, J. M. (1979). *Decomposition in Terrestrial Ecosystems*. University of California Press.
27. Bohlen, P. J., Scheu, S., Hale, C. M., McLean, M. A., Migge-Kleian, S., Groffman, P. M. (2004). Non-native invasive earthworms as agents of change in northern temperate forest ecosystems. *Frontiers in Ecology and the Environment*, 2(8), 427–435.
28. Blouin, M., Hodson, M. E., Delgado, E. A., Baker, G., Brussaard, L., Butt, K. R., Dai, J., Dendooven, L., Pérès, G., Tondoh, E. J., Cluzeau, D., Brun, J. J. (2013). A review

of earthworm impact on soil function and ecosystem services. *European Journal of Soil Science*, 64(2), 161–182.
29. Lavelle, P., Bignell, D., Lepage, M., Wolters, V., Roger, P., Ineson, P., Heal, O. W., Sousa, J. P. (1997). Soil function in a changing world: The role of invertebrate ecosystem engineers. *European Journal of Soil Biology*, 33(4), 159–193.
30. Nair, P. K. R. (2011). Agroforestry for ecosystem services and environmental benefits: An overview. *Agroforestry Systems*, 81(1), 1–10.
31. Bardgett, R. D., van der Putten, W. H. (2014). Belowground biodiversity and ecosystem functioning. *Nature*, 515(7528), 505–511.
32. Silva, R. B., Oliveira, L. J., Brown, G. G. (2018). The role of detritivores in agroforestry systems: A review. *Agroforestry Systems*, 92(2), 265–275.
33. Mafongoya, P. L., Giller, K. E. (1999). Decomposition and nutrient release patterns of tree prunings and litter. *Agroforestry Systems*, 46(3), 229–250.
34. Lauber, C. L., Hamady, M., Knight, R., Fierer, N. (2009). Pyrosequencing-based assessment of soil pH as a predictor of soil bacterial community structure at the continental scale. *Applied and Environmental Microbiology*, 75(15), 5111–5120.
35. García-Gil, J. C., Plaza, C., Soler-Rovira, P., Polo, A. (2004). Long-term effects of municipal solid waste compost application on soil enzyme activities and microbial biomass. *Soil Biology and Biochemistry*, 36(11), 1767–1775.
36. Wu, H., Chen, Y., Wu, Y. (2016). Effects of organic manure on soil microorganisms and enzyme activities during maize (Zea mays L.) cultivation. *Applied Soil Ecology*, 108, 207–215.
37. Mbagwu, J. S. C., Abalu, G. I. O., Iwuafor, E. N. O. (2004). Effects of straw mulch on soil properties, growth, and yield of maize in a humid ultisol in southeastern Nigeria. *Plant and Soil*, 263(1–2), 151–160.
38. Ngwira, A. R., Thierfelder, C., Lambert, D. M., Kassie, G. T. (2014). Conservation agriculture systems have contrasting effects on maize and cowpea residue decomposition. *Soil and Tillage Research*, 144, 184–194.
39. Lachat, T., Bütler, R., Duelli, P. (2018). Snags and retention trees are keystone structures for saproxylic beetles and other insects in Swiss lowland agricultural forests. *Forest Ecology and Management*, 417, 180–188.
40. Gundale, M. J., Metlen, K. L., Fiedler, C. E. (2017). Restoring arbuscular mycorrhizal fungal communities in an annual crop: Impact on plant growth and soil microbial communities. *Ecological Applications*, 27(3), 858–871.
41. Marshall, E. J. P., West, T. M., Kleijn, D. (2006). Impacts of an agri-environment field margin prescription on the flora and fauna of arable farmland in different landscapes. *Agriculture, Ecosystems & Environment*, 113(1–4), 36–44.
42. Mäder, P., Fliessbach, A., Dubois, D., Gunst, L., Fried, P., Niggli, U. (2002). Soil fertility and biodiversity in organic farming. *Science*, 296(5573), 1694–1697.
43. Gliessman, S. R. (2014). *Agroecology: The Ecology of Sustainable Food Systems*. CRC Press.
44. Taberlet, P., Coissac, E., Pompanon, F., Brochmann, C., Willerslev, E. (2012). Towards next-generation biodiversity assessment using DNA metabarcoding. *Molecular Ecology*, 21(8), 2045–2050.
45. Pettorelli, N., Safi, K., Turner, W. (2014). Satellite remote sensing for applied ecologists: Opportunities and challenges. *Journal of Applied Ecology*, 51(4), 839–848.

46. Schmidt, O., Dyckmans, J. (2015). The use of stable isotope probing to identify key microbial genera driving biologically mediated soil stabilization. *Global Change Biology*, 21(11), 3358–3368.
47. Díaz, S., Pascual, U., Stenseke, M., Martín-López, B., Watson, R. T., Molnár, Z., ... & Shirayama, Y. (2019). Assessing nature's contributions to people. *Science*, 366(6462), 1040–1042.
48. Woodcock, B. A., Heard, M. S., Vanbergen, A. J., Roy, D. B. (2016). Assessing the impacts of crop protection on non-target invertebrate communities. *Annals of Applied Biology*, 168(2), 143–152.
49. Mouillot, D., Bellwood, D. R., Baraloto, C., Chave, J., Galzin, R., Harmelin-Vivien, M., Kulbicki, M., Lavergne, S., Lavorel, S., Mouquet, N., Timothy Paine, C. E., Renaud, J., Thuiller, W. (2013). Rare species support vulnerable functions in high-diversity ecosystems. *PLoS Biology*, 11(5), e1001569.
50. Dawson, I. K., Vinceti, B., Weber, J. C., Neufeldt, H., Russell, J., Lengkeek, A., Antoine, K., Kindt, R., Barnekow Lillesø, J.-P., Kindt, R. (2019). Climate change and tree genetic resource management: Maintaining and enhancing the productivity and value of smallholder tropical agroforestry landscapes. A review. *Agroforestry Systems*, 93(2), 519–541.
51. Pretty, J. (2018). Intensification for redesigned and sustainable agricultural systems. *Science*, 362(6417).
52. Asare, R., Afari-Sefa, V., Penczykowski, R. M. (2012). The role of biodiversity for agroforestry options in cocoa landscapes. *Agroforestry Systems*, 84(3), 393–407.
53. Iñiguez-Mendoza, L., Ku-Vera, J. C., Solorio-Sánchez, F. J., Torres-Acosta, J. F. J., Aguilar-Caballero, A. J. (2017). Forage quality and its voluntary intake by goats in a silvopastoral system with high and low densities of Leucaena leucocephala. *Agroforestry Systems*, 91(3), 485–496.
54. Mhango, W. G., Snapp, S., Kanyama-Phiri, G. Y. (2020). Biological nitrogen fixation of pigeonpea and groundnut: Quantifying response across 18 farm sites in northern Malawi. In *Just Enough Nitrogen: Perspectives on How to Get There for Regions with Too Much and Too Little Nitrogen* (pp. 139–153). Cham: Springer International Publishing.
55. Nicholls, C. I., Parikh, S. J., Altieri, M. A. (2010). The effects of a polyculture cover crop on vineyard pest populations. *Journal of Sustainable Agriculture*, 34(2), 210–233.
56. Hairiah, K., Sulistyani, H., Roshetko, J. M., van Noordwijk, M. (2011). Options for smallholder farmers in Southeast Asia for integrating food into landscapes while reducing greenhouse gas emissions. *Mitigation and Adaptation Strategies for Global Change*, 16(1), 111–127.
57. Thierfelder, C., Rusinamhodzi, L., Ngwira, A. R. (2017). Conservation agriculture in southern Africa: Advances in knowledge. *Renewable Agriculture and Food Systems*, 32(3), 200–218.
58. Lal, R. (2015). Restoring soil quality to mitigate soil degradation. *Sustainability*, 7(5), 5875–5895.
59. Nair, P. K. R. (2012). Climate change mitigation: A low-hanging fruit of agroforestry. In: *Agroforestry—The Future of Global Land Use* (pp. 211–234). Springer.

60. Perfecto, I., Vandermeer, J. (2010). The agroecological matrix as an alternative to the land-sparing/agriculture intensification model. *PNAS*, 107(13), 5786–5791.
61. Bhoi, T. K., Samal, I., Majhi, P. K., Komal, J., Mahanta, D. K., Pradhan, A. K., Saini, V., Nikhil Raj, M., Ahmad, M. A., Behera, P. P., Ashwini, M. (2022). Insight into aphid mediated Potato Virus Y transmission: A molecular to bioinformatics prospective. *Frontiers in Microbiology*, 13, 1001454.
62. Mahanta, D. K., Jangra, S., Ghosh, A., Sharma, P. K., Iquebal, M. A., Jaiswal, S., Baranwal, V. K., Kalia, V. K., Chander, S. (2022). Groundnut bud necrosis virus modulates the expression of innate immune, endocytosis, and cuticle development-associated genes to circulate and propagate in its vector, Thrips palmi. *Frontiers in Microbiology*, 13, 773238.
63. Bhatnagar, S., Mahanta, D. K., Vyas, V., Samal, I., Komal, J., Bhoi, T. K. (2023). Storage pest management with nanopesticides incorporating silicon nanoparticles: A novel approach for sustainable crop preservation and food security. *Silicon*, 16(2), 1–13.
64. Mahanta, D. K., Komal, J., Samal, I., Bhoi, T. K., Dubey, V. K., Pradhan, K., Nekkanti, A., Gouda, M. N. R., Saini, V., Negi, N., Bhateja, S., Jat, H. K., Jeengar, D. (2023a). Nutritional aspects and dietary benefits of "Silkworms": Current scenario and future outlook. *Frontiers in Nutrition*, 10, 1121508.
65. Samal, I., Bhoi, T. K., Vyas, V., Majhi, P. K., Mahanta, D. K., Komal, J., Singh, S., Dinesh Kumar, P. V., Acharya, L. K. (2023a). Resistance to fungicides in entomopathogenic fungi: Underlying mechanisms, consequences, and opportunities for progress. *Tropical Plant Pathology*, 49, 5–17.
66. Mahanta, D. K., Bhoi, T. K., Komal, J., Samal, I., Nikhil, R. M., Paschapur, A. U., Singh, G., Kumar, P. V. D., Desai, H. R., Ahmad, M. A., Singh, P. P., Majhi, P. K., Mukherjee, U., Singh, P., Saini, V., Shahanaz, Srinivasa, N., Yele, Y. (2023b). Insect-pathogen crosstalk and the cellular-molecular mechanisms of insect immunity: Uncovering the underlying signaling pathways and immune regulatory function of non-coding RNAs. *Frontiers in Immunology*, 14, 1169152.
67. Samal, I., Bhoi, T. K., Majhi, P. K., Murmu, S., Pradhan, A. K., Kumar, D., Saini, V., Paschapur, A. U., Raj, M. N., Ankur, Manik, S., Behera, P. P., Mahanta, D. K., Komal, J., Alam, P., Balawi, T. A. (2023b). Combatting insects mediated biotic stress through plant associated endophytic entomopathogenic fungi in horticultural crops. *Frontiers in Plant Science*, 13, 1098673.
68. Komal, J., Desai, H. R., Samal, I., Mastinu, A., Patel, R. D., Kumar, P. D., Majhi, P. K., Mahanta, D. K., Bhoi, T. K. (2023). Unveiling the genetic symphony: Harnessing CRISPR-Cas genome editing for effective insect pest management. *Plants*, 12(23), 3961.
69. Samal, I., Bhoi, T. K., Mahanta, D. K., Komal, J. (2023c). Establishing the role of silicon (Si) in plant resistance to insects: A bibliometric approach. *Silicon*, 16, 2119–2128.
70. Majhi, P. K., Samal, I., Bhoi, T. K., Pattnaik, P., Pradhan, C., Gupta, A., Mahanta, D. K., Senapati, S. K. (2023). Climate-smart agriculture: An integrated approach to address climate change and food security. In: *Climate Change and Insect Biodiversity* (pp. 221–245). CRC Press.

71. Mahanta, D. K., Samal, I., Komal, J., Bhoi, T. K., Majhi, P. K., Ahmad, M. A. (2023). Understanding anthropogenic climate change, its consequences on insect pests, and techniques in forecasting and monitoring pest dynamics: A contemporary scenario. In: *Climate Change and Insect Biodiversity* (pp. 44–67). CRC Press.
72. Bhoi, T. K., Samal, I., Mahanta, D. K., Komal, J., Jinger, D., Sahoo, M. R., Achary, G. C., Nayak, P., Sunani, S. K., Saini, V., Raghuraman, M., Singh, S. (2023). Understanding how silicon fertilization impacts chemical ecology and multitrophic interactions among plants, insects and beneficial arthropods. *Silicon*, 15(6), 2529–2549.
73. Bhoi, T. K., Samal, I., Mahanta, D. K., Komal, J., Majhi, P. K., Ankur. (2024). Exploring Melatonin's potential as an alternative strategy for protecting plants from biotic stresses. In: *Melatonin in Plants: A Pleiotropic Molecule for Abiotic Stresses and Pathogen Infection* (pp. 223–242). Springer Nature Singapore: Singapore.
74. Samal, I., Bhoi, T. K., Pradhan, A. K., Mahanta, D. K. (2023d). Plantix app: A success story of artificial intelligence in plant protection. *Van Sangyan*, 24.
75. Mahanta, D. K., Bhoi, T. K., Komal, J., Samal, I., Mastinu, A. (2024). Spatial, spectral and temporal insights: Harnessing high-resolution satellite remote sensing and artificial intelligence for early monitoring of wood boring pests in forests. *Plant Stress*, 11, 100381.
76. Samal, I., Mahanta, D. K., Bhoi, T. K., Komal, J., Jatav, H. S., Jatav, S. S., Sathyanarayana, E. (2024). *Approaches of Biochar in Ecosystem Management: Current Scenario and Future Perspectives*. IntechOpen.
77. Vyas, V., Bhoi, T. K., Samal, I., Singh, S., Mahanta, D. K. (2024). Revitalizing degraded soils with agroforestry interventions: Opportunities, challenges, and future direction. In: *Agroforestry to Combat Global Challenges: Current Prospects and Future Challenges* (pp. 529–549).
78. Samal, I., Bhoi, T. K., Mahanta, D. K., Komal, J., Singh, S. (2024). Biorational pest management: Potentials, unintended consequences, and future concerns. *Biorationals and Biopesticides: Pest Management* (pp. 47–75).

CHAPTER 5

Fostering Microbe Synergy in Agroforestry: Unveiling the Power of Microbial Partnerships for Sustainable Agricultural Practices

LELLAPALLI RITHESH,[1] POOJA,[2*] PRASHANT CHAUHAN,[2] and ABHISHEK KUMAR[3]

[1]*Department of Plant Pathology, Kerala Agricultural University, Thiruvananthapuram, Kerala, India*

2Department of Plant Pathology, CCS Haryana Agricultural University, Hisar, Haryana, India

[3]*Department of Agriculture, Maharishi Markandeshwar (Deemed to be University), Mullana, Ambala, Haryana, India*

*Corresponding author

ABSTRACT

Agroforestry integrates crops and trees within the same landscape to maximize ecological interactions and benefits. A key advantage is biodiversity conservation, which is vital amid ongoing deforestation and soil degradation. Microbial biodiversity, functioning as the core of agroforestry ecosystems, supports sustainable production of food, fiber, and nontimber forest products. Microorganisms play a crucial role in improving soil health, facilitating nutrient cycling, and enhancing plant resilience. Symbiotic and free-living nitrogen-fixing bacteria, phosphate-solubilizing organisms, and stress-tolerant microbes are particularly significant for sustaining productivity under both biotic and abiotic stresses. Microbial diversity

Agroforestry and Climate Change Mitigation: Harnessing Multitrophic Interactions for Sustainable Plant Protection and Ecosystem Resilience. Hanuman Singh Jatav, Vishnu D. Rajput, Ipsita Samal, Tanmaya Kumar Bhoi (Eds.)

DOI: 10.1201/9781779641397-5

in agroforestry systems is shaped by plant species composition, habitat conditions, and management practices, which together promote ecosystem services essential for sustainable development. Research increasingly highlights the potential of microbial consortia to strengthen plant growth and soil fertility while reducing environmental stress. However, despite abundant knowledge, microbial innovations remain underutilized in practical agroforestry. Harnessing the functional potential of the plant microbiome offers a promising pathway to improve agroforestry health and resilience. This chapter therefore emphasizes the importance of understanding plant–microbe, plant–plant, and microbe–microbe interactions to enhance soil health and achieve sustainable agroforestry systems..

5.1 INTRODUCTION

Agroforestry, a sustainable land management practice that integrates trees and shrubs with agricultural crops and/or livestock, has emerged as a multifaceted approach to address the challenges of modern agriculture [12]. By combining the benefits of both forestry and agriculture, agroforestry contributes significantly to biodiversity conservation, soil health improvement, and overall ecosystem resilience. Agroforestry systems promote biodiversity conservation by creating diverse ecological niches. The integration of trees and crops provides habitat for various flora and fauna, fostering a balanced ecosystem. This is particularly crucial in the face of global biodiversity loss, where traditional agricultural practices often contribute to habitat destruction and monoculture-related ecological imbalances [32]. Agroforestry mitigates these issues by supporting a variety of habitats that accommodate a wide array of plant and animal species.

Moreover, the presence of trees in agroforestry systems enhances biodiversity above and below ground. Tree canopies offer nesting sites for birds, while their roots create microenvironments conducive to the proliferation of beneficial soil organisms. This dynamic interaction between trees, crops, and associated organisms underscores the intricate web of life within agroforestry landscapes [26]. Soil health is a cornerstone of sustainable agriculture, and agroforestry plays a pivotal role in enhancing soil fertility and structure. The roots of trees in agroforestry systems anchor the soil, preventing erosion, and promoting water retention. As the tree roots penetrate the soil, they create channels for water infiltration, enhancing overall soil structure [6]. The organic matter from fallen leaves and other plant residues further contributes to nutrient cycling, fostering a nutrient-rich environment for crops [10].

Microorganisms, particularly bacteria and fungi, constitute a vital component of soil biodiversity. The role of microbial diversity in sustainable agroforestry is profound. Microbes contribute to nutrient cycling, disease suppression, and overall soil fertility [4]. In agroforestry systems, the intricate relationships between plants and microbes become especially significant. Plant roots release exudates that attract specific microbial communities, forming symbiotic partnerships that enhance nutrient uptake and disease resistance [19]. Understanding the plant–microbe synergy in agroforestry is crucial for optimizing the benefits of microbial partnerships. Certain tree species in agroforestry systems form mycorrhizal associations, where the plant roots and mycorrhizal fungi engage in a mutualistic exchange of nutrients. This not only enhances nutrient absorption by plants but also contributes to soil structure improvement [31]. Additionally, the microbial communities associated with agroforestry plants play a pivotal role in suppressing pathogens and promoting overall plant health [3].

Agroforestry stands as a sustainable approach that harmonizes agriculture, forestry, and biodiversity conservation. Microbial diversity, acting as a linchpin in agroecosystems, forms intricate partnerships with plants, contributing to soil health and overall ecosystem resilience. Recognizing and harnessing the synergies between plants and microbes in agroforestry is paramount for realizing the full potential of this integrated land management strategy. This chapter provides an overview of microbial communities in various plant habitats within agroforestry and highlights the pivotal importance of microbial diversity for promoting plant growth and maintaining soil health.

5.2 FOUNDATIONS OF MICROBIAL ECOLOGY

Microbial ecology is a fascinating field that delves into the intricate relationships between microorganisms and their environments. At its core, microbial ecology seeks to understand the basics of microbial communities, examining the dynamics and interactions among various microorganisms. The foundations of microbial ecology encompass the exploration of microbial communities, their roles in soil health, and their significance in agroforestry systems. These insights not only contribute to our understanding of fundamental ecological processes but also provide valuable knowledge for developing sustainable practices that benefit both the environment and human societies.

5.2.1 BASICS OF MICROBIAL COMMUNITIES

Microbial communities are complex networks of microscopic organisms, including bacteria, archaea, fungi, and viruses (Figure 5.1). These communities exist virtually everywhere, from the depths of oceans to the soil beneath our feet. Understanding the basics of microbial communities involves exploring the diversity of microorganisms, their interactions, and the factors that shape their composition. The roles of these communities extend beyond simple coexistence, as they actively contribute to ecological processes such as nutrient cycling, organic matter decomposition, and symbiotic relationships with higher organisms [8]. Microbial communities are not static entities; they adapt and respond to environmental changes. Factors like temperature, pH, nutrient availability, and the presence of other microorganisms can significantly influence the structure and function of microbial communities [17]. Researchers use advanced techniques, such as metagenomics and molecular biology tools, to unravel the complexities of microbial interactions and identify key players within these communities.

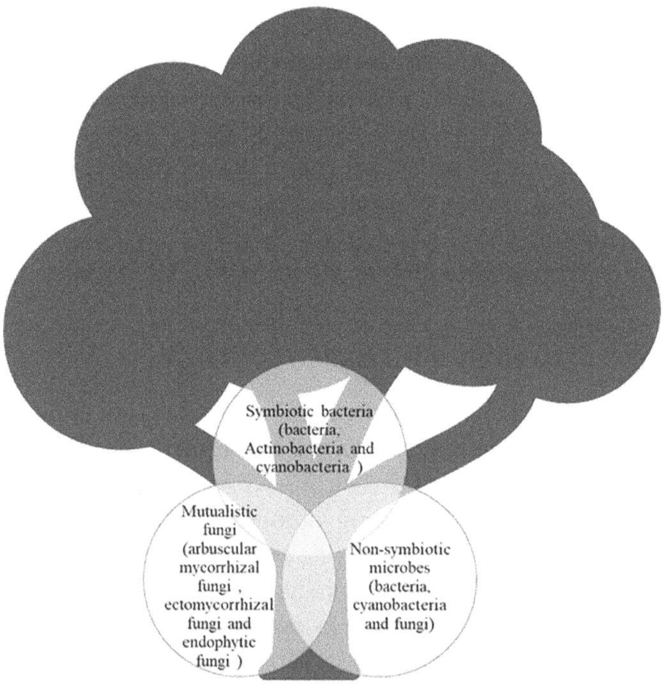

FIGURE 5.1 The diverse microbial community associated with agroforestry system.

5.2.2 ROLE OF MICROORGANISMS IN SOIL HEALTH

One of the critical domains of microbial ecology is the study of microorganisms in soil ecosystems. Microorganisms play a pivotal role in maintaining soil health, influencing its fertility, structure, and overall productivity. Bacteria and fungi, for example, contribute to nutrient cycling by decomposing organic matter and releasing essential nutrients for plant growth. Some microorganisms form symbiotic relationships with plant roots, aiding in nutrient uptake and enhancing the plants' resilience to environmental stressors. Understanding the dynamics of microbial communities in soil allows scientists and agricultural experts to develop sustainable practices that promote soil health. This knowledge is crucial for addressing global challenges such as food security and climate change, as healthy soils contribute to increased crop yields, carbon sequestration, and overall ecosystem stability.

5.2.3 KEY MICROBIAL PLAYERS IN AGROFORESTRY

Agroforestry, an innovative land use system that integrates trees and shrubs with crops and livestock, relies heavily on microbial interactions for success. Microorganisms in the soil contribute to nutrient cycling and facilitate the establishment of beneficial relationships between plants and microbes. Nitrogen-fixing bacteria, for instance, play a key role in providing plants with essential nutrients, nitrogen, promoting the growth of both trees and crops. Moreover, mycorrhiza form symbiotic associations with plant roots, enhancing nutrient absorption and water uptake. These microbes contribute to the overall resilience of agroforestry systems, making them more adaptable to changing environmental conditions. Studying the key microbial players in agroforestry systems allows researchers and practitioners to optimize land management practices, improving the sustainability and productivity of these integrated systems.

5.3 MICROBIAL DIVERSITY IN AGROFORESTRY

The intricate relationships between plants and their microbial communities play a crucial role in shaping the agroforestry systems. Agroforestry systems encompass diverse plant habitats, ranging from tree canopies to soil rhizospheres. Each of these habitats harbors unique microbial communities that contribute to nutrient cycling, disease suppression, and overall ecosystem resilience [22]. In the tree canopy, for instance, epiphytic microbes play a vital role in nutrient acquisition and protection against pathogens [38]. The rhizosphere, on the other hand, is rich in root-associated microbes that form symbiotic relationships with plants, enhancing nutrient uptake and promoting plant growth [41].

Recent studies employing advanced molecular techniques, such as metagenomics and next-generation sequencing, have provided deeper insights into the microbial diversity within agroforestry systems. These methods enable the identification of specific microbial taxa and their functional roles, elucidating the complex interactions between plants and microbes [25]. The intricate relationships between plants and microbial communities within agroforestry systems are essential for promoting plant growth and maintaining soil health. Understanding the microbial diversity in various plant habitats allows for the optimization of agroforestry practices to harness the full potential of these interactions. As research in this field progresses, the integration of microbial considerations into agroforestry management strategies holds promise for enhancing the sustainability and productivity of agroecosystems.

Microbial diversity is a cornerstone in agroforestry systems, playing a multifaceted and critical role in promoting plant growth, maintaining soil health, and ensuring the overall sustainability of agricultural ecosystems. The implications of microbial diversity are as follows:

5.3.1 NUTRIENT CYCLING AND AVAILABILITY

Microbes in agroforestry systems are key players in nutrient cycling. They contribute significantly to the decomposition of organic matter, breaking down complex compounds into simpler forms and releasing essential nutrients such as nitrogen, phosphorus, and potassium. This process enhances soil fertility and nutrient availability, providing plants with the building blocks necessary for their growth and development. The intricate relationship between microbes and nutrient cycling in agroforestry systems fosters a dynamic and well-balanced nutrient environment, ultimately promoting healthier and more robust plant growth.

5.3.2 DISEASE SUPPRESSION

Diverse microbial communities serve as natural allies in disease suppression within agroforestry systems. Through mechanisms such as competitive exclusion and the production of antimicrobial compounds, microbes act as a frontline defense against harmful pathogens. In the integrated environment of agroforestry, microbial diversity contributes to a balanced ecosystem that reduces the prevalence of plant diseases. This not only benefits individual plants but also contributes to the overall resilience of the agroforestry system, minimizing the need for synthetic interventions.

5.3.3 SOIL STRUCTURE AND EROSION CONTROL

Microbes are instrumental in maintaining soil structure and preventing erosion, critical considerations in agroforestry systems. Fungal hyphae, for example, form intricate networks that bind soil particles together, creating stable aggregates. This network enhances soil structure, making it more resistant to erosion and promoting water retention. In agroforestry, where diverse vegetation is integrated, microbial contributions to soil stability become especially important for mitigating erosion risks. The combined effect of roots and microbial networks helps anchor the soil, preventing nutrient loss and ensuring a stable substrate for plant growth.

5.3.4 CLIMATE RESILIENCE

Microbial diversity in agroforestry systems contributes significantly to climate resilience. The adaptability of plants to changing environmental conditions is facilitated by the microbial community, which helps to enhance stress tolerance. In the face of climate change-related challenges such as drought, extreme temperatures, or fluctuations in precipitation, the diverse microbial population collaborates with plant roots to mitigate stress impacts. This microbial-assisted adaptability is crucial for maintaining the productivity and sustainability of agroforestry systems under dynamic climatic conditions.

5.4 SYMBIOTIC RELATIONSHIPS IN AGROFORESTRY

One of the key factors contributing to the success of agroforestry systems is the role played by nitrogen-fixing bacteria. These bacteria play a crucial role in nitrogen cycling, both in symbiotic and nonsymbiotic relationships, impacting plant health and soil fertility. Understanding the mechanisms and implications of nitrogen fixation in agroforestry is essential for optimizing sustainable agricultural practices [18].

5.4.1 SYMBIOTIC NITROGEN-FIXING BACTERIA

Symbiotic nitrogen-fixing bacteria form mutualistic relationships with certain plants, forming nodules on their roots. The most well-known example of this symbiosis is the partnership between leguminous plants and Rhizobia bacteria. The bacteria infect the root nodules, where they convert atmospheric nitrogen (N_2) into a form (ammonia, NH_3) that the plant can use for its growth [37]. This process not only provides the host plant with a direct source of nitrogen but also contributes to the overall nitrogen content of the soil. Leguminous trees, such as Acacia and Robinia, commonly used in agroforestry, can enhance soil fertility through this symbiotic relationship.

5.4.2 NONSYMBIOTIC NITROGEN-FIXING BACTERIA

In addition to symbiotic relationships, nonsymbiotic nitrogen-fixing bacteria, such as *Azotobacter* and *Clostridium*, also play a crucial role in nitrogen fixation [1]. These free-living bacteria are not confined to root nodules but exist in the soil, fixing atmospheric nitrogen directly. They contribute to the nitrogen pool in the soil, making it available for plant uptake. In agroforestry systems, the presence of these nonsymbiotic bacteria helps maintain a steady supply of nitrogen to support the growth of both trees and associated crops.

5.4.3 ROLE IN NITROGEN CYCLING

Nitrogen-fixing bacteria contribute significantly to nitrogen cycling in agroforestry systems. The symbiotic and nonsymbiotic bacteria fix atmospheric nitrogen, converting it into forms that plants can utilize. This process enriches the soil with nitrogen, enhancing nutrient availability for plants. The nitrogen-fixed by these bacteria is released into the soil through root exudates, decomposing plant residues, and microbial activity, further promoting nutrient cycling within the agroecosystem.

5.4.4 IMPLICATIONS FOR PLANT HEALTH AND SOIL FERTILITY

The presence of nitrogen-fixing bacteria in agroforestry has profound implications for plant health and soil fertility. The nitrogen supplied by these bacteria serves as a natural fertilizer, reducing the need for external nitrogen inputs. This sustainable approach contributes to the long-term health of the agroforestry system by maintaining soil fertility and promoting robust plant growth. Improved nitrogen availability positively affects crop yields, making agroforestry a resilient and ecologically sound agricultural practice. Nitrogen-fixing bacteria, both symbiotic and nonsymbiotic, are integral to the success of agroforestry systems [21]. By enhancing nitrogen cycling, these bacteria contribute to improved plant health and soil fertility. The sustainable and mutually beneficial relationships formed between nitrogen-fixing bacteria and plants make agroforestry a promising approach for meeting the growing demand for food production while minimizing environmental impact. Implementing and optimizing agroforestry practices that harness the potential of nitrogen-fixing bacteria can play a vital role in sustainable agriculture.

5.5 PHOSPHATE-SOLUBILIZING ORGANISMS (PSOS) AND SOIL HEALTH

PSOs play a pivotal role in enhancing nutrient availability in soil, thereby contributing significantly to soil health and promoting plant growth. In the context of agroforestry systems, understanding the significance of these organisms becomes crucial for sustainable and productive agricultural practices.

5.5.1 SIGNIFICANCE OF PSOS

Phosphorus is an essential nutrient for plant growth and development, but it often exists in forms that are unavailable to plants. PSOs, including bacteria and fungi, possess the unique ability to convert insoluble phosphates into soluble forms through various enzymatic and metabolic processes. This solubilization makes phosphorus more accessible to plants, leading to improved nutrient uptake and utilization. Numerous studies highlight the positive impact of PSOs on nutrient availability.

For instance, a study by Tang et al. [34] demonstrated the phosphate-solubilizing activity of rhizospheric bacteria, emphasizing their potential to enhance soil fertility. This increased availability of phosphorus positively influences other nutrient cycles, creating a cascade effect that improves overall soil health.

5.5.2 CONTRIBUTIONS TO SOIL HEALTH

In agroforestry systems, which integrate trees and shrubs with crops, the role of PSOs becomes even more critical. Trees and plants in these systems often have diverse nutrient requirements, and the symbiotic relationship between PSOs and plant roots aids in nutrient acquisition. The exploration of PSOs reveals their crucial role in improving nutrient availability, soil health, and plant growth in agroforestry systems. As we continue to seek sustainable agricultural practices, understanding and harnessing the potential of PSOs offer a promising avenue for optimizing nutrient use efficiency and promoting ecosystem resilience. This collaboration contributes to the overall health of the soil in the following ways.

5.5.2.1 IMPROVED NUTRIENT CYCLING

PSOs play a pivotal role in nutrient cycling within agroforestry systems. Through their ability to release phosphorus from organic matter, they make this essential nutrient more accessible to plants. This improved cycling of phosphorus contributes to a more balanced nutrient supply in the soil, creating an environment conducive to plant growth and overall soil health. The enhanced availability of phosphorus, a vital nutrient for plant development, fosters a healthier soil ecosystem and reduces the need for external phosphorus inputs.

5.5.2.2 ENHANCED SOIL STRUCTURE

The activity of PSOs has a positive correlation with improved soil structure in agroforestry systems. As plants efficiently utilize nutrients, they contribute to robust root development and stimulate microbial activity in the rhizosphere. This increased root growth and microbial presence, facilitated by PSOs, enhance soil aggregation and stability. The root's penetration and the secretions from microbial activity contribute to the formation of soil aggregates, creating a more porous and well-structured soil. Improved soil structure benefits water infiltration, root exploration, and nutrient movement, ultimately contributing to a healthier and more productive agroforestry soil environment [7].

5.5.2.3 DISEASE SUPPRESSION

In addition to their role in nutrient availability, PSOs exhibit a capacity for disease suppression in plants within agroforestry systems. By colonizing the rhizosphere, these organisms establish a competitive advantage over pathogenic microorganisms. This competitive exclusion limits the growth and activity of harmful pathogens, providing a natural defense mechanism for plants against diseases. The presence of PSOs in the root zone creates a protective barrier, reducing the susceptibility of plants to soil-borne diseases. This not only contributes to the overall health of individual plants but also supports the resilience of the entire agroforestry system [23].

5.6 MICROBIAL RESPONSES TO BIOTIC AND ABIOTIC STRESS

Microbes play a crucial role in shaping the resilience of ecosystems, particularly in agroforestry, where the interaction between plants, pests, and environmental stressors is complex. This chapter explores the microbial responses to both biotic and abiotic stress, focusing on their contributions to resilience against pests and environmental challenges in the context of agroforestry. Additionally, case studies in Table 5.1 will illustrate successful microbial interventions in agroforestry systems (Figure 5.2). Microbial responses to biotic and abiotic stress are fundamental to building resilience in agroforestry systems. Harnessing the potential of beneficial microbes offers sustainable solutions for pest management and environmental stress mitigation. By promoting microbial diversity and understanding microbial interactions, agroforestry can thrive in the face of diverse stressors, contributing to global food security and environmental sustainability.

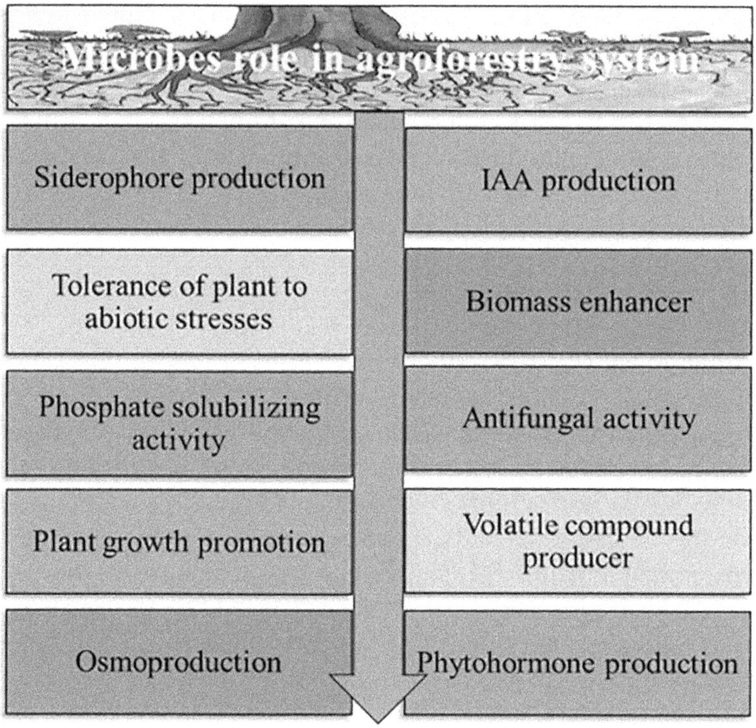

FIGURE 5.2 Role of microbial community in the agroforestry system.

5.6.1 MICROBIAL CONTRIBUTIONS TO RESILIENCE

Microorganisms, including bacteria, fungi, and viruses, are integral components of the soil microbiome, contributing significantly to plant health and ecosystem resilience. In response to biotic stressors such as pests, certain microbes exhibit antagonistic activities against pathogens, acting as natural biopesticides. For example, mycorrhizal fungi establish symbiotic relationships with plant roots, enhancing nutrient uptake and conferring resistance to soilborne pathogens. These microbial interactions can mitigate the impact of pests on crops, promoting a healthier and more resilient agroforestry system.

Moreover, microbes also play a crucial role in alleviating abiotic stressors, such as drought, salinity, and nutrient deficiency. Some bacteria have been identified for their ability to produce stress-tolerant compounds or induce systemic resistance in plants, enhancing their adaptability to

TABLE 5.1 Role of Beneficial Microbial Community in the Agroforestry System

S. No.	Microorganism	Source	Activity	Reference
1	*Stenotrophomonas* strain 169	Aerial parts of poplar	PGP activities, siderophore production, phosphate mobilization, IAA production, tolerance of plant to abiotic stresses	[36]
2	*Enterobacter* sp. strain 638	Willow and poplar trees	Biomass enhancer	[33]
3	*Burkholderia* sp.	Roots and Rhizospheric soil from apple trees	Antagonistic activity against C. gloeosporioides	[27]
4	Burkholderia, Pseudomonas, Serratia, Enterobacter, Pantoea, Klebsiella, Acinetobacter, and Bacillus	Soil from Pine-forest	Phosphate solubilizing activity, antifungal activity against fusarium species, siderophore production, phytase production, IAA production	[24]
5	*Stenotrophomonas maltophilia* Sm97	*Pistacia atlan-tica* L.	Siderophore production	[14]
			Plant growth promotion	
6	*Paenibacillus* sp. NEB	*Morinda citrifolia* L.	antagonistic activity against smut pathogen	[20]
7	*Phialocephala sphareoides*	Norway spruce	Inhibit all phytopathogens, Promoted the root shoot ratio of Norway spruce seedlings	[35]
8	*Stenotrophomonas maltophilia*	*Equisetum arvense*	Ability to synthesize IAA-like compounds, Highest metabolic activity	[39]
9	*Paenibacillus* sp. RRB2, *Azospirillum brasilense* RRAK5, *Bacillus subtilis* subsp. subtilis RRD8, *Burholdria kururiensis* RRAK1, and *Pseudomonas stutzeri* RRAN2	Nanmangalam reserve forest	Biomass enhancer	[29]

TABLE 5.1 *(Continued)*

S. No.	Microorganism	Source	Activity	Reference
10	*Burkholderia territorii* A63, *Burkholderia metallica* A53, *Pseudomonas geniculata* 95, and *Bacillus pumilus* 104	Citrus rhizosphere	Antimicrobial potential against citrus pathogen under dual culture test, Phosphate solubilizer, Siderophore producer, Volatile compound producer, Osmoproduction, Phytohormone production	[30]
11	*Stenotrophomonas maltophilia RMC6*	*Mucuna utilis* var. *capitata* L. (Safed Kaunch)	Plant growth promoter (IAA, phosphate solubilization and ACC deaminase, siderophore production)	[2]

adverse environmental conditions. Understanding the intricate network of microbial interactions in the soil can aid in the development of sustainable agroforestry practices that promote resilience against a range of stressors.

5.6.2 MYCORRHIZAL FUNGI IN AGROFORESTRY SYSTEMS

A study conducted by Fahad et al. [15] investigated the role of mycorrhizal fungi in enhancing agroforestry resilience. The researchers found that the presence of specific mycorrhizal species improved nutrient cycling, increased drought resistance, and reduced the incidence of soilborne diseases in agroforestry systems. This highlights the potential of mycorrhizal fungi as a microbial tool to enhance agroforestry sustainability.

5.6.3 BIOCONTROL AGENTS FOR PEST MANAGEMENT

In a case study by Da Silva Folli-Pereira et al. [11], biocontrol agents were applied to agroforestry systems to combat pest infestations. The results demonstrated a significant reduction in pest populations and associated damage to crops. This microbial intervention showcased the potential for environmentally friendly pest management strategies in agroforestry, emphasizing the importance of harnessing microbial biodiversity.

5.7 MICROBIAL AND PLANT SPECIES INTERACTIONS

Microbial consortia, defined as complex communities of diverse microorganisms, play a crucial role in the functioning of ecosystems, particularly in the context of plant–microbe and microbe–microbe interactions. In agroforestry systems, understanding these interactions is essential for promoting sustainable practices and optimizing the health of the system. This discussion delves into how microbial diversity increases through these interactions and explores their implications for the health and management of agroforestry systems.

5.7.1 PLANT–MICROBE INTERACTIONS: A NEXUS OF DIVERSITY

Plant–microbe interactions are pivotal for nutrient cycling, disease resistance, and overall ecosystem health. The rhizosphere, the region of soil influenced by root secretions, is a hotspot for these interactions [28].

Plants release a myriad of compounds, including root exudates, shaping the microbial communities around their roots. In return, microbes contribute to nutrient acquisition, protect plants from pathogens, and modulate the plant's response to environmental stressors. The interaction between mycorrhiza and plants exemplifies this synergy. Mycorrhizal associations enhance nutrient uptake for plants, and in turn, plants provide carbon compounds to the fungi [40]. This mutualistic exchange fosters microbial diversity as different plant species engage with a variety of microbial partners, creating a complex web of interactions in the rhizosphere.

5.7.2 MICROBE–MICROBE INTERACTIONS: SHAPING MICROBIAL COMMUNITIES

Microbes seldom operate in isolation; instead, they engage in intricate interactions with other microorganisms. Competition, predation, and mutualism among microbes further contribute to microbial diversity. For instance, antibiosis, where one microbe produces substances inhibiting the growth of another, is a common mechanism. This dynamic interplay among microbes shapes the composition and function of microbial communities in agroforestry systems. Biofilm formation is another example of microbe–microbe interaction. Biofilms are microbial communities attached to surfaces, and they play a crucial role in nutrient cycling and the establishment of stable microbial consortia. The interactions within these biofilms create microenvironments that support diverse microbial life, contributing to overall ecosystem resilience [9].

Understanding microbial consortia and their interactions has profound implications for the health and management of agroforestry systems. The intricate relationships between plants and microbes, as well as among different microbial species, play a pivotal role in shaping the dynamics of agroforestry ecosystems. Recognizing and harnessing these interactions can lead to more sustainable and resilient agricultural practices, with far-reaching consequences for soil health, nutrient cycling, and the overall well-being of agroforestry environments. The implications of understanding and harnessing microbial interactions in agroforestry systems are far-reaching. From nutrient cycling to disease suppression, climate resilience, and biodiversity conservation, the integration of microbial knowledge offers a holistic approach to sustainable and resilient agricultural practices. Agroforestry systems stand to benefit not only from increased productivity but also from the creation of environmentally harmonious and diverse ecosystems.

5.8 CURRENT RESEARCH AND INNOVATIONS

Agroforestry, an integrated land-use management system that combines trees and shrubs with crops and/or livestock, has gained significant attention as a sustainable approach to agricultural production. Within the realm of agroforestry, the intricate relationships between plants and microorganisms play a crucial role in shaping ecosystem dynamics. Recent research has delved into understanding and harnessing these plant–microbe synergies to enhance agricultural productivity while promoting sustainability. This chapter provides an overview of the current state of microbial research in agroforestry, highlighting key studies and innovations that showcase the potential of microbial partnerships.

Recent studies have unveiled the complex and dynamic interactions between plants and microbes in agroforestry systems. One notable area of focus is the exploration of mycorrhiza, which form symbiotic associations with plant roots. These fungi enhance nutrient uptake by plants, especially phosphorus, and contribute to increased resilience against environmental stresses. Research by Dollinger and Jose [13] demonstrated that incorporating mycorrhizal fungi into agroforestry practices led to improved nutrient cycling and enhanced soil fertility. Additionally, advancements in molecular biology techniques have enabled researchers to unravel the diversity and functions of microbial communities in agroforestry soils. Metagenomic studies, such as those conducted by Bharati et al. [5], have revealed the presence of beneficial bacteria with the capacity to promote plant growth and suppress pathogens. Understanding the roles of specific microbial species in nutrient cycling and disease suppression allows for targeted interventions to enhance agroforestry system resilience.

The application of microbial partnerships in agroforestry has yielded tangible benefits for sustainable development. One notable success story involves the use of nitrogen-fixing bacteria in association with leguminous trees. This symbiotic relationship enhances soil fertility by converting atmospheric nitrogen into a form accessible to plants, reducing the need for synthetic fertilizers. Research by Kim and Isaac [18] demonstrated that integrating nitrogen-fixing bacteria with agroforestry systems not only increased crop yields but also contributed to a reduction in greenhouse gas emissions associated with nitrogen fertilizer production. Furthermore, the promotion of beneficial microbial communities has shown promise in disease management. Inoculating agroforestry systems with antagonistic microorganisms can suppress soil-borne pathogens, reducing the reliance on chemical pesticides. This approach not only safeguards crop health but also mitigates the environmental impact of agrochemical use. Studies by Greff et al. [16] highlight the potential of using biocontrol agents to manage soil-borne diseases in agroforestry, paving the way for more sustainable pest management strategies.

The evolving field of microbial research within agroforestry holds immense promise for sustainable agriculture. As we delve deeper into the intricate relationships between plants and microbes, innovative applications are emerging, reshaping the landscape of agroforestry practices. The success stories presented here underscore the potential of harnessing plant–microbe synergies to enhance productivity, reduce environmental impact, and contribute to the broader goal of sustainable development. Continued research in this area is essential for unlocking further insights and optimizing the utilization of microbial partnerships in agroforestry systems.

5.9 CHALLENGES

The challenges in implementing microbial innovations in agroforestry are substantial, but addressing them offers promising opportunities for sustainable and resilient agricultural systems. Future research should strive to unravel the complexities of plant–microbe interactions, optimize microbial consortia, and integrate precision agriculture approaches. Simultaneously, efforts to engage and educate local communities are crucial for the successful adoption of microbial technologies in agroforestry, paving the way for a more sustainable and productive future. While the challenges in implementing microbial innovations in agroforestry are substantial, they present exciting opportunities for research and development. Future efforts should focus on unraveling the complexities of plant–microbe interactions, optimizing microbial consortia, addressing environmental sensitivities, and ensuring seamless integration with existing agricultural practices. Engaging local communities in the process is essential for the successful adoption of microbial technologies, ultimately contributing to the creation of more sustainable and productive agroforestry systems.

5.9.1 DIVERSITY AND SPECIFICITY

A central challenge lies in the intricate and diverse nature of plant–microbe interactions. The vast array of plant species interacts with a multitude of microbial partners, each with specific functions. This diversity poses a significant hurdle in translating research findings into practical applications for agroforestry. To overcome this challenge, future research must delve into the complexities of these interactions, identifying key microbial players and understanding their roles in different contexts. Developing microbial consortia tailored to specific agroforestry systems requires a nuanced understanding of the functional diversity within microbial communities. Researchers should explore how different plant species interact with unique microbial assemblages and how these interactions influence plant health and productivity. This knowledge is essential for designing microbial interventions that are not only effective but also adaptable to the diversity of plant–microbe associations in agroforestry landscapes.

5.9.2 ENVIRONMENTAL SENSITIVITY

Microbial partnerships are highly sensitive to environmental variations, including soil conditions, climate, and land management practices. Agroforestry systems often span diverse landscapes, from tropical to temperate regions, each with its unique environmental characteristics. This diversity makes it challenging to develop microbial solutions that remain consistently effective across different conditions. Researchers must focus on understanding how environmental factors influence the dynamics of microbial communities and their interactions with plants. This includes investigating the impact of varying soil types, climate conditions, and land management practices on the stability and efficacy of microbial interventions. Additionally, developing microbial formulations that can adapt to different environmental conditions is crucial for the widespread and successful implementation of these innovations in diverse agroforestry systems.

5.9.3 INTEGRATION WITH EXISTING PRACTICES

The adoption of microbial innovations in agroforestry must seamlessly integrate with existing agricultural practices to gain acceptance from farmers. Resistance to change and disruption of traditional methods are common barriers. Research efforts should, therefore, address how to integrate microbial solutions into established agroforestry systems without causing significant disruptions. This requires understanding the socioeconomic and cultural aspects of farming communities. Researchers and extension services should work closely with farmers to identify points of synergy between microbial innovations and existing practices. Developing practical guidelines and demonstrating the compatibility of microbial solutions with traditional agroforestry methods can facilitate a smoother transition and increase the likelihood of widespread adoption.

5.10 FUTURE RESEARCH AND PRACTICAL APPLICATIONS

5.10.1 MICROBIAL CONSORTIA RESEARCH

Future research endeavors in the realm of microbe synergy in agroforestry should be geared toward a more nuanced understanding and manipulation of microbial consortia. It is increasingly evident that the synergistic actions of multiple microbial species can potentially surpass the efficacy of individual strains in fostering plant growth and ensuring overall plant health. This necessitates a departure from the traditional focus on isolated microbial strains toward a more holistic approach that acknowledges the complex interplay within microbial communities. Researchers should delve into deciphering the intricate dynamics of microbial consortia, exploring how different species interact and contribute to plant–microbe partnerships. This could involve investigating the optimal combinations of microbes that exhibit complementary functions, such as nitrogen fixation, phosphate solubilization, and disease suppression. Furthermore, understanding the temporal and spatial dynamics of microbial consortia in different agroforestry systems will be crucial for developing tailored and effective microbial interventions.

5.10.2 PRECISION AGRICULTURE APPROACHES

The integration of precision agriculture techniques holds immense promise in optimizing the application of microbial synergies in agroforestry.

Advanced molecular tools and remote sensing technologies can be leveraged to precisely tailor microbial interventions to specific agroforestry contexts. This approach allows for real-time monitoring of environmental conditions and the microbial community, enabling adaptive management strategies. Through precision agriculture, researchers and farmers can gain insights into the dynamic interactions between microbes and plants, adjusting microbial applications based on the evolving needs of the agroforestry system. For instance, if a particular microbial consortium is thriving under specific soil conditions, precision agriculture tools can identify this success and inform future interventions. This responsive and data-driven approach enhances the efficiency and sustainability of microbe-based agroforestry practices.

5.10.3 COMMUNITY ENGAGEMENT AND EDUCATION

The successful implementation of microbial innovations in agroforestry necessitates active engagement with local communities and comprehensive education initiatives targeted at farmers. Bridging the knowledge gap surrounding microbial technologies is crucial for fostering acceptance and ensuring the widespread adoption of these innovations. Researchers, extension services, and agricultural educators should collaborate to develop outreach programs that not only disseminate information but also engage farmers in a dialogue. Emphasizing the tangible benefits of microbial synergies in agroforestry, such as increased crop yield, improved soil health, and reduced dependence on synthetic inputs, can help garner support from local communities. Moreover, involving farmers in the research process, such as field trials and demonstrations, can provide them with firsthand experience and build trust in these innovative approaches. The future of microbe synergy in agroforestry lies in a comprehensive understanding of microbial consortia, the application of precision agriculture techniques, and the active involvement of communities through education and engagement efforts. This holistic approach has the potential to revolutionize sustainable agriculture practices by harnessing the power of microbial partnerships.

5.11 CONCLUSION

The symbiotic relationship between plants and microbes in agroforestry is crucial for ecosystem health and productivity. This chapter reveals the multifaceted roles of microbes in fostering a synergistic relationship with plants, such as mycorrhizal fungi that enhance nutrient uptake and improve soil structure. The microbiome balance contributes to the resilience of agroforestry ecosystems, with beneficial microbes acting as guardians and suppressing soil-borne diseases. The microbial community also affects the above-ground components of agroforestry systems, modulating plant defense mechanisms and promoting stress tolerance. Integrating microbial ecology into curricula and training programs will empower the next generation of farmers and practitioners. Policymakers must recognize the importance of microbial diversity in agroforestry and allocate resources for research and development. Fostering outreach programs and collaborations between research institutions, agricultural extension services, and local communities can facilitate knowledge exchange and the co-creation of solutions tailored to diverse agroforestry systems.

KEYWORDS

- **agroforestry**
- **ecosystem**
- **endophytic microbes**
- **microbial biodiversity**
- **plant microbiota**

REFERENCES

1. Aasfar, A., Bargaz, A., Yaakoubi, K., Hilali, A., Bennis, I., Zeroual, Y., Kadmiri, I. M. (2021). Nitrogen fixing azotobacter species as potential soil biological enhancers for crop nutrition and yield stability. *Frontiers in Microbiology*, 12(February), 628379. https://doi.org/10.3389/fmicb.2021.628379.
2. Aeron A., Dubey R. C., Maheshwari D. K. (2020). Characterization of a plant-growth-promoting non-nodulating endophytic bacterium (Stenotrophomonas maltophilia) from the root nodules of Mucuna utilis var. capitata L. (Safed Kaunch). *Canadian Journal of Microbiology*, 66, 670–677.
3. Afridi, M. S., Ali, S., Salam, A., César Terra, W., Hafeez, A., Sumaira, Ali, B., et al. (2022). Plant microbiome engineering: Hopes or hypes. *Biology*, 11(12), 1782. https://doi.org/10.3390/biology11121782.

4. Aguilar-Paredes, A., Valdés, G., Nuti, M. (2020). Ecosystem functions of microbial consortia in sustainable agriculture. *Agronomy*, 10(12), 1902. https://doi.org/10.3390/agronomy10121902.
5. Bharati, A., Anuragi, H., Rajarajan, K., Kumar Handa, A., Arunachalam, A. (2022). Soil microbial metagenomics in agroforestry system: Tools and techniques. In: Kumar, A., Choudhury, B., Dayanandan, S., Khan, M. L. (Eds.), *Molecular Genetics and Genomics Tools in Biodiversity Conservation* (pp. 171–183). Springer Nature Singapore: Singapore. https://doi.org/10.1007/978-981-16-6005-4_8.
6. Bodner, G., Mentler, A., Keiblinger, K. (2021). Plant roots for sustainable soil structure management in cropping systems. In: Rengel, Z., Djalovic, I. (Eds.), *The Root Systems in Sustainable Agricultural Intensification* (pp. 45–90), 1st edn, Wiley. https://doi.org/10.1002/9781119525417.ch3.
7. Brown, C. M. B., Nepomuceno, R. A., Anarna, J. A., Brown, M. B. (2022). Plant growth promoting rhizobacteria—Advances and future prospects. In: *New and Future Developments in Microbial Biotechnology and Bioengineering* (pp. 1–28). Elsevier. https://doi.org/10.1016/B978-0-323-85577-8.00003-2.
8. Cao, T., Luo, Y., Shi, M., Tian, X., Kuzyakov, Y. (2024). Microbial interactions for nutrient acquisition in soil: Miners, scavengers, and Carriers. *Soil Biology and Biochemistry*, 188(January), 109215. https://doi.org/10.1016/j.soilbio.2023.109215.
9. Caruana, J. C., Walper, S. A. (2020). Bacterial membrane vesicles as mediators of microbe—Microbe and microbe—Host community interactions. *Frontiers in Microbiology*, 11(March), 432. https://doi.org/10.3389/fmicb.2020.00432.
10. Coonan, E. C., Kirkby, C. A., Kirkegaard, J. A., Amidy, M. R., Strong, C. L., Richardson, A. E. (2020). Microorganisms and nutrient stoichiometry as mediators of soil organic matter dynamics. *Nutrient Cycling in Agroecosystems*, 117(3), 273–298. https://doi.org/10.1007/s10705-020-10076-8.
11. Da Silva Folli-Pereira, M., Bertolazi, A. A., Garlet, J., Passamani, L. Z., De Souza, S. B., Costa, F. S. F., Ramos, A. C. (2022). Microorganisms as biocontrol agents of pests and diseases. In: Arora, N. K., Bouizgarne, B. (Eds.), *Microbial Biotechnology for Sustainable Agriculture Volume 1*, (33, pp. 143–184). Microorganisms for Sustainability. Springer Nature Singapore: Singapore. https://doi.org/10.1007/978-981-16-4843-4_4.
12. Dagar, J. C., Sileshi, G. W., Akinnifesi, F. K. (2020). Agroforestry to enhance livelihood security in Africa: Research trends and emerging challenges. In: Dagar, J. C., Gupta, S. R., Teketay, D. (Eds.), *Agroforestry for Degraded Landscapes* (pp. 71–134). Springer: Singapore. https://doi.org/10.1007/978-981-15-4136-0_3.
13. Dollinger, J., Jose, S. (2018). Agroforestry for soil health. *Agroforestry Systems*, 92(2), 213–219. https://doi.org/10.1007/s10457-018-0223-9.
14. Etminani F., Harighi, B. (2018). Isolation and identification of endophytic bacteria with plant growth promoting activity and biocontrol potential from wild pistachio trees. *Plant Pathology Journal (Faisalabad)*, 34, 208–217.
15. Fahad, S., Bhanudas Chavan, S., Ravindra Chichaghare, A., Ramani Uthappa, A., Kumar, M., Kakade, V., Pradhan, A., et al. (2022). Agroforestry systems for soil health improvement and maintenance. *Sustainability*, 14(22), 14877. https://doi.org/10.3390/su142214877.

16. Greff, B., Sáhó, A., Lakatos, E., Varga, L. (2023). Biocontrol activity of aromatic and medicinal plants and their bioactive components against soil-borne pathogens. *Plants*, 12(4), 706. https://doi.org/10.3390/plants12040706.
17. Kang, E., Li, Y., Zhang, X., Yan, Z., Wu, H., Li, M., Yan, L., Zhang, K., Wang, J., Kang, X. (2021). Soil pH and nutrients shape the vertical distribution of microbial communities in an alpine wetland. *Science of the Total Environment*, 774(June), 145780. https://doi.org/10.1016/j.scitotenv.2021.145780.
18. Kim, D.-G., Isaac, M. E. (2022). Nitrogen dynamics in agroforestry systems. A review. *Agronomy for Sustainable Development*, 42(4), 60. https://doi.org/10.1007/s13593-022-00791-7.
19. Kumar, N., Haldar, S., Saikia, R. (2023). Root exudation as a strategy for plants to deal with salt stress: An updated review. *Environmental and Experimental Botany*, 216(December), 105518. https://doi.org/10.1016/j.envexpbot.2023.105518.
20. Liu, Y., Bai, F., Li, T., Yan, H. (2018). An endophytic strain of genus Paenibacillus isolated from the fruits of Noni (*Morinda citrifolia* L.) has antagonistic activity against a Noni's pathogenic strain of genus Aspergillus. *Microbial Pathogenesis*, 125, 158–163.
21. Mahmud, K., Makaju, S., Ibrahim, R., Missaoui, A. (2020). Current progress in nitrogen fixing plants and microbiome research. *Plants*, 9(1), 97. https://doi.org/10.3390/plants9010097.
22. Meena, M., Yadav, G., Sonigra, P., Nagda, A., Mehta, T., Swapnil, P., Harish, Marwal, A., Kumar, S. (2023). Multifarious responses of forest soil microbial community toward climate change. *Microbial Ecology*, 86(1), 49–74. https://doi.org/10.1007/s00248-022-02051-3.
23. Mitra, D., Anđelković, S., Panneerselvam, P., Senapati, A., Vasić, T., Ganeshamurthy, A. N., Chauhan, M., Uniyal, N., Mahakur, B., Radha, T. K. (2020). Phosphate-solubilizing microbes and biocontrol agent for plant nutrition and protection: Current perspective. *Communications in Soil Science and Plant Analysis*, 51(5), 645–657. https://doi.org/10.1080/00103624.2020.1729379.
24. Mittal, D., Shukla, R., Verma, S., Sagar, A., Verma, K. S., Pandey, A., Negi, Y. S., Saini, R. V., Saini, A. K. (2019). Fire in pine grown regions of Himalayas depletes cultivable plant growth promoting beneficial microbes in the soil. *Applied Soil Ecology*, 139, 117–124.
25. Nam, N. N., Do, H. D. K., Trinh, K. T. L., Lee, N. Y. (2023). Metagenomics: An effective approach for exploring microbial diversity and functions. *Foods*, 12(11), 2140. https://doi.org/10.3390/foods12112140.
26. Ollinaho, O. I., Kröger, M. (2021). Agroforestry transitions: The good, the bad and the ugly. *Journal of Rural Studies*, 82(February), 210–221. https://doi.org/10.1016/j.jrurstud.2021.01.016.
27. Passos, J. F. M. D., Costa, P. B. D., Costa, M. D., Zaffari, G. R., Nava, G., Boneti, J. I., de Oliveira, A. M., Passaglia, L. M. (2014). Cultivable bacteria isolated from apple trees cultivated under different crop systems: Diversity and antagonistic activity against Colletotrichum gloeosporioides. *Genetics and Molecular Biology*, 37, 560–572.
28. Pathan, S. I., Ceccherini, M. T., Sunseri, F., Lupini, A. (2020). Rhizosphere as hotspot for plant-soil-microbe interaction. In: Datta, R., Meena, R. S., Pathan, S. I.,

Ceccherini, M. T. (Eds.), *Carbon and Nitrogen Cycling in Soil* (pp. 17–43). Springer Singapore: Singapore. https://doi.org/10.1007/978-981-13-7264-3_2.

29. Radhapriya, P., Ramachandran, A., Palani, P. (2018). Indigenous plant growth-promoting bacteria enhance plant growth, biomass, and nutrient uptake in degraded forest plants. *3 Biotech*, 8, 1–10.
30. Riera, N., Handique, U., Zhang, Y., Dewdney, M. M., Wang, N. (2017). Characterization of antimicrobial-producing beneficial bacteria isolated from Huanglongbing escape citrus trees. *Frontiers in Microbiology*, 8, 2415.
31. Selim, M. M. (2020). Introduction to the integrated nutrient management strategies and their contribution to yield and soil properties. *International Journal of Agronomy*, 2020(May), 1–14. https://doi.org/10.1155/2020/2821678.
32. Suarez, A., Gwozdz, W. (2023). On the relation between monocultures and ecosystem services in the global South: A review. *Biological Conservation*, 278(February), 109870. https://doi.org/10.1016/j.biocon.2022.109870.
33. Taghavi, S., Garafola, C., Monchy, S., Newman, L., Hoffman, A., Weyens, N., Barac, T., Vangronsveld, J., van der Lelie, D. (2009). Genome survey and characterization of endophytic bacteria exhibiting a beneficial effect on growth and development of poplar trees. *Applied and Environmental Microbiology*, 75(3), 748–757.
34. Tang, A., Haruna, A. O., Muhamad Ab. Majid, N., Jalloh, M. B. (2020). Potential PGPR properties of cellulolytic, nitrogen-fixing, phosphate-solubilizing bacteria in rehabilitated tropical forest soil. *Microorganisms*, 8(3), 442. https://doi.org/10.3390/microorganisms8030442.
35. Terhonen, E., Sipari, N., Asiegbu, F. O. (2016). Inhibition of phytopathogens by fungal root endophytes of Norway spruce. *Biological Control*, 99, 53–63.
36. Ulrich K., Kube M., Becker R., Schneck V., Ulrich A. (2021). Genomic analysis of the endophytic Stenotrophomonas strain 169 reveals features related to plant-growth promotion and stress tolerance. *Frontiers in Microbiology,* 12, 687463.
37. Vats, S., Srivastava, P., Saxena, S., Mudgil, B., Kumar, N. (2021). Beneficial effects of nitrogen-fixing bacteria for agriculture of the future. In: Cruz, C., Vishwakarma, K., Choudhary, D. K., Varma, A. (Eds.), *Soil Nitrogen Ecology* (62, pp. 305–325). Soil Biology. Springer International Publishing: Cham. https://doi.org/10.1007/978-3-030-71206-8_15.
38. Verma, S.K., Sahu, P.K., Kumar, K., Pal, G., Gond, S.K., Kharwar, R.N., White, J.F. (2021). Endophyte roles in nutrient acquisition, root system architecture development and oxidative stress tolerance. *Journal of Applied Microbiology*, 131(5), 2161–2177. https://doi.org/10.1111/jam.15111.
39. Woźniak M., Gałązka A., Tyśkiewicz R., Jaroszuk-ściseł J. (2019). Endophytic bacteria potentially promote plant growth by synthesizing different metabolites and their phenotypic/physiological profiles in the biolog gen iii microplate™ test. *International Journal of Molecular Sciences*, 20, 5283.
40. Xie, K., Ren, Y., Chen, A., Yang, C., Zheng, Q., Chen. J., Wang, D., Li, Y., Hu, S., Xu, G. (2022). Plant nitrogen nutrition: The roles of arbuscular mycorrhizal fungi. *Journal of Plant Physiology*, 269(February), 153591. https://doi.org/10.1016/j.jplph.2021.153591.

41. Xiong, Q., Hu, J., Wei, H., Zhang, H., Zhu, J. (2021). Relationship between plant roots, rhizosphere microorganisms, and nitrogen and its special focus on rice. *Agriculture*, 11(3), 234. https://doi.org/10.3390/agriculture11030234.

CHAPTER 6

Multitrophic Interaction in Agroforestry: An Overview

CHANDNI[1*] and VICTOR THINGUJAM[2]

[1]*ICAR-Central Institute of Temperate Horticulture, Regional Station, Mukteshwar, Uttarakhand, India*

[2]*ICAR-Research Complex for Eastern Region, Regional Station, Palandu, Ranchi, Jharkhand, India*

*Corresponding author

ABSTRACT

Agroforestry systems are increasingly valued for enhancing agricultural sustainability through complex multitrophic interactions. By integrating trees with crops and/or livestock, these systems create ecological networks where plants, herbivores, predators, and decomposers interact to maintain biodiversity, regulate pests, support nutrient cycling, and improve productivity. Such interactions strengthen ecosystem resilience and stability, offering a pathway toward sustainable food systems and environmental conservation. Trees in agroecosystems enhance resource use efficiency, ecological balance, and habitat diversity, thereby contributing to food security, rural livelihoods, and climate change mitigation. Maximizing these benefits requires careful design and management. Key factors include appropriate species selection, spatial arrangements, and compatible tree–crop–livestock combinations. Thoughtful planning ensures that positive interactions outweigh competition, enabling long-term productivity and ecosystem health. A deeper understanding of multitrophic relationships is essential to harness their potential, supported by ongoing research and

Agroforestry and Climate Change Mitigation: Harnessing Multitrophic Interactions for Sustainable Plant Protection and Ecosystem Resilience. Hanuman Singh Jatav, Vishnu D. Rajput, Ipsita Samal, Tanmaya Kumar Bhoi (Eds.)

DOI: 10.1201/9781779641397-6

adaptive management. Agroforestry, by leveraging ecological processes, offers a practical strategy to address global challenges such as biodiversity loss, land degradation, and climate change. However, realizing its full potential requires supportive policies, knowledge dissemination, and wider adoption. Strengthening research and policy frameworks will help scale agroforestry systems as resilient, sustainable solutions for future agriculture.

6.1 INTRODUCTION TO AGROFORESTRY SYSTEMS

6.1.1 INTRODUCTION

The practice of cultivating agricultural crops and trees in close proximity to each other is an ancient tradition observed by farmers worldwide. Agroforestry is a recently coined term for a long-standing practice. It can be described as the practice of cultivating trees in agriculture. Agroforestry refers to a range of land-use systems and technologies that intentionally combine the cultivation of woody perennial plants (such as trees, shrubs, palms, and bamboos) with crops and/or animals on the same land. This integration can occur in several ways, either spatially or temporally [1]. Agroforestry is a sustainable land-use system that involves combining food crops (annuals) with tree crops (perennials) and/or livestock on the same unit of land. This can be done either alternately or simultaneously, with the goal of maintaining or increasing total yields. Agroforestry practices are tailored to the social and cultural characteristics of the local people, as well as the economic and ecological conditions of the area [2]. Agroforestry has several advantages, such as decreased nutrient runoff, the storage of carbon, enhanced soil quality, erosion prevention, improved wildlife habitat, less reliance on fossil fuels, and increased ability to withstand an uncertain agricultural future [3–9]. Therefore, incorporating trees and other perennial plants into a land-use system can mitigate numerous detrimental impacts of agriculture. In addition to its positive impact on the environment, agroforestry has the potential to provide a wide range of products, including timber, crops, fruits, nuts, mushrooms, forages, cattle, biomass, and herbal medicine [10]. The three basic components of agroforestry systems are crops, trees, and livestock.

Agroforestry systems can be classified according to the following sets of criteria:

1.	Structural basis	Nature of components	Agrisilvicultural systems Silvopastoral systems Agrosilvopastoral systems Other systems
		Arrangement of components	Spatial arrangement Temporal arrangement
2.	Functional basis	Productive function Protective function	
3.	Socioeconomic basis	Low input, high input, or subsistence, commercial or intermediate	
4	Ecological basis	Arid, semiarid, tropical, lowland humid tropics, etc	

Soil developing under the natural forest is known to be well structured, with good water holding capacity, and contains higher organic matter, and farmers, since time immemorial, were well aware of it, and they started clearing the forestland for cultivation. Once the soil quality is depleted, they will move on to clearing new patches for cultivation, leaving the former patches barren and wasted. The tree ecosystem is a relatively closed, self-sufficient system with continuous nutrient transfer, storage, and cycling. Therefore, if trees are incorporated in this land alongside the cultivation of annual crops, the microclimate will be restored, making the land favorable for sustainable cultivation. Microclimate amelioration due to the shade and windbreak effect of the trees regulates the temperature, humidity, and movement of air as well as the temperature and moisture of the soil.

Shading effect by the trees causes a reduction in the temperature and temperature fluctuations, which in turn reduces loss of soil organic matter by oxidation. Shading can also suppress some of the light-demanding weeds. The presence of a mulch or litter layer under shade trees also reduces evaporation loss but increases infiltration of rainwater. Trees serving as wind breaks control wind erosion, reducing the loss of the fertile top layers of soil. The different layers of tree canopies also reduce the impact of raindrops on the soil, thereby reducing rainwater erosion. Leaf litter from woody perennials increases the organic matter content of the soil. Approximately 20–25% of the total living biomass of the tree, consisting of roots, is constantly dying and decaying and is being added as organic matter to the soil. These will increase the activity of the favorable microorganism present in the root zone. Biological Nitrogen fixation and solubilization of relatively unavailable nutrients like phosphate through the activity of mycorrhiza and phosphate-solubilizing bacteria. Nitrogen fixation by the leguminous trees improves the fertility status of the agricultural lands through the addition of nitrogen in large amounts.

Trees are generally assumed to have a deep and spreading root system and hence have the capability to exploit more soil depth. Therefore, they take up water and nutrients from the depths of the soil, which are beyond the reach of the annual crops. These nutrients are returned to the surface layers through litter fall from the trees. Even the benefits of including woody perennials can be extended from agricultural land to the adjoining water bodies by reducing the siltation rate.

Agroforestry should therefore be considered more of a dynamic process than a system toward a more sustainable natural system of production.

6.2 THE MULTITROPHIC INTERACTION CONCEPT

6.2.1 MULTITROPHIC INTERACTIONS IN THE CONTEXT OF AGROFORESTRY

The success of an agroforestry system relies heavily on the exploitation of component interactions. In the context of agroforestry, multitrophic interactions relate to the intricate relationships and interactions between several trophic levels within an agroecosystem that includes agricultural crops and trees (forestry). It refers to the influence of one component of a system on the performance of the other components as well as the system as a whole.

The presence of trees next to crops in agroforestry systems generates a dynamic and varied ecosystem that affects a variety of creatures, including insects, plants, microorganisms, and other wildlife. The links between these many organisms are both direct and indirect in multitrophic interactions. Multitrophic interactions can either be positive or negative interactions. Key elements of agroforestry's multitrophic interactions include the following:

Cooperation and competition: Crops and trees may face competition for the same resources, such as sunlight, nutrients, and water. They can also coexist in mutualistic interactions, in which some trees help crops by fixing nitrogen or offering shade [11].

Allelopathy: Allelopathy refers to the inhibition of one plant by chemical compounds that are released into the soil from neighboring plants [12].

Microclimate amelioration: The primary cause of microclimate amelioration pertaining to soil moisture and temperature relations is the utilization of trees for shade, as well as for live support, live fences, windbreaks, and shelterbelts. Offering shade results in a complicated web of interactions that go far beyond just blocking off light and heat [13].

Pest and disease control: In agroforestry systems, the presence of trees can affect the number of pests. Certain tree species have the potential to function as natural enemies of pests by giving predators a place to live or by creating substances that repel or discourage bugs. This may help with organic pest management. Reduced temperature and humidity fluctuations under shade can also suppress disease and pests [14].

Pollination: Bees and other pollinators drawn to trees can improve fruit development and pollinate surrounding crops. Via this beneficial relationship, agroforestry systems may produce higher yields. Three main benefits of agroforestry methods for pollinators are as follows: (1) providing habitat, such as food supplies and places to nest or lay eggs; (2) improving the connectivity between individual sites and the landscape; and (3) reducing exposure to pesticides [15].

Microbial interactions: Crop and tree roots interact with soil microbes to affect soil health and nutrient cycling. Because the roots of crops and trees are mixed together in agroforestry, their interactions and the resulting exudates have the potential to change the pH and organic matter content of the bulk soil as well as the rhizosphere, which could lead to a substantial variation in the microbial community in the soil [16]. Certain trees and mycorrhizal fungi can develop symbiotic connections that improve the trees' and the related crops' ability to absorb nutrients.

Biodiversity: Agroforestry can improve habitat and landscape variability, increase structural complexity, and aid in the conservation of biodiversity in agricultural and forest settings. [17]. Compared to monoculture systems, agroforestry systems are naturally able to support a higher diversity of plant and animal species. Complex ecological interactions, such as those involving herbivores, predators, and decomposers, may result from this diversity.

Understanding and being able to control these multitrophic interactions are necessary for designing robust and sustainable agricultural systems in agroforestry, as well as for selecting the right species for a given system. By enhancing positive interactions and avoiding negative ones within their systems, farmers can enhance the overall productivity and ecological health of their agroforestry systems.

6.2.2 SIGNIFICANCE OF UNDERSTANDING MULTITROPHIC INTERACTIONS IN AGROFORESTRY

It is important to understand the multitrophic interactions that occur in agroforestry because they directly affect the resilience, productivity, and sustainability of agroecosystems. It is essential to comprehend these interactions for the following reasons:

Biological control of pests: The presence of trees in agricultural systems and the complexity of the landscape can both positively impact the biological management of particular pests [18–21]. It can be achieved through multitrophic interactions in natural pest management. Farmers can recognize and promote the existence of pests' natural enemies by knowing the interactions between trees, crops, and other species. This promotes sustainable and ecologically friendly pest management by lowering the need for chemical pesticides.

Enhanced output and pollination: Fruit set and crop pollination are significantly impacted by the pollinators that trees in agroforestry systems can draw. Agroforestry can improve habitat quality, pollination services, crop quality, and yields by diversifying agricultural landscapes and giving pollinators additional food and shelter [22]. Higher yields and better-quality agricultural products are the results of this. Farmers can create agroforestry systems that maximize pollination services by having a thorough understanding of these interactions.

Nutrient cycling, soil health, and microbial interactions: Through its capacity to modify the microclimate, the integration of trees with agricultural crops, along with appropriate species selection and management techniques, helps to improve the biological, chemical, and physical aspects of the degraded soil as well as its structure [23]. Some tree species may encourage the development of advantageous microorganisms, which would increase soil health and nutrient cycling. This knowledge can help guide management strategies that improve the fertility and structure of the soil.

Habitat for beneficial organisms: Agroforestry systems support a wide range of creatures by virtue of their varied vegetation structure. This includes pollinators, other helpful species, and natural enemies of pests. Resilience, ecosystem services, and the creation and maintenance of a biodiverse environment are all aided by an understanding of multitrophic interactions.

Ecosystem resilience: Agroforestry aids farmers in addressing climate change by helping them create resilient microclimates for their crops and livestock, as well as preparing for extreme weather events [24]. Climate resilience refers to the ability of a system or community to withstand and recover from the impacts of climate change, such as extreme weather events and rising temperatures. Agroforestry systems that incorporate a greater diversity of species often exhibit greater resilience to environmental stresses, such as fluctuations in climate. The intricate web of multitrophic connections enhances the system's ability to withstand disturbances and adapt to changing external conditions.

Sustainable utilization of resources: In agroforestry systems, the process by which mycorrhiza and phosphate-solubilizing bacteria fix nitrogen in the environment and solubilize nutrients that are comparatively inaccessible helps related crops. The soil organic matter content is increased by the leaf litter generated by woody perennials. The roots of the tree, which are continuously decomposing, add organic matter to the soil. The ecosystem of trees is essentially a closed, self-sufficient system that cycles, stores, and transfers nutrients continuously. The selection and management of tree species that support the agroecosystem's overall nutrient balance are made possible by an understanding of these relationships.

Economic benefits: The ability of the components of the system to optimize resource use while simultaneously preserving "complementary" interactions is largely responsible for the success of intercropping. This frequently increases production per unit of land area, which raises economic rewards. A system of intercropping that combines extremely diverse components (woody and nonwoody, for example) may maximize productivity per unit land area and stagger the demand for scarce resources throughout time and place [25]. Diversification of sources of income is encouraged by agroforestry. Farmers can deliberately integrate trees and crops to increase total economic returns and lower the hazards associated with monoculture by knowing how various components interact.

Effects on the environment over time: The provision of ecosystem services, such as carbon sequestration, habitat provision, and water purification, is facilitated by multitrophic interactions. Recognizing these relationships reduces adverse environmental effects and promotes the long-term viability of agroforestry systems. In conclusion, a thorough grasp of multitrophic interactions in agroforestry empowers farmers to make well-informed choices that result in agricultural systems that are more robust, sustainable, and fruitful. By assisting in the reconciliation of ecological preservation and agricultural output, it advances the general welfare of the farming community and the environment.

6.3 PLANT–SOIL INTERACTIONS IN AGROFORESTRY

6.3.1 TREE–CROP INTERACTIONS ON SOIL HEALTH AND NUTRIENT CYCLING

The health of the soil and the cycling of nutrients are significantly impacted by tree–crop interactions in agroforestry systems. The intricate relationships between soil microbes, crop and tree root systems, and the nutrient cycle in the ecosystem are all a part of these interactions. The following are a few ways that crop–tree interactions affect the health of the soil and the cycling of nutrients:

Root System Dynamics: Competition and complementarity crop roots and tree roots may compete with one another for subsurface resources such as nutrients and water. On the other hand, certain tree species might have root systems that work well with particular crops. Shallow-rooted crops might profit from nutrients in the upper soil horizons, whereas deep-rooted trees can reach minerals in the deeper soil layers. The perennial nature of the trees allows their root systems to explore deeper soil volumes laterally and vertically, as well as to take advantage of zones with concentrated, rich supply of nutrients and/or water. The resources needed by the trees are met if these zones are primarily inaccessible to annual crops, and trees can raise the system's overall production [26].

Mycorrhizal Associations: Mycorrhizal fungi and crops frequently coexist in symbiotic relationships. These fungi form a network that helps nutrients, phosphorus in particular, move from the soil to the plants. For both trees and crops, this connection improves nutrient uptake efficiency.

Nitrogen Fixation: With the aid of nitrogen-fixing bacteria found in their root nodules, several tree species are able to fix atmospheric nitrogen. By increasing the amount of nitrogen in the soil that is accessible for plant uptake, the nitrogen-fixing trees help to benefit both the trees and the crops that are related to them.

Organic Matter Decomposition: The presence of trees facilitates the decomposition of organic matter by providing a source of organic matter in the form of leaf litter and other plant debris. Microorganisms decompose organic materials, leading to the release of nutrients into the soil. The process of tree litter decomposition can enhance the quantity of carbon and other essential elements that become accessible to crops and soil organisms. The primary source of organic material in forest soil is the vegetation that accumulates as a layer (litter) on the soil surface and is partially distributed into the soil [27].

Root Binding: Tree roots aid in the binding of soil particles, which strengthens the structure of the soil. This increases soil stability, lessens erosion, and increases water infiltration. Maintaining nutrient-rich topsoil and halting nutrient loss depend heavily on preventing soil erosion. The highest stability on hillslopes and river banks is achieved by a mixture of tree species with deep roots and grasses with intense fine roots [28].

Water-Use Efficiency: Trees tend to absorb water at a different rate than crops, with trees frequently reaching deeper soil layers. This may help agroforestry systems use water more effectively. Given the tight relationship between water availability and nutrient transport in the soil, the increased water-use efficiency may have an impact on nutrient availability and cycling.

Allelopathy: Certain tree species release chemicals through their leaves or roots that might affect the growth of plants in the vicinity. Through altering plant interactions within the environment, this allelopathic effect can either impede or stimulate the growth of specific crops, thus disrupting the cycle of nutrients.

Buffering Effects: Trees stabilize nitrogen availability in the soil and can act as a buffer against nutrient cycling. When crop demand is high or the environment is unfavorable, trees can serve as a reservoir for nutrients, releasing them gradually as needed. The application of agroforestry techniques can influence the physical, chemical, and biological properties of soil and stimulate plant growth by acting as a substantial source of soil organic matter. Agroforestry can help restore soil health by stabilizing soils, promoting aggregate formation, storing carbon in soils, improving nutrient availability, and maintaining and promoting a healthy soil biota [29].

In order to develop and manage agroforestry systems that maximize nutrient cycling and preserve soil health, it is imperative to comprehend the tree–crop interactions. These interactions can be strengthened by using appropriate species selection, spatial arrangement, and management techniques, which will promote a productive and sustainable agroecosystem.

6.3.2 IMPACT OF ROOT EXUDATES AND LITTER DECOMPOSITION

Agroforestry systems and other ecosystems depend on root exudates and litter decomposition for healthy soil and nutrient cycling. Both activities are crucial for forming the soil environment, affecting plant development, and maintaining the ecosystem's general health.

6.3.2.1 ROOT EXUDATES

Nutrient Mobilization and Organic Compounds: Through their roots, plants discharge a wide range of organic substances into the soil. Root exudates are a group of chemicals that contain organic acids, sugars, amino acids, and other compounds. These exudates have the ability to increase microbial activity and speed up the decomposition of organic materials, which increases the availability of nutrients for plants [30].

Microbial Interactions in the Rhizosphere: Microbial communities have a significant impact on the physiology and growth of plants, which is essential to their functioning. Plant growth benefits from the presence of several members of the rhizosphere microbiome [31]. Because root exudates are present, the rhizosphere zone around plant roots sees an increase in microbial activity. Plant roots and soil microbes interact to produce a dynamic environment that is favorable to microorganism growth, improving soil structure, reducing disease, and cycling nutrients.

Symbiotic Mycorrhizal Associations: Certain root exudates help mycorrhizal fungi and plants form symbiotic partnerships. In turn, these fungi help plants absorb more nutrients, particularly phosphorus and other minerals. By functioning as an extension of the plant root system, the mycorrhizal network enhances nutrient acquisition and transfer for the plant.

Allelopathic Effects: Exudates from roots can affect neighboring plants' growth and development. Certain substances secreted by plant roots prevent rival plants from germinating or growing, which helps the dominant plant take control of its immediate environment. Allelopathy can be useful as a management tool in agricultural systems to suppress weed growth, fend off pests and diseases, and enhance the relationship between soil nutrients and microorganisms [32].

Soil Aggregation: Some root exudates contain chemicals that promote soil agglomeration by acting as glue-like agents. Securing soil particles together, this promotes root penetration, water infiltration, and general soil stability, all of which contribute to improved soil structure.

6.3.2.2 LITTER DECOMPOSITION

Leaf Litter and Plant Debris: The organic matter that trees shed, including their leaves and other plant components, accumulates in the layer of surface litter in the soil. The addition of organic matter to the soil is a vital source of nutrients. Carbon, phosphorus, and nitrogen are nutrients that are released into the soil during the decomposition of litter.

Microbial Decomposition: Microorganisms, such as fungi and bacteria, are essential to the breakdown of litter. Through enzymatic reactions, they decompose complex organic substances into simpler forms, liberating nutrients that plants can absorb.

Soil Organic Carbon: Soil organic carbon is formed through the breakdown of plant litter. This is a crucial part of the organic matter in the soil, which enhances the soil's structure, capacity to hold water, and availability of nutrients. More than 40% of the total organic carbon of the terrestrial ecosystems is stored in forest soils, making them an important carbon reservoir [33]. Increasing the carbon contained in forest soils may help to mitigate climate change by sequestering more carbon dioxide (CO_2) from the atmosphere [34].

Continuous Nutrient Supply: Plant litter decomposition and its impact on soil nutrients are significant components of the ecosystem material cycle. An understanding of these processes is essential for species selection and allocation to support the efficient use of plant litter in agroforestry [35]. The constant decomposition of litter gives the soil a steady supply of nutrients. Particularly in environments with soils deficient in nutrients, the release of nutrients promotes plant growth.

Decomposer Organisms: Soil fauna, including insects, earthworms, and other detritivores, contribute to the breakdown of litter in addition to microorganisms. Their actions promote the soil nutrition cycle and further break down of organic materials.

An effective understanding of the effects of root exudates and litter decomposition is necessary for agroforestry system management. In addition to improving nitrogen cycling and soil fertility, appropriate management techniques that assist these processes also increase the agro-ecosystem's overall productivity and sustainability.

6.4 BIODIVERSITY AND PEST MANAGEMENT

6.4.1 INFLUENCE OF DIVERSE PLANT SPECIES IN AGROFORESTRY SYSTEMS ON THE PEST POPULATIONS

In agroforestry systems, a variety of plant species can have a significant impact on insect populations, affecting their distribution, abundance, and behavior. Plant diversity, or the interactions between diverse plant species, affects insect populations via a variety of ecological processes. Several plant species found in agroforestry systems have the following effects on insect populations:

Habitat Complexity: The presence of multiple plant species in agroecosystems generates a complex and varied environment. The complexity of the environment may hinder pests in their ability to locate and establish themselves on host plants. Vegetation diversity can help reduce the impact of pests and diseases through various mechanisms. These include diverting resources and using stimulo-deterrent diversion, disrupting spatial and temporal cycles, allelopathy effects, soil suppressiveness, crop physiological resistance, conserving natural enemies, and facilitating their action against aerial pests, as well as direct and indirect architectural/physical effects [36].

Natural Enemy Attraction: A wide range of insects, including predators and parasitoids, which are natural enemies of pests, are drawn to diverse plantings. Certain plants generate resources beneficial to insects, such as pollen, nectar, or other material. One such substitute food is extra floral nectar, which is produced by plants that are known to be less vulnerable to herbivory [37]. The increased predation and parasitism of pest populations brought about by the presence of these natural enemies can aid in biological control.

Allelopathic Effects: Some plant species emit chemicals into the air or soil that might affect pests' survival and behavior. This phenomenon is known as allelopathy. Certain compounds possess the ability to repel or poison bugs, thereby serving as a natural means of controlling pest populations. Diversity of plant-produced compounds in Agroforestry systems can result in a dynamic and adaptable pest management environment.

Trap Cropping: By offering insect pests an alternate, preferred option, trap cropping is a unique form of companion planting technique that is typically used for insect pest management through vegetation diversification [38]. In agroforestry systems, certain plant species can serve as trap crops, drawing pests away from the primary crops. In order to maintain the health of the primary crops while managing and controlling pest populations, these sacrificial plants attract pests to a concentrated area.

Dilution Effect: Diverse plantings may offer a wider range of resources for natural enemies, including nonpest prey species and resources such as pollen and nectar from the plant itself, thereby creating communities of natural enemies and enhancing their effects on pests [39]. The density of the host plant preferred by pests might be decreased by diverse plantings. In situations where pests come into contact with both host and nonhost plants, their chances of successfully reproducing and surviving may be reduced. Preventing pests from accessing suitable host plants can aid in the management of pest populations.

Interference with Pest Behavior: By erecting physical barriers, obscuring host plant appearance, or changing the environment, diverse plantings can impede insect behavior. For instance, plants with dense or higher foliage may produce shade, which reduces the environmental suitability for particular pests.

Ecosystem Resilience: General health and resilience of agroforestry systems are bolstered by a variety of plant species. Plants in good health are frequently more resistant to insect invasions. Diverse plant species can also improve soil health and nutrient cycling, which indirectly affects a plant's ability to fend off pests.

Biological Diversity and Stability: Elevated plant diversity is a factor in the overall stability of an ecosystem. Pest outbreaks are less likely to cause harm in stable ecosystems due to the availability of a variety of natural enemies, plant-to-plant competition, and other ecological checks and balances.

In order to develop and manage agroforestry systems that support sustainable pest management approaches, it is imperative to comprehend these mechanisms and capitalize on the advantages of plant diversity. Farmers can increase resilience and lessen their need for chemical pest management by using the ecological complexity of a variety of plants.

6.4.2 THE ROLE OF NATURAL ENEMIES AND PREDATORS IN PEST CONTROL

Because agroforestry systems integrate trees, crops, and other vegetation to form a complex and diversified landscape, the involvement of natural enemies and predators in pest control is very important. A rich and dynamic environment that sustains a diversified community of natural enemies is made possible by the interactions between various plant species and the existence of multiple habitats. In agroforestry systems, natural enemies and predators help manage pests in the following ways:

Diverse Habitats: Agroforestry systems diversify habitats, which are created by the presence of different plant species, and draw a variety of natural enemies. Varied plants sustain a range of useful species by providing varied resources, such as nectar, pollen, and shelter. Agroforestry can help in increasing pollinator diversity, which is necessary for both food production and preservation of wild plant populations [40].

Promoting Biodiversity: By creating niches for diseases, parasitoids, and predators, agroforestry encourages biodiversity at several trophic levels. This biodiversity helps keep some pests under control and strengthens the ecosystem's resilience.

Natural Enemy Abundance and Diversity: Agroforestry systems give natural enemies food and cover, which boosts the number and variety of predators and parasitoids. The biological control of pests is improved by this higher population density.

Enhanced Trophic Interactions: Intricate trophic interactions are fostered by the intricacy of the agroforestry landscape. By interacting with pests as well as one another, predators and parasitoids help to create an environment that is more self-regulating and in balance. The combination of trees and crops provides greater niche diversity and complexity in both time and area than the polyculture of annual crops, so agroforestry holds promise for enhancing insect diversity and lowering pest concerns [41].

Corridors for Movement: Agroforestry systems can serve as pathways that make it easier for natural enemies to migrate around the terrain. Because of this connection, parasitoids and predators can target pest populations across the entire region by traveling to different regions of the agroecosystem. Pollinators and other beneficial insects can find refuge in windbreaks, hedgerows, and riparian buffers provided they are adequately shielded from pesticides. These agroforestry refugia offer additional feed during pesticide applications in the vicinity of surrounding crop fields or orchards, annual crop harvesting, and other disturbances [42].

Feeding Resources for Beneficial Insects: Food sources for beneficial insects, such as nectar and pollen-producing plants and shrubs, are common in agroforestry systems. These nutrients give adults of natural enemy, the sustenance they need to live long lives, reproduce successfully, and generally suppress pests.

Biological Monitoring: Indicators of the existence of pests may come from natural enemies. As a biological monitoring technique, tracking the number and activity of natural enemies in agroforestry systems can assist farmers in identifying and mitigating pest pressures.

Suppression of Pest Populations: By preying on or parasitizing pest species, predators and parasitoids effectively reduce pest populations. Reducing the need for synthetic pesticides can be one benefit of using this natural method of pest management.

6.4.3 INTEGRATED PEST MANAGEMENT (IPM)

Holistic Approach: IPM techniques work best in agroforestry systems. In order to manage pests in a sustainable and environmentally responsible way, IPM integrates biological, cultural, mechanical, and chemical control strategies. One of the main components of IPM's biological control component is the existence of natural enemies.

Habitat Conservation: The principles of conservation biological control are aligned with agroforestry activities that improve habitat variety and protect natural enemy populations. Farmers can aid in the preservation of natural enemies by avoiding the degradation of habitat and using fewer broad-spectrum insecticides.

In conclusion, agroforestry systems that incorporate both crops and trees produce a diverse landscape that encourages the presence and activities of natural enemies and predators. Thus, a balanced and resilient agroecosystem is promoted through efficient and sustainable pest control.

6.5 NUTRIENT CYCLING AND PRODUCTIVITY

6.5.1 ENHANCE NUTRIENT CYCLING AND SOIL FERTILITY BY AGROFORESTRY SYSTEMS

Through a number of methods, agroforestry systems greatly improve soil fertility and nutrient cycling. In these systems, the coexistence of trees and crops and/or livestock fosters a variety of interactions that have a beneficial impact on the dynamics of nutrients.

Input of Organic Material: There are a variety of organic material inputs that trees in agroforestry systems contribute, including leaves, twigs, and other plant detritus. This organic debris serves as a source of nutrients and provides substrates for the decomposition process that is carried out by microorganisms. Litter falling in the soil has a substantial impact on the preservation of the hydrological cycles, in addition to having an effect on the ecology and the features of the soil. The types of trees, the way the land is managed, the amount of litter, and the quality of litter all have an impact on the degree to which the soil is enriched. There are a variety of criteria that determine the amount of nutrients that are released, including temperature, moisture content, microbial activity, and the nature of the litter [43]. Nutrients are converted through the process of breakdown of various components of litter.

Microbial Decomposition: The organic material in tree litter is broken down by microorganisms such as bacteria and fungi. Through the release of nutrients such as carbon, phosphorus, and nitrogen into the soil, this decomposition process improves soil fertility and enriches the nutrient content. A greater variety of decomposer and other species, which offer ecosystem services like pest control, is encouraged by the growth of litter shade trees [44]. Decomposers also play a crucial role in the ecosystems' cycles of phosphorus and nitrogen.

Nitrogen Fixation: In agroforestry systems, several tree species (legumes) fix atmospheric nitrogen by coexisting symbiotically with bacteria that fix nitrogen. Trees and related crops both gain from this biological nitrogen fixation, which enriches the soil with accessible nitrogen. Two species of Rhizobium can nodulate with tropical legumes: the fast-growing Rhizobium (sensu stricto) strains and the slow-growing Rhizobium (now called Bradyrhizobium) strains that form the cowpea miscellany [45]. A particular class of trees, such as *Sesbania grandiflora* and *Leucaena leucocephala* (with a few exceptions), only nodulates with Rhizobium (sensu stricto). Another type, such as *Acacia mearnsii* or *Acacia albida*, only nodulates in conjunction with Bradyrhizobium. Because it nodulates with both Rhizobium and Bradyrhizobium, a third category, such as *Acacia seyal*, is more promiscuous.

Root Exudates: Organic substances, such as sugars and organic acids, are released into the soil by trees and crops. By affecting the availability and solubility of soil nutrients, these exudates promote microbial activity and aid in the cycling of nutrients.

Mycorrhizal Fungi: Nearly all land-dwelling plants develop symbiotic connections with arbuscular mycorrhizal fungi, which promote growth and productivity, particularly in times of abiotic stress. The fungus releases a number of hormones that help the plant grow an extensive root system, which enhances the plant's ability to absorb nutrients, including phosphorus, water, and minerals. Plant tolerance and resilience to abiotic stresses such as salt, drought, and heavy metal toxicity are also enhanced by the fungus. Plants, in return, provide sugar to the fungus, which is the primary source of energy and is essential for survival [46]. By facilitating nitrogen transport between trees and crops, the mycorrhizal network encourages nutrient cycling.

Access to Deeper Soil Layers: Since trees frequently have deep root systems, they can reach nutrients in the soil that crops with shallow roots would not be able to. Litter falling from the trees returns these nutrients to the upper layers of the soil, which raises nutrients such as potassium and phosphorus to the surface layers where other plants can benefit from them.

Competition for Resources: Although crops and trees may compete with one another for sunlight, water, and nutrients, this competition may actually improve crop nutrient utilization. Effective resource allocation is encouraged by the competitive interactions, which stop one species from absorbing too many nutrients at the expense of other species. The right choice of species in a system can reduce competition to a great extent.

Complementary: Some tree species may help related crops meet their nutrient requirements. For example, the tree component in a system can raise the amount of nutrients in the soil, which is beneficial for crops through biological Nitrogen fixation and solubilization of relatively unavailable nutrients like phosphate.

Windbreaks and Erosion Control: Windbreaks are a common feature of agroforestry systems, and they aid in lowering soil erosion. Agroforestry systems help to minimize soil deterioration, lessen soil and nutrient erosion, and buffer the effects of rainfall [23]. The trees maintain soil fertility by retaining more nutrients in the root zone and preventing nutrient-rich topsoil from being lost due to erosion.

Water-Use Efficiency and Nutrient Availability: Trees can help increase water-use efficiency because of their extensive root system. Since nutrient transport in the soil is intimately related to water availability, this in turn affects nutrient availability.

Manure as a Nutrient Source: Animal dung serves as an extra source of nutrients in agroforestry systems that include livestock. Manure enriches the soil with minerals and organic materials, increasing its fertility.

Agroforestry Practices and Nutrient Management: In agroforestry systems, nutrient cycling can be maximized by strategic planting design and thoughtful species selection. For example, combining phosphorus-available species with nitrogen-fixing species can have a synergistic effect on nutrient cycling.

In conclusion, agroforestry systems improve soil fertility and nutrient cycling by using a mix of nitrogen fixation, efficient nutrient use, organic matter input, and mycorrhizal relationships. These systems improve overall ecosystem resilience, preserve soil health, and lower nutrient runoff, all of which support sustainable agriculture practices.

6.5.2 IMPACT ON CROP YIELDS AND OVERALL PRODUCTIVITY

Agroforestry systems may provide a resilient and diverse agricultural environment, which can have a substantial impact on crop yields and overall production. Agroforestry systems, which combine trees with crops and/or livestock, provide a number of advantages that raise productivity.

Diversification of Income Streams: Agroforestry provides a source of nourishment and diversifies food sources, and so is essential in maintaining food availability. The pillar of food availability is also supported by the increased and diverse income obtained from agroforestry products, which makes it possible for people to buy food [47]. By diversifying, revenue may be stabilized, and the hazards of relying just on one crop can be decreased. Furthermore, the presence of medicinal plants and the availability of fuelwood for cooking inside agroforestry systems improve human health and digestive capacity. The components of agroforestry systems offer a year-round source of food and revenue.

Enhanced Resource-Use Efficiency: In agroforestry systems, the complementary resource uses of crops and trees can increase the overall efficiency of resource usage. For instance, deep-rooted trees may have less competition from shallow-rooted crops because they can reach water and nutrients in deeper soil layers.

Nutrient Cycling: The addition of trees to a cropping system improves the cycling of nutrients. Decomposition of leaf litter, nitrogen fixation, and mycorrhizal relationships all increase the amount of nutrients available in the soil, which enhances soil fertility and supports crop growth.

Reduced Water Stress: By altering the water balance in the system, shade trees can help the crop of interest. Even though they use water, shade trees have no impact on the quantity of water required by crops in shade systems [48]. By lowering evaporation, offering shade, and enhancing water infiltration, trees in agroforestry systems can aid in water conservation. As a result, crops may experience less water stress, particularly during dry spells, and produce more consistently.

Protection From Wind and Harsh Weather: Agroforestry systems frequently incorporate windbreaks and shelterbelts to shield crops from wind and severe weather. Windbreaks also help to regulate the microclimate. This defense can lessen crop damage from physical impacts, stop soil erosion, and improve the microclimate that supports crop growth.

Natural Enemies: In agroforestry systems, a variety of flora serves as a food source for populations of predators and parasitoids, which are important biological pest controllers. By assisting in the management of insect populations, these natural enemies can lessen crop loss and the need for chemical pesticides.

Attracting Pollinators: In agroforestry systems, trees that bloom well can draw in pollinators such as butterflies and bees. Increased productivity can result from improved pollination services, which can raise the output of crops that depend on pollination.

Carbon Sequestration and Climate Resilience: Agroforestry is frequently regarded as a financially viable method of mitigating climate change. In addition to storing carbon in soils and woody biomass, agroforestry systems lower soil greenhouse gas emissions. The majority of agroforestry systems have the capacity to store carbon; however, this capacity varies depending on the tree species and the management practices [49]. Furthermore, the presence of trees might increase the system's resistance to harsh weather events and climatic fluctuations.

Enhanced Productivity: The availability of tree fodder and the possibility of agro-pastoral integration can both lead to increased livestock productivity in agroforestry systems that incorporate livestock. Within the same system, this integrated strategy benefits cattle as well as crops.

Social and Environmental Benefits: Numerous ecosystem services, such as better water quality, wildlife habitat, and increased aesthetics, can be obtained from agroforestry systems. These services support the general health of the environment and community.

Although agroforestry systems have many advantages, their precise effects on crop yields and productivity can differ based on a number of variables, including the system's design, the species of crop and tree used, and the local environment. All things considered, agroforestry provides an all-encompassing and sustainable method of agricultural production that takes into account social, economic, and environmental objectives.

6.6 CARBON SEQUESTRATION AND CLIMATE MITIGATION

Carbon sequestration and climate mitigation have emerged as topics of utmost importance in discussions taking place all over the world in response to the growing worries regarding climate change. These phrases encompass a variety of approaches that are targeted at lowering the levels of greenhouse gases that are present in the atmosphere, with a specific emphasis on CO_2. In the context of climate change mitigation efforts, carbon sequestration, which refers to the act of absorbing and storing carbon from the atmosphere, is an essential component.

At the core of carbon sequestration is the recognition that elevated levels of CO_2 contribute significantly to the greenhouse effect, exacerbating global warming and its associated impacts. As trees and other plants undergo photosynthesis, they absorb atmospheric CO_2 and convert it into organic carbon, releasing oxygen in the process. Forests, oceans, and soil are natural carbon sinks, acting as repositories for this sequestered carbon.

Forests, in particular, are indispensable in the context of carbon sequestration. Large expanses of trees serve as carbon reservoirs, with the capacity to absorb substantial amounts of CO_2 from the atmosphere [26]. This has led to initiatives such as afforestation and reforestation, emphasizing the importance of preserving existing forests and strategically planting new ones.

Agroforestry, a sustainable land-use practice, also contributes significantly to carbon sequestration. By integrating trees and shrubs with agricultural activities, agroforestry systems enhance the sequestration potential of both ecosystems. Trees, through their biomass, sequester carbon, while the agroforestry model fosters a balanced and integrated approach to land use, promoting biodiversity and resilience.

Beyond natural sinks, technological solutions such as carbon capture and storage (CCS) have gained attention. CCS involves capturing CO_2 emissions from industrial sources and power plants, transporting them, and securely storing them underground. While these technologies show promise, challenges such as cost, scalability, and potential environmental risks necessitate careful consideration.

The significance of carbon sequestration extends beyond merely reducing atmospheric CO_2 levels. It directly aligns with broader climate mitigation goals, aiming to limit the global temperature increase and mitigate the impacts of climate change. By sequestering carbon, we not only mitigate the greenhouse effect but also contribute to the restoration of ecosystems, fostering biodiversity and supporting sustainable land management practices.

Despite the evident benefits, challenges persist in the pursuit of effective carbon sequestration and climate mitigation. Deforestation, land-use changes, and the degradation of natural carbon sinks counteract sequestration efforts. Addressing these challenges requires a multipronged approach that combines conservation, sustainable land management, and global cooperation.

In conclusion, carbon sequestration and climate mitigation are integral components of the collective effort to combat climate change. Whether through the preservation of natural carbon sinks, the implementation of agroforestry practices, or the development of innovative technologies, the journey toward a sustainable future requires a steadfast commitment to reducing carbon emissions and enhancing our planet's capacity to sequester carbon. In these endeavors lies the promise of a more resilient and harmonious coexistence with the environment.

6.6.1 THE CONTRIBUTION OF AGROFORESTRY TO CARBON SEQUESTRATION AND MITIGATING CLIMATE CHANGE

Agroforestry, the integrated approach of combining trees and shrubs with agricultural crops, stands as a significant contributor to carbon sequestration and plays a vital role in mitigating climate change. This practice addresses the dual challenge of enhancing food production while promoting environmental sustainability.

One of the key contributions of agroforestry lies in its ability to sequester CO_2 from the atmosphere. Trees, through the process of photosynthesis, absorb CO_2 and convert it into organic carbon, storing it in their biomass. The carbon sequestration potential of trees in agroforestry systems is substantial, as they accumulate biomass over time, serving as long-term carbon sinks. The diverse range of species in agroforestry also enhances biodiversity, fostering resilient ecosystems that contribute to overall environmental health.

Furthermore, agroforestry mitigates climate change by promoting sustainable land-use practices. Traditional agriculture often involves deforestation, leading to the release of stored carbon into the atmosphere. In contrast, agroforestry systems maintain or enhance carbon stocks in the soil and vegetation, acting as a counterforce to deforestation. The integration of trees within agricultural landscapes also improves soil structure, fertility, and water retention, enhancing the overall resilience of the ecosystem.

The impact of agroforestry on climate change mitigation extends beyond carbon sequestration. By providing shade, windbreaks, and reducing soil erosion, agroforestry helps create microclimates conducive to crop growth. This can contribute to increased agricultural productivity, creating a positive feedback loop by simultaneously addressing food security and climate resilience.

Agroforestry emerges as a practical and sustainable solution in the fight against climate change. Its dual benefits of carbon sequestration and improved land-use practices position it as a valuable strategy for mitigating the impacts of climate change while supporting livelihoods and food production. As the world grapples with the challenges of a changing climate, agroforestry stands out as a holistic approach that aligns environmental conservation with agricultural sustainability.

6.7 MULTITROPHIC INTERACTIONS INFLUENCING CARBON DYNAMICS

Multitrophic interactions, involving various levels of the food chain and ecosystem, play a crucial role in shaping carbon dynamics within ecosystems. These interactions influence the cycling and distribution of carbon through intricate relationships among plants, herbivores, predators, decomposers, and microorganisms [47].

At the foundation of multitrophic interactions are the primary producers—plants. Through photosynthesis, plants capture CO_2 from the atmosphere and convert it into organic carbon compounds. However, the fate of this captured carbon is not solely determined by the plant itself; it is profoundly influenced by interactions with other trophic levels.

Herbivores, the primary consumers in the trophic hierarchy, play a significant role in carbon dynamics. When herbivores graze on plants, they influence the growth and allocation of carbon within the vegetation. Herbivory-induced stress responses in plants can affect photosynthetic rates, altering the amount of carbon sequestered [26]. Additionally, the deposition of herbivore feces and urine contributes to nutrient cycling and influences soil microbial activity, affecting the decomposition of organic matter and the release of carbon into the soil.

Predators, the secondary consumers, further shape carbon dynamics by regulating herbivore populations. The presence of predators can indirectly enhance carbon sequestration by preventing overgrazing and allowing vegetation to thrive. This predator-mediated control of herbivores creates a cascade effect on the entire ecosystem, influencing the abundance and distribution of carbon across trophic levels.

Decomposers, including bacteria and fungi, represent another critical component of multitrophic interactions. They break down organic matter, releasing CO_2 back into the atmosphere during the process of decomposition. However, the efficiency of decomposition is influenced by various factors, including the quality of plant litter, microbial activity, and the presence of detritivores. The interactions among decomposers, detritivores, and plants influence the rate of carbon turnover in the ecosystem.

In conclusion, multitrophic interactions intricately weave the story of carbon dynamics within ecosystems. From plants to herbivores, predators to decomposers, the relationships among these trophic levels influence carbon sequestration, nutrient cycling, and overall ecosystem functioning.

Understanding these interactions is essential for comprehending how ecosystems respond to environmental changes and for implementing effective strategies for carbon management and climate change mitigation.

6.8 CASE STUDIES

Agroforestry systems worldwide showcase the successful integration of trees, crops, and livestock, exemplifying the positive impact of multi-trophic interactions on sustainable agriculture, biodiversity, and carbon dynamics. Here are a few compelling case studies that illustrate the efficacy of agroforestry:

6.8.1 TAUNGYA SYSTEM IN SOUTHEAST ASIA

Southeast Asia has witnessed the success of the Taungya system, where fast-growing trees are intercropped with food crops. Initially developed in Myanmar, this system has been adopted in various countries such as Indonesia and Thailand. Farmers plant trees in the early stages of the crop cycle, providing shade and windbreaks. The trees enhance carbon sequestration, and their leaf litter contributes to nutrient cycling, benefiting both the crops and the overall ecosystem.

6.8.2 TYPES OF TAUNGYA

Sl. No.	Type	Description
1	Departmental	Under this, the forest department raises plantations and agricultural crops. The primary goal of growing agricultural products and maintaining plantations is to keep the land clear of undesired vegetation by hiring daily paid laborers.
2	Leased	The individual who offers the most money for growing crops for a predetermined period of years and ensures maintenance of the tree plantation is granted a lease on the plantation land.
3	Village	The most successful among the three Taungya systems. In this method, the individuals who have established residence in a hamlet located within the forest cultivate crops for this specific objective. Typically, each family possesses between 0.8 to 1.7 hectares of land on which they engage in tree cultivation and crop farming for a duration of 3 to 5 years.

6.8.3 SILVOPASTORAL SYSTEMS IN LATIN AMERICA

Silvopastoral systems integrate trees with livestock grazing. In Colombia, for instance, farmers have embraced this approach to improve forage quality, provide shade for livestock, and sequester carbon. Trees like *L. leucocephala* contribute nitrogen to the soil, enhancing grass growth. The system's multitrophic interactions support a diverse array of wildlife, promoting ecological balance [17].

6.8.4 ALLEY CROPPING IN SUB-SAHARAN AFRICA

In Sub-Saharan Africa, alley cropping has proven successful in addressing soil fertility and enhancing food production. Farmers plant rows of trees in between rows of food crops, forming "alleys." The pruned branches of the trees serve as valuable green manure, enriching the soil with organic matter and nutrients. This system exemplifies how multitrophic interactions enhance soil fertility, benefitting both the trees and the associated crops.

6.8.5 HOME GARDENS IN SOUTHEAST ASIA

Home gardens, prevalent in countries such as Indonesia and Malaysia, represent diverse and integrated agroforestry systems around households. Fruit trees, timber species, and various crops are strategically planted to maximize space and resources. The diverse plantings support a multitude of interactions—from beneficial insect pollinators to soil microorganisms. The organic waste generated is often recycled, contributing to nutrient cycling and minimizing environmental impact.

6.8.6 AGROFORESTRY IN THE UNITED STATES—WALNUT-APPLE SYSTEMS

In parts of the United States, agroforestry is practiced with walnut and apple orchards. The trees provide timber and fruit, and the leaf litter acts as a natural mulch, reducing the need for synthetic inputs. The root exudates from trees enhance microbial activity in the soil, influencing nutrient availability for the crops. This multitrophic interaction fosters a more resilient and sustainable agricultural system.

6.8.7 NERALE (AMLA) AGROFORESTRY IN KARNATAKA

In Karnataka, farmers have adopted agroforestry systems centered around the cultivation of Nerale (Indian gooseberry or Amla) trees. Nerale is a hardy tree that thrives in arid and semiarid regions. Farmers intercrop Nerale with various crops such as millets, pulses, and oilseeds. The trees provide shade, improve soil fertility, and contribute to water conservation. The leaves and organic matter from Nerale trees enhance nutrient cycling, fostering a more sustainable and productive agricultural landscape. This agroforestry practice has not only improved the livelihoods of farmers but also contributed to increased resilience in the face of climate variability.

6.8.8 SUBABUL AGROFORESTRY IN ANDHRA PRADESH

Subabul (*L. leucocephala*) is a fast-growing tree species that has found widespread use in agroforestry systems in Andhra Pradesh. Farmers intercrop Subabul with crops such as groundnut, maize, and sorghum. The nitrogen-fixing properties of Subabul improve soil fertility, benefiting the associated crops. The pruned branches of Subabul serve as valuable fodder for livestock, creating an integrated and sustainable system that addresses both food and fodder needs. This agroforestry model has demonstrated positive outcomes in terms of increased crop yields, improved soil health, and enhanced fodder availability for livestock, contributing to the overall well-being of farming communities.

These case studies underscore the versatility of agroforestry systems and the positive outcomes derived from multitrophic interactions. The success stories highlight the potential for increased agricultural productivity, enhanced ecosystem services, and climate change mitigation through the integration of trees with traditional farming practices. As global agriculture faces the challenges of climate change and sustainability, these examples provide valuable insights into the practical application of agroforestry principles for the benefit of both farmers and the environment.

6.9 CHALLENGES AND MANAGEMENT STRATEGIES FOR MULTITROPHIC INTERACTIONS IN AGROFORESTRY

Agroforestry systems, characterized by the integration of trees with crops and often livestock, rely on intricate multitrophic interactions that involve various levels of the food chain and ecosystem. While these interactions contribute to the resilience and sustainability of agroforestry, managing them effectively presents unique challenges. Here, we explore potential challenges and propose strategies to optimize multitrophic interactions for sustainable agriculture [26].

6.9.1 CHALLENGES

1. **Competition for Resources:** One significant challenge in agroforestry is the competition for resources among trees, crops, and sometimes livestock. Root competition for water and nutrients can occur, affecting the growth and productivity of both trees and crops. Striking a balance that meets the needs of all components in the system is crucial.
2. **Pest and Disease Dynamics:** Multitrophic interactions can influence pest and disease dynamics. While some tree species may act as hosts for certain pests or diseases, they can also harbor natural enemies that control pests. Finding a balance that minimizes pest damage while preserving beneficial organisms is a delicate task.
3. **Complexity in Ecosystem Dynamics:** Agroforestry systems are inherently complex, with multiple species interacting simultaneously. Understanding the dynamics of these interactions is challenging, as they can be influenced by factors such as climate, soil conditions, and management practices. The complexity makes predicting outcomes and managing interventions more intricate.
4. **Biodiversity and Ecosystem Services:** While agroforestry aims to enhance biodiversity and ecosystem services, improper management may lead to unintended consequences. Excessive tree cover may reduce light availability for crops, affecting their growth. Striking the right balance to optimize biodiversity while maximizing agricultural productivity is a persistent challenge.

5. **Temporal and Spatial Dynamics:** Multitrophic interactions can vary over time and space. The effectiveness of natural enemies in pest control, for instance, may fluctuate seasonally. Managing these temporal and spatial dynamics requires adaptive management strategies that consider the dynamic nature of agroforestry systems.

6.10 MANAGEMENT STRATEGIES

1. **Comprehensive Planning and Design:** A holistic approach to planning and designing agroforestry systems is crucial. Considering the specific needs and interactions of each species, farmers can strategically place trees, crops, and livestock to minimize competition and maximize synergies. This involves selecting tree species that complement rather than compete with crops for resources.
2. **Species Selection and Diversity:** Careful selection of tree and crop species is essential for successful multitrophic interactions. Choosing tree species that provide ecosystem services, such as nitrogen fixation or pest control, enhances the overall system. Incorporating diverse species increases biodiversity, creating a resilient system that can adapt to changing conditions.
3. **IPM:** Implementing IPM practices is vital for managing pest and disease dynamics. Monitoring pest populations, using resistant crop varieties, and promoting the presence of natural enemies through habitat management contribute to sustainable pest control. Using biological control methods, such as introducing beneficial insects, can help maintain a balance in the ecosystem.
4. **Agroecological Practices:** Applying agroecological principles, such as crop rotation and polyculture, can enhance the resilience of agroforestry systems. These practices reduce the risk of pests and diseases while optimizing resource use. Integrating cover crops into the system contributes to soil health and nutrient cycling.
5. **Education and Capacity Building:** Farmers play a central role in managing agroforestry systems. Providing education and

capacity-building initiatives can empower farmers with the knowledge and skills needed to understand and optimize multi-trophic interactions. Training programs on sustainable practices and adaptive management contribute to the long-term success of agroforestry.

6. **Adaptive Management and Research:** Recognizing the dynamic nature of agroforestry systems, an adaptive management approach is crucial. Regular monitoring and research to understand changing interactions and ecosystem dynamics can inform adaptive strategies. Farmer participation in research initiatives facilitates the co-creation of knowledge and the development of context-specific solutions.
7. **Policy Support:** Governments and policymakers can contribute to the success of agroforestry by implementing supportive policies. Incentives for farmers adopting agroforestry practices, research funding, and policies that promote sustainable land management can create an enabling environment for successful multitrophic interactions.

Therefore, managing multitrophic interactions in agroforestry requires a thoughtful and adaptive approach. While challenges exist, the potential benefits in terms of increased resilience, biodiversity, and sustainability make agroforestry a compelling solution for the future of agriculture [17]. Through careful planning, diverse species selection, and ongoing education and research, farmers can optimize multitrophic interactions to create productive and ecologically sound agroforestry systems.

6.10.1 FUTURE DIRECTIONS AND RESEARCH NEEDS IN AGROFORESTRY

As agroforestry continues to gain recognition for its potential in promoting sustainable agriculture, there are several gaps in current research that warrant further investigation. Addressing these gaps and advancing our understanding of multitrophic interactions within agroforestry systems can pave the way for more effective and resilient agricultural practices. Here, we explore future directions and research needs in agroforestry to optimize multitrophic interactions.

6.10.2 IDENTIFIED GAPS AND RESEARCH NEEDS

1. **Quantifying Ecosystem Services:** While the potential benefits of agroforestry in terms of ecosystem services are acknowledged, quantifying these services remains a challenge. Research is needed to develop standardized methods for assessing and quantifying the various services provided by agroforestry systems, including carbon sequestration, water regulation, and biodiversity enhancement. Understanding the economic value of these services can further incentivize adoption.
2. **Climate Change Resilience:** With climate change posing new challenges to agriculture, there is a need for research on how agroforestry systems can enhance climate change resilience. This includes investigating the role of diverse tree–crop combinations in buffering against extreme weather events, improving water-use efficiency, and contributing to climate change adaptation strategies for farmers.
3. **Interactive Effects of Climate and Multitrophic Interactions:** The interactive effects of climate change and multitrophic interactions in agroforestry systems are not fully understood. Research should focus on how changing climate conditions influence the dynamics of pests, diseases, and natural enemies within agroforestry systems. This knowledge can inform adaptive management strategies to mitigate potential risks.
4. **Innovative Agroforestry Models:** While traditional agroforestry models have proven successful, there is a need for research into innovative and context-specific agroforestry models. This includes exploring novel combinations of tree and crop species, assessing the potential for integrating nontraditional crops, and understanding the implications of agroforestry in different agro-ecological zones.
5. **Long-Term Studies:** Many existing studies on agroforestry are of short duration, limiting our understanding of the long-term dynamics and sustainability of these systems. Long-term studies are essential to monitor changes in soil health, tree–crop interactions, and overall system performance over extended periods.

This will provide insights into the durability and resilience of agroforestry practices.

6. **Social and Economic Dimensions:** Understanding the social and economic dimensions of agroforestry adoption is critical for its successful implementation. Research should explore the socio-economic factors influencing farmer decision-making, the impact of agroforestry on livelihoods, and the role of community engagement in the adoption and management of agroforestry systems.
7. **Policy and Institutional Support:** Research is needed to assess the effectiveness of existing policies and institutional frameworks in supporting agroforestry adoption. Identifying barriers and incentives for farmers, evaluating the impact of subsidies and extension services, and recommending policy adjustments to promote sustainable agroforestry practices are essential research areas.

6.10.3 POTENTIAL FOR IMPROVING AGROFORESTRY SYSTEMS

1. **Enhanced Tree–Crop Interactions:** Research can delve into understanding specific tree–crop interactions that lead to mutual benefits. This includes exploring root interactions, nutrient exchange mechanisms, and allelopathic effects that positively influence both tree and crop growth. Optimizing these interactions can enhance overall system productivity.
2. **Precision Agroforestry:** Advances in technology can be harnessed for precision agroforestry. Research on precision tree planting, using remote sensing for monitoring tree and crop health, and developing decision support tools can contribute to more efficient and precise agroforestry management.
3. **Integration of Agroforestry into Digital Agriculture:** Research should explore the integration of agroforestry practices into digital agriculture platforms. Using data analytics, sensors, and modeling tools can enhance the monitoring and management of agroforestry systems. This can empower farmers with real-time information for decision-making.

4. **Agroforestry in Urban and Periurban Areas:** Investigating the potential for agroforestry in urban and peri-urban areas is a promising avenue. Research can explore how small-scale agroforestry systems can contribute to urban food security, green spaces, and climate resilience. Understanding the social dynamics of urban agroforestry is also crucial.
5. **Multistakeholder Collaboration:** Research should focus on the development of collaborative models involving researchers, farmers, policymakers, and local communities. This interdisciplinary approach can facilitate the co-creation of knowledge, ensuring that research outcomes are relevant, accessible, and applicable in diverse contexts.
6. **Agroforestry for Ecosystem Restoration:** Investigating the potential of agroforestry in ecosystem restoration efforts is essential. Research can explore how agroforestry practices can contribute to reforestation, soil rehabilitation, and biodiversity conservation, aligning with global goals such as the Bonn Challenge and the UN Decade on Ecosystem Restoration.
7. **Capacity Building and Extension Programs:** Research should assess the effectiveness of capacity-building programs and extension services in promoting agroforestry adoption. Understanding the learning needs of farmers, developing targeted training modules, and evaluating the impact of extension efforts are critical for scaling up agroforestry practices.

Addressing these research needs and exploring future directions in agroforestry can unlock its full potential as a sustainable and resilient agricultural system. As global challenges such as climate change, food security, and biodiversity loss intensify, agroforestry stands as a dynamic solution that, with continued research and innovation, can contribute significantly to building more sustainable and regenerative agricultural landscapes.

6.11 CONCLUSION

The intricate dance of multitrophic interactions within agroforestry systems paints a compelling picture of sustainable and resilient agriculture. As we navigate the complexities of a changing climate and increasing demands on our agricultural landscapes, understanding and harnessing these interactions become paramount. The synergy between trees, crops, and the myriad of organisms within these ecosystems holds the key to unlocking a harmonious balance that benefits both the environment and agricultural productivity.

Agroforestry emerges not only as a land-use strategy but also as a holistic paradigm that integrates ecological principles with the practical needs of farmers. From carbon sequestration to pest control, nutrient cycling to enhanced biodiversity, the multifaceted benefits of multitrophic interactions contribute to the overarching goal of sustainable agriculture. The success stories from diverse regions, whether the rolling hills of Uttarakhand or the vibrant landscapes of Southeast Asia, exemplify the adaptability and versatility of agroforestry systems.

As we look toward the future, it is imperative to address the gaps in our knowledge through continued research and innovative practices. Long-term studies, precision agroforestry, and a deeper understanding of socioeconomic dynamics will be pivotal. Agroforestry's potential extends beyond the confines of traditional agriculture, offering solutions for climate change adaptation, ecosystem restoration, and community resilience.

In the tapestry of agroforestry, where trees stand as pillars of sustainability and crops weave the fabric of nourishment, multitrophic interactions form the intricate patterns that define a regenerative and harmonious landscape. It is a call to action for researchers, policymakers, and farmers alike to embrace the richness of these interactions, fostering a future where agriculture is not merely a means of sustenance but a stewardship of the land for generations to come.

KEYWORDS

- **multitrophic interaction**
- **agroforestry ecosystems**
- **land-use system**
- **crop pollination**
- **tree-crop interactions**

REFERENCES

1. Lungdren, B. O., Raintree, J. B. (1982). Sustained agroforestry. In: *Agricultural Research for Development. Potentials and Challenges in Asia* (pp. 37–49).
2. Chundawat, B. S., Gautam, S. K. (1993). *Textbook of Agroforestry*, Oxford and IBH Publishing.
3. Davis, A. S., Hill, J. D., Chase, C. A., Johanns, A. M., Liebman, M. (2012). Increasing cropping system diversity balances productivity, profitability and environmental health. *PLoS One*, 7(10), 7149.
4. Winans, K. S., Tardif, A. S., Lteif, A. E., Whalen, J. K. (2015). Carbon sequestration potential and cost-benefit analysis of hybrid poplar, grain corn and hay cultivation in southern Quebec, Canada. *Agroforestry System*, 89, 421–433.
5. Dixon, R. K., Winjum, J. K., Andrasko, K. J., Lee, J. J., Schroeder, P. E. (1994). Integrated land-use systems: Assessment of promising agroforest and alternative land-use practices to enhance carbon conservation and sequestration. *Climate Change*, 27, 71–92.
6. Jordan, N. R., Davis, A. S. (2015). Middle-way strategies for sustainable intensification of agriculture. *Bioscience*, 6, 513–519.
7. Baah-Acheamfour, M., Carlyle, C. N., Bork, E. W., Chang, S. X. (2014). Trees increase soil carbon and its stability in three agroforestry systems in central Alberta, Canada. *Forest Ecology and Management*, 328, 131–139.
8. Caudill, S. A., DeClerck, F. J. A., Husband, T. P. (2015). Connecting sustainable agriculture and wildlife conservation: Does shade coffee provide habitat for mammals. *Agriculture, Ecosystems & Environment*, 199, 85–93.
9. Brandle, J. R., Schoeneberger, M. M. (2014). Working trees: Supporting agriculture and healthy landscapes. *Journal of Tropical Forest Science*, 26, 305–308.
10. Gold, M., Cernusca, M., Hall, M. (2006). *Training Manual for Applied Agroforestry Practices*, University of Missouri.
11. Etherington, J. R. (1975). *Environment and Plant Ecology*.
12. Putnam, A. R., Tang, C. S. (1986). *The Science of Allelopathy*. Wiley Inter-Science.
13. Willey, R. W. (1975). The use of shade in coffee, cocoa and tea. *Horticultural Abstracts*, 45, 791–7898.
14. Lass, R. A. (1985). Diseases, In: Wood, G. A. R., Lass, R. A. (Eds.), *Cocoa* (4th edn.), Longman: London, UK.
15. Bentrup, G., Hopwood, J., Adamson, N. L., Vaughan, M. (2019). Temperate agroforestry systems and insect pollinators: A review. *Forests*, *10*(11), 981.
16. Duchene, O., Vian, J. F., Celette, F. (2017). Intercropping with legume for agroecological cropping systems: Complementarity and facilitation processes and the importance of soil microorganisms. A review. *Agriculture, Ecosystems & Environment*, 240, 148–161.
17. Torralba, M., Fagerholm, N., Burgess, P. J., Moreno, G., Plieninger, T. (2016). Do European agroforestry systems enhance biodiversity and ecosystem services? A meta-analysis. *Agriculture, Ecosystems & Environment*, 230, 150–161.
18. Perfecto, I., Vandermeer, J. H., Bautista, G. L., Nuñez, G. I., Greenberg, R., Bichier, P., Langridge, S. (2004). Greater predation in shaded cofee farms: The role of resident neotropical birds. *Ecology*, 85, 2677–2681.

19. Bianchi, F. J. J. A., Goedhart, P. W., Baveco, J. M. (2008). Enhanced pest control in cabbage crops near forest in The Netherlands. *Landscape Ecology*, 23, 595–602.
20. Tscharntke, T., Clough, Y., Bhagwat, S. A., Buchori, D., Faust, H., Hertel, D., Hölscher, D., Juhrbandt, J., Kessler, M., Perfecto, I., Scherber, C., Schroth, G., Veldkamp, E., Wanger, T. C. (2011). Multifunctional shade-tree management in tropical agroforestry landscapes—A review. *Journal of Applied Ecology*, 48, 619–629.
21. Karp, D. S., Mendenhall, C. D., Sandí, R. F., Chaumont, N., Ehrlich, P. R., Hadly, E. A., Daily, G. C. (2013). Forest bolsters bird abundance, pest control and coffee yield. *Ecology Letters*, 16, 1339–1347.
22. Smagghe, G., Boecking, O., Maccagnani, B., Mand, M., Kevan, P. G. (2020). *Entomovectoring for Precision Biocontrol and Enhanced Pollination of Crops*, Springer Nature.
23. Fahad, S., Chavan, S. B., Chichaghare, A. R., Uthappa, A. R., Kumar, M., Kakade, V., Pradhan, A., Jinger, D., Rawale, G., Yadav, D. K., Kumar, V., Farooq, T. H. , Ali, B., Sawant, A. V., Saud, S., Chen, S., Poczai, P. (2022). Agroforestry systems for soil health improvement and maintenance. *Sustainability*, 14(22), 14877.
24. BogaleG. A., Bekele, S. E. (2023). Sustainability of agroforestry practices and their resilience to climate change adaptation and mitigation in Sub-Saharan Africa: A review. *Ekologia*, 42(2), 179–192.
25. Thevathasan, N. V., Gordon, A. M. (2003). Refining agroforestry systems for sustainable agro-ecosystems in temperate North America. *XII World Forestry Congress*.
26. Noordwijk, M. V., Lawson, G., Soumare, A., Groot, J. J. R., Hairiah, K. (2015). Root distribution of trees and crops: Competition and/or complementarity. *Tree-crop Interactions: Agroforestry in a Changing Climate*, 2(8), 221–257.
27. Santa, R. I., Tarazona, T. (2001). Nutrient cycling in a natural beech forest and adjacent planted pine in northern Spain. *Forestry*, 74, 11–28.
28. Hairiah, K., Widianto, W., Suprayogo, D., Noordwijk, M. V. (2020). Tree roots anchoring and binding soil: Reducing landslide risk in Indonesian agroforestry. *Land*, *9*(8), 256.
29. Du, X., Jian, J., Du, C., Stewart, R. D. (2022). Conservation management decreases surface runoff and soil erosion. *International Soil and Water Conservation Research*, 10, 188–196.
30. Ma, W., Tang, S., Dengzeng, Z., Zhang, D., Zhang, T., Ma, X. (2022). Root exudates contribute to belowground ecosystem hotspots: A review. *Frontiers in Microbiology*, 13, 937940.
31. Mendes, R., Garbeva, P., Raaijmakers, J. M. (2013). The rhizosphere microbiome: Significance of plant beneficial, plant pathogenic, and human pathogenic microorganisms. *FEMS Microbiology Reviews*, 37(5), 634–663.
32. Xie, Y., Tian, L., Han, X., Yang, Y. (2021). Research advances in allelopathy of Volatile Organic Compounds (VOCs) of plants. *Horticulturae*, 7(9), 278.
33. Wei, X., Shao, M., Gale, W., Li, L. (2014). Global pattern of soil carbon losses due to the conversion of forests to agricultural land. *Scientific Reports*, 4, 4062.
34. Lal, R. (2005). Forest soils and carbon sequestration. *Forest Ecology and Management*, 220, 242–258.

35. Liu, S., Yang, R., Peng, X., Hou, C., Ma, J., Guo, J. (2022). Contributions of plant litter decomposition to soil nutrients in ecological tea gardens. *Agriculture*, 12(7), 957.
36. Ratnadass, A., Fernandes, P., Avelino, J., Habib, R. (2011). Plant species diversity for sustainable management of crop pests and diseases in agroecosystems: A review. *Agronomy for Sustainable Development*, 32, 273–303.
37. Rezende, M. Q. Venzon, M., Perez, A. L., Cardoso, I. M., Janssen, A. (2014). Extrafloral nectaries of associated trees can enhance natural pest control. *Agriculture, Ecosystems & Environment*, 118, 198–203.
38. Sarkar, S. C., Wang, E., Wu, S., Lei, Z. (2018). Application of trap cropping as companion plants for the management of agricultural pests: A review. *Insects*, 9(4), 128.
39. Parker, J. E., Snyder, W. E., Hamilton, G. C., Saona, S. R. (2012). *Companion Planting and Insect Pest Control*, IntechOpen.
40. Udawatta, R. P., Rankoth, L., Jose, S. (2019). Agroforestry and biodiversity. *Sustainability*, 11(10), 2879.
41. Stamps, W. T., Linit, M. J. (1997). Plant diversity and arthropod communities: Implications for temperate agroforestry. *Agroforestry Systems*, 39(1), 73–89.
42. Vaughan, M., Adamson, N., MacFarland, K. (2017). *Using Agroforestry Practices to Reduce Pesticide Risks to Pollinators & Other Agriculturally Beneficial Insects*. USDA.
43. Ronisha, B., Solanki, Arora, S. (2015). Leaf litter dynamics in Agroforestry system affecting microbial activity in Saline Soils. *Journal of Soil & Water Conservation*, 14(4), 333–339.
44. Aldana, P., Rahman, M. M., Lezama, P., Cruz, I. (2019). Litter decomposition process in coffee agroforestry systems. *Journal of Forest and Environmental Science*, 35(2), 121–139.
45. Elkan, G. H. (1984). Taxonomy and metabolism of Rhizobium and its genetic relationships. In: Alexander, M. (Ed.), *Biological Nitrogen Fixation, Ecology, Technology, and Physiology*. Plenum Press: New York.
46. Wahab, A., Muhammad, M., Munir, A., Abdi, G., Zaman, W., Ayaz, A., Khizar, C., Reddy, S. P. P. Role of arbuscular mycorrhizal fungi in regulating growth, enhancing productivity, and potentially influencing ecosystems under abiotic and biotic stresses. *Plants*, *2*(17), 3102.
47. Datta, P., Behera, B., Rahut, D. B. (2023). Assessing the role of agriculture-forestry-livestock nexus in improving farmers' food security in South Asia: A systematic literature review. *Agricultural Systems*, 213, 103807.
48. Lin, B. B. (2010). The role of agroforestry in reducing water loss through soil evaporation and crop transpiration in coffee agroecosystems. *Agricultural and Forest Meteorology*, 150(4), 510–518.
49. Dhyani, S. K., Ram, A., Dev, I. (2016). Potential of agroforestry systems in carbon sequestration in India. *Indian Journal of Agricultural Sciences*, 86(9), 1103–1112.

CHAPTER 7

Multitrophic Pollination Networks for Biodiversity and Crop Yield Enhancement

SUNAULLAH BHAT, NUTAN KARKI, and AMIT UMESH PASCHAPUR*

Crop Protection Division, ICAR-Vivekananda Parvatiya Krishi Anusandhan Sansthan, Almora, Uttarakhand, India

*Corresponding author

ABSTRACT

In the face of biodiversity loss and increasing global food demand, multitrophic pollination networks offer a crucial strategy for sustainable agriculture and ecosystem health. These networks extend beyond the traditional plant–pollinator interactions, encompassing a web of relationships among plants, pollinators, herbivores, predators, parasites, and competitors. Their combined activities influence not only plant reproduction and crop yields but also biodiversity conservation, pest regulation, and ecological stability. This chapter explores the significance of multitrophic pollination networks by highlighting their role in enhancing crop productivity and maintaining ecosystem resilience. Effective pollination outcomes often depend on the presence of multiple interacting species, which collectively support ecosystem services. Biodiversity conservation benefits from such networks as they sustain species richness and functional diversity. In crop pollination, multitrophic interactions improve yield and quality by ensuring diverse and effective pollinator services, contributing to global food security. Pest regulation is strengthened when natural enemies recruited through these networks help control herbivores, reducing reliance on chemical pesticides. Moreover, the ecological stability provided by these

Agroforestry and Climate Change Mitigation: Harnessing Multitrophic Interactions for Sustainable Plant Protection and Ecosystem Resilience. Hanuman Singh Jatav, Vishnu D. Rajput, Ipsita Samal, Tanmaya Kumar Bhoi (Eds.)

DOI: 10.1201/9781779641397-7

networks ensures the long-term maintenance of essential ecosystem functions. Recognizing and managing the complexity of multitrophic pollination networks is therefore vital for achieving sustainable agriculture and conserving biodiversity, fostering resilience in a changing world.

7.1 INTRODUCTION

One of the biggest challenges in ecology is keeping biodiversity in place, which is extremely important for ecosystem health [1]. Plants and animals interact to maintain biodiversity throughout the world, which is why this chapter is focused on understanding how these interactions contribute to biodiversity [2]. Pollinators play a crucial role in global biodiversity, serving as vital components of ecosystems [3]. Their significance lies in ensuring the continuity of essential ecological processes, particularly pollination. In agriculture, the pollination of crop plants by animals, primarily insects, is a crucial trophic interaction, supporting the formation of yields in 75% of all crops [4, 5]. This process significantly contributes to the reproductive success of both native and cultivated plants, highlighting the interconnectedness of pollinators with the overall health and functioning of ecosystems. The improvement of pollination service in agriculture is directly related to high yield [6–8].

As more and more crops around the world need insect-mediated pollination, bee has a bigger job for the same [9]. Global studies indicate a decline in wild pollinator diversity, particularly bees. Concerns usually revolve around the impact on crop yield and global food security. Surprisingly, about 80% of global pollination services come from just 2% of pollinator species. The vast majority of the approximately 20,000 bee species globally, including those occasionally visiting crops, lack direct economic importance due to mismatches in foraging ranges and crop locations or flowering times. Most crops benefit from pollination, yet they represent only a small portion of flowering plant biodiversity, insufficient to support many pollinator species.

Some plant species play crucial roles in providing nesting and alternative flower resources for crop pollinators, while others have no direct or indirect contribution to ecosystem services. Bees play a crucial role in pollinating many of the cross-pollinated crops, vegetables, fruits, and nuts that make our diets healthier. They provide important nutrients that go along with our regular foods, helping to balance our diets and prevent nutrient shortages [10–12]. In general, pollinator diversity leads to improved pollination [13] and also enhanced quality as well as quantity of crop [5–7, 14]. Intensification of agriculture, characterized by heightened agrochemical usage and landscape simplification, is identified as a significant threat to pollination services. This is attributed to the resultant decline in both pollinator abundance and diversity [15].

7.2 POLLINATION NETWORKS: A MULTITROPHIC PERSPECTIVE

Pollination networks serve as visual representations of the interactions between animals, particularly insects, and plants within a specific locale where these coexisting species are found (Figure 7.1). Ecologists have extensively recognized and characterized the interactions among pollinators [16, 17]. In instances involving eusocial insects such as honey bees and bumble bees, the presence of additional species has been shown to induce resource partitioning through active avoidance [18, 19], modify pollinator foraging behaviors [20], and promote short-term floral specialization [21, 22]. Despite these findings, significant knowledge gaps persist concerning the outcomes of pollinator–pollinator interactions in natural communities. The persistence of these gaps is primarily attributable to the intricate challenge of establishing a connection between the effects of such interactions and population dynamics [17].

FIGURE 7.1 Pollination network: interaction of insects with flowering plants.

Understanding the interactions among pollinators presents a challenge based on real-world observations because environmental conditions can significantly influence how pollinators behave with others of their kind and different species. Specifically, the relationships between plants and pollinators rely on the external conditions faced by the pollinators [23, 24]. When there is a high availability of resources, measured in terms of flower density, pollen, or nectar volume, the duration of foraging trips for bumble bees tends to decrease, and there is an increase in the frequency of floral visits [20, 25].

Additionally, changes in the interactions between insects and plants are influenced by variations in temperature. Elevated temperatures have been noted to reduce the time spent on individual flowers, as seen in some bees exposed to low temperatures [26], and to encourage specialization in floral preferences along an elevation gradient [27]. Therefore, investigating the context in which these interactions unfold is just as significant as examining the interactions themselves.

In addition to these interactions, the abundance and behavior of pollinators can be influenced by indirect ecological interactions with various nonpollinator species across different trophic guilds, including predators and parasitoids (Figure 7.2). This also involves leaf herbivores and soil fungi and bacteria that impact plant biomass, architecture, and physiology [28]. To summarize, crop and natural systems can mutually influence each other through numerous indirect pathways, many of which have not been thoroughly studied.

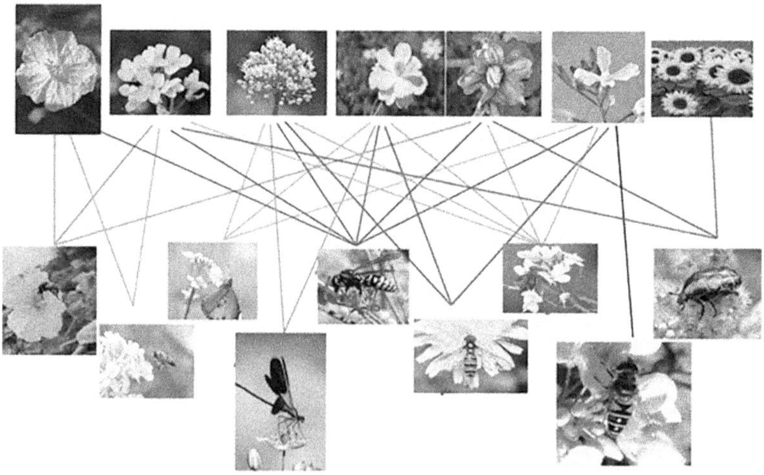

FIGURE 7.2 Interaction of nonpollinators that influence dominant species.

Source: Adapted from Ref. [29]. https://creativecommons.org/licenses/by/4.0/

7.3 BIODIVERSITY CONSERVATION AND MULTITROPHIC INTERACTIONS

Global change (climate change, use of pesticides, etc.) is triggering species loss across several trophic groups, with most potential effects on the services that ecosystems provide to humans. Anticipating the practical outcomes of biodiversity reduction in authentic, multilevel communities continues to be a complex task. The current biodiversity crisis has prompted over two decades of research into the consequences of species loss on both the functioning of the ecosystem as well as societal well-being [30, 31]. Now one of the widely accepted scientific consensuses recognizes that the presence of greater diversity within trophic levels, encompassing the functional traits, genes, and species, leads to greater efficiency in resource capture. This diversity ultimately contributes to heightened biomass production [32], improves the nutrient cycling, and accelerates the decomposition rate [33]. This understanding aligns with the well-established positive and saturating relationship observed within biodiversity and the functioning of ecosystems. The most robust indication supporting these connections often arises from controlled manipulative studies. However, identifying clear and reliable links between the loss of biodiversity and ecological function in the complex natural systems remains challenging [34]. This challenge endures, partly due to the fact that the impacts of diversity in real-world systems can be obscured by a multitude of factors.

Initially, a significant portion of existing studies has concentrated on loss of biodiversity among a single trophic level, commonly known as monotrophic extinction [35, 36]. However, biodiversity loss typically spans multiple trophic levels due to the direct coextinction of the dependent species [37] or through indirect extinction of cascading effects [38]. Functional outcomes of the extinction within two or more trophic levels can exhibit qualitative differences from the predictions as given by monotrophic studies [34]. This distinction arises because the species in different trophic levels influence the function of ecosystem both independently and interactively [38, 39]. Additionally, the properties of the food web significantly impact the order as well as the severity of the secondary extinctions [40, 41].

There is mounting evidence indicating that environmental changes are negatively impacting several species and groups of pollinators, as well as the plants they pollinate, in many locations around the world. These pressures encompass a range of factors, including climate change (Figure 7.3), intensified agricultural practices, diseases, pests, pathogens, and invasive species.

Changes in climate directly affect the phenology and distribution of pollinator species, causing disruptions in the temporal and spatial alignment of plant–pollinator ecological interactions. If the response to climate change leads to alterations in the distribution and blooming periods of flowering plants, the spatial arrangement of pollinators and the timing of their nectar requirements must also adapt. This adjustment can result in a shortage of nectar resources for pollinators during crucial periods. The resulting mismatch holds the potential for cascading effects on ecosystem functions (Figure 7.3), particularly for plants and pollinators with specialized relationships [42].

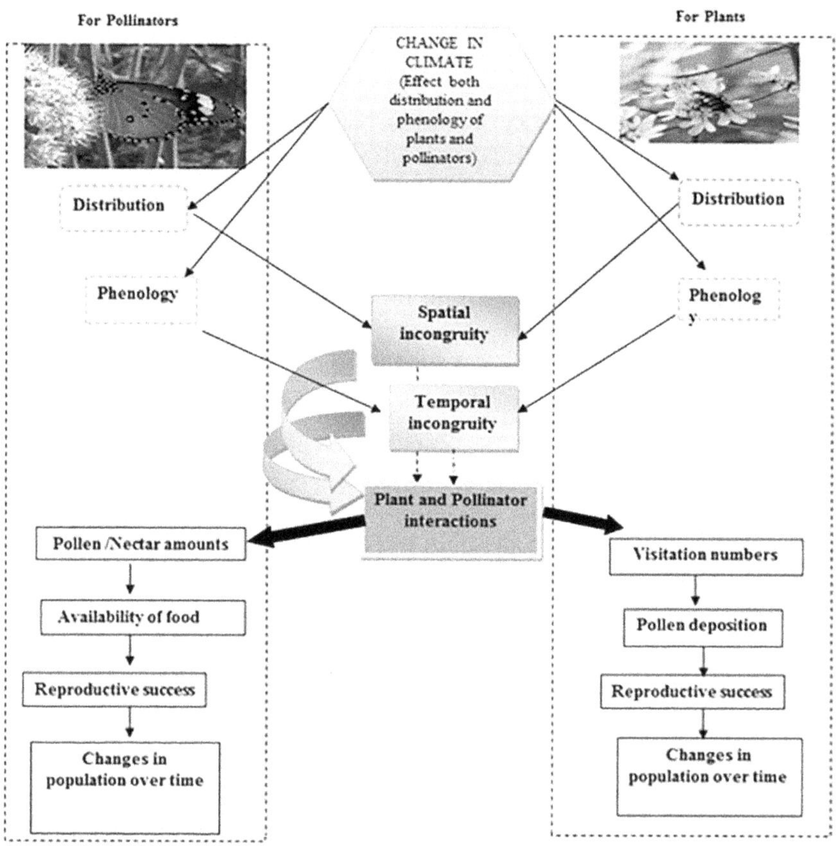

FIGURE 7.3 Impact of climate change on plant–pollinator interactions.

7.4 CROP YIELD ENHANCEMENT THROUGH MULTITROPHIC POLLINATION

Addressing the escalating need for agricultural products while simultaneously mitigating adverse effects on biodiversity and ecosystem health stands out as one of the paramount global challenges [43]. Intensive agricultural practices and the simplification of agroecosystems pose a significant threat to farmland biodiversity and the associated ecosystem services on a global scale [44].

Strategies for yield enhancement in agriculture encompass a diverse range of approaches aimed at optimizing crop productivity while ensuring sustainability [45]. These strategies include the integration of advanced technologies, precision agriculture techniques, and the adoption of innovative farming practices and conserving natural pollinators. Emphasizing the efficient use of resources, such as water and fertilizers, and leveraging genetic advancements through crop breeding and biotechnology are key components of yield enhancement efforts. By embracing a holistic approach that considers environmental impact and resource conservation, farmers can achieve increased yields without compromising long-term ecosystem health and biodiversity [46].

Enhancing yields through pollination entails optimizing the natural pollination process, facilitating the transfer of pollen from male to female plant parts to boost fertilization and increase crop production. This optimization is achieved by promoting the presence and activity of pollinators, such as bees, butterflies, and birds, within agricultural environments [6, 7]. Employing practices that safeguard and enhance pollinator habitats [47], utilizing managed pollination services [6, 7], and selecting crop varieties that attract pollinators are vital strategies for maximizing the beneficial effects of pollination on agricultural yields.

7.5 PEST REGULATION IN MULTITROPHIC NETWORKS

In multitrophic pollination systems, the regulation of pests is intricately connected to the interactions among multiple trophic levels, involving not only pollinators and plants but also natural enemies of pests [48, 49]. This integrated approach to pest regulation relies on the ecological balance within a diverse ecosystem [50, 51]. By fostering a biodiverse environment that supports a variety of species, including pollinators and their natural predators, a more resilient and self-regulating system emerges. This, in turn, contributes to effective pest control, reducing the need for external interventions such as pesticides [52]. Implementing and preserving such multitrophic interactions can be a sustainable strategy for managing pests in agricultural ecosystems while maintaining the health of pollination networks [14].

Most insects as well as other arthropods that are found in the agroecosystems do not feed on or harm the plants. Some of them feed on and destroy insect pests only. In most cases, the biological activity of these beneficial organisms can prevent or greatly limit pest problems. In a multitrophic system, some of the farmer's friends can be categorized broadly as predators, parasites, and parasitoids (Figure 7.4). In the field of entomology, biological control is described as the utilization of insect predators, parasitoids, entomopathogenic nematodes, or microbial pathogens to overcome the populations of different insect pests [53, 54]. Biological control is more unpredictable, time taking, and somewhat less effective than pesticides, that is, chemical control. Though biological control is a safer means of pest control, its consequences are to be studied properly before employing it [55].

The economically important pest species attacking vegetable crops are *Macrosiphum euphorbiae*, *Aphis gossypii*, *Myzus persicae*, and *Aulacorthum solani* [56], whereas the insect pests of ornamentals are *M. euphorbiae*, *M. persicae*, *A. gossypii*, *A. solani*, *Brachycaudus helichrysi*, *Macrosiphoniella sanborni*, and *Macrosiphum rosae* [57]. Studies conducted regarding the effectiveness of ladybird beetles as predators of aphids in greenhouses and glasshouses have found two highly effective ladybird beetles, (1) *Coccinella septempunctata* and (2) *Hippodamia convergens*, to control aphid populations [58]. The aphid gall midge, *Aphidoletes aphidimyza*, serves as an effective control for the outbreak of aphids. The midges are reported for their effectiveness in controlling aphid infestation, controlling more aphids than they actually intake. The larvae of midges established in the aphid population rapidly eradicate the adult aphids [59]. *Tuta absoluta* (the tomato leafminer) is one of the key insect pests of tomato in greenhouses. Its larvae feed both on foliage and tomato fruit. Numerous predatory species, such as *Macrolophus pygmaeus*, *Dicyphus bolivari*, *Dicyphus errans*, *Nabis pseudoferus*, and *Nesidiocoris tenuis*, have been evaluated as the natural enemies of *T. absoluta* [60]. Among parasitoids, *Bracon nigricans*, as tested in lab condition, illustrates good behavioral plasticity to parasitize the *T. absoluta*.

Necremnus tutae, a eulophid parasitoid, was determined to show high mortality of *T. absoluta* larvae [61].

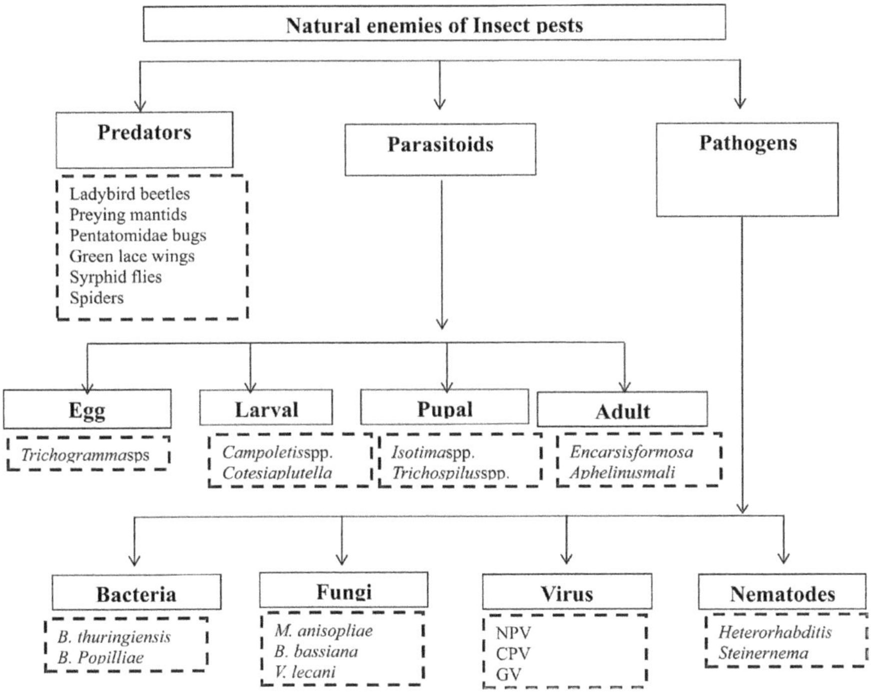

FIGURE 7.4 Natural enemies of insect pests.

7.6 ECOLOGICAL STABILITY IN MULTITROPHIC POLLINATION

Ecological stability in multitrophic pollination systems refers to the resilience and balance maintained within the ecosystem [62], where multiple trophic levels, including plants, pollinators, and natural enemies of pests, interact harmoniously [63]. The intricate networks of mutualistic interactions between pollinator and plant species underscore the interdependence within ecological systems [64]. The decline in pollinator populations can reverberate, disrupting the dynamics and stability of entire plant–pollinator communities [65]. These mutualistic relationships, characterized by diverse and specialized interactions, play a vital role in the reproductive success of plants. As pollinator abundance or diversity diminishes, the repercussions extend beyond individual species, affecting the resilience and health of the broader ecosystem [66]. Recognizing the far-reaching consequences of pollinator decline emphasizes the urgency of conservation efforts to safeguard these intricate ecological relationships. Plant–pollinator networks demonstrate a nested structure, wherein specialist species tend to form connections with subsets of generalist species.

The impact of ecosystem complexity, associated with biodiversity and connectance, on the stability and resilience of plant–pollinator systems is not well understood. While biodiversity is generally acknowledged as a contributor to ecosystem functioning, the effects on stability remain diverse, ranging from positive to negative. Network structure, including nestedness and connectance, also plays a crucial role in ecological stability. For instance, in random interaction networks, increased connectance and/or size may reduce system resilience, leading to instability. The dynamics of predator–prey interactions enhance stability, while mutualistic dynamics characterized by nestedness may favor instability [67].

The interplay between biodiversity and network structure in influencing the stability of plant–pollinator communities requires further investigation. Ecosystems, an intricate system, involve various interactions among diverse species [68]. Traditional ecological theory suggests that increased complexity, in terms of the number and intensity of interactions, destabilizes ecosystems [69]. Mutualistic interactions, such as those between plants and pollinators, have been viewed as potentially destabilizing [70]. However, despite theoretical predictions, mutualisms are prevalent and play a crucial role in maintaining biodiversity, particularly in agricultural ecosystems vital for human well-being [71]. Addressing the gap between theory and observation, we introduce a consumer–resource theory that integrates food webs and mutualistic networks into "multiplex" networks with different interaction types [72]. This multiplex model explores how mutualism influences the diversity, stability, and function of complex ecosystems.

7.7 MANAGEMENT STRATEGIES FOR ENHANCING MULTITROPHIC POLLINATION

Strategies to enhance multitrophic pollination include cultivating diverse floral resources, preserving natural habitats, and implementing Integrated Pest Management (IPM) practices for sustainable pest control.

7.7.1 DIVERSE FLORAL RESOURCES

Cultivate a variety of flowering plants to attract a broad range of pollinators. Diverse flora supports different trophic levels, creating a resilient ecosystem. In recent studies, a positive correlation has been observed between the richness and abundance of floral resources and the diversity and activity of pollinators at the landscape level [4, 73, 74]. These functional relationships, where species from different trophic levels reciprocally influence each other, play an essential role in conserving populations and sustaining the ecosystem services provided by the pollinators. Since pollinating insects are generally more specialized than the plants they pollinate, a decrease in plant species richness is likely to negatively impact pollinator richness. Moreover, the presence of other floral resources, such as natural habitat patches or attractive plant species [75], in the vicinity can enhance pollinator activity toward plants, potentially improving the quality and quantity of pollination services for both wild and crop plant species [4]. Revitalizing the natural diversity of plants by introducing native flowering species within the flight ranges of bees in landscapes with limited resources can offer the essential pollen, nectar, and shelter required by pollinators throughout the growing season. This approach aids in sustaining the populations and diversity of wild bees [76, 77].

7.7.2 NATURAL HABITAT PRESERVATION

Protect and preserve the natural habitats surrounding agricultural areas. These habitats provide nesting sites and resources for pollinators and their natural enemies. Pollinators of different insect orders need nesting for survival as well as to reproduce their future generation.

Most of the bees can be classified into three main categories: (1) solitary, (2) social, and (3) parasitic. Over 75% of known bee species fall under the solitary type, where individual females work independently to construct nests and care for their offspring [78]. The sociality of bees forms a spectrum from solitary nesting to advanced eusociality, with blurred distinctions between each category and instances of social polymorphism [79, 80]. Eusocial species, marked by overlapping generations and a permanent reproductive division of labor with a queen at the helm, include primitively eusocial and socially polymorphic bees, particularly within the ground-nesting Halictidae family [81]. In these species, queens establish nests on their own before handing over responsibilities to subsequent generations of workers. Solitary bees may share communal nests but maintain independent behavior, producing offspring in their own brood cells. In temperate climates, queens of primitively eusocial species emerge at the start of the growing season, select nest locations, and initiate construction. This process continues through generations of workers, males, and the next season's queens [82]. Therefore, conservation strategies such as providing trap nests, conservation of natural habitats where these bees construct nests is the need of hours.

7.7.3 INTEGRATED PEST MANAGEMENT

Implement IPM practices to control pests. This reduces the need for chemical pesticides, promoting a healthier environment for pollinators and their associated species. The concept of IPM was introduced in the late 1950s and widely adopted during the 1970s and 1980s. IPM is a long-term management approach, which combines biological, cultural, and chemical strategies to combat pests up to a tolerable level, with growers spending minimum cost and also safeguarding our environment [83]. IPM also offers applied techniques for the effective control of pests in the greenhouses. With the implementation of comprehensive cultural techniques and monitoring methods, precise identification of problems, and timely application as well as assessment of suitable management approaches, farmers can improve crop production while reducing their dependence on monotonous pesticide applications [56].

7.7.4 HEDGEROWS AND GREEN CORRIDORS

Hedgerows and green corridors link fragmented landscapes, providing travel pathways for pollinators and habitats for beneficial insects. While initial studies examined the direct impact of hedges on agricultural production, recent attention has shifted toward their role in pollination, pest control, biodiversity conservation, and soil health. Urban hedges, although less connected to production, are gaining importance as refuges for nature in cities, particularly significant for the 55% of the global population residing in urban areas.

7.7.5 CONSERVATION OF NATURAL ENEMIES

Encourage the presence of natural enemies such as predatory insects and birds. They contribute to pest control and create a balanced, multitrophic environment. IPM relies on biological control as a key element, involving the regulation of pest populations through natural enemies such as predators, parasitoids, and pathogens. Conservation of natural enemies is achieved through various agroecosystem manipulations, including habitat- and semiochemical-based approaches. Habitat manipulation strategies encompass vegetational diversity, supplemental food sources, shelters, artificial open rearing systems, farmscaping, and selective use of pesticides. Plants play a crucial role by providing nectar, pollen, and sap as food resources and emitting herbivore-induced plant volatiles (HIPVs). Methyl salicylate, a natural enemy attractant, is highlighted for its potential in attracting beneficial insects [84]. This chapter focuses exclusively on the methods employed to conserve natural enemies in diverse crop ecosystems.

7.7.6 REDUCED PESTICIDE USE

Minimize the use of broad-spectrum pesticides. Targeted application of pesticides and opting for bee-friendly alternatives help protect pollinators. To conserve natural enemies effectively, it is crucial to employ active ingredients with minimal nontarget activity in pesticides. Mitigating undesired side effects of pesticides on natural enemies can be achieved by adjusting the time, place, and mode of application. This strategic adaptation helps minimize the impact on beneficial organisms and ensures the targeted conservation of natural enemies.

7.7.7 CONSERVATION OF NATIVE POLLINATORS

Support native pollinator species by preserving their habitats and avoiding the introduction of nonnative species that may disrupt existing ecological balances. The conservation of native pollinators is a critical undertaking to safeguard biodiversity and maintain ecological balance. Strategies for their conservation involve creating and preserving habitats rich in diverse native plants, minimizing pesticide use, and establishing protected areas.

Additionally, raising awareness about the importance of pollinators, implementing sustainable agricultural practices, and supporting research on pollinator biology and behavior contribute to their effective conservation. By adopting these measures, we can help ensure the well-being and sustainability of native pollinator populations, which play a vital role in pollination and ecosystem health.

7.7.8 RESEARCH AND MONITORING

Conduct ongoing research to understand local pollinator communities and their interactions. Understanding the life cycle and pollination effectiveness must be done at the local level. Regular monitoring helps adapt management strategies based on evolving ecological dynamics. Adopting these management strategies fosters a harmonious multitrophic pollination system, benefiting crops, biodiversity, and overall ecosystem health.

7.8 CHALLENGES AND FUTURE DIRECTIONS

- **Complexity and connectivity:** Understanding and managing multitrophic pollination networks is challenging due to their intricate complexity and dynamic nature. Identifying and mapping interactions across multiple trophic levels is a complex task.
- **Species interactions:** Unraveling the intricate interactions among various species within the network poses a challenge. Deciphering how different organisms influence each other in multitrophic systems requires comprehensive research.
- **Environmental variability:** Pollination networks are susceptible to environmental changes, such as climate fluctuations and habitat alterations. Adapting to and mitigating the impact of these changes on multitrophic interactions is a key challenge.
- **Invasive species:** The introduction of invasive species can disrupt existing pollination networks. Managing and mitigating the impact of invasive species on native flora and fauna is crucial for maintaining biodiversity.
- **Land use changes:** Anthropogenic activities, including changes in land use, can alter the structure of multitrophic networks.

Balancing agricultural practices with conservation efforts is essential to sustain both biodiversity and crop yield.

- **Data collection and analysis:** Gathering comprehensive data on diverse species interactions and effectively analyzing this information is a persistent challenge. Advanced technologies and interdisciplinary collaboration are crucial for addressing this issue.

7.8.1 FUTURE DIRECTIONS

- **Technological advances:** Leveraging emerging technologies such as advanced sensors, DNA barcoding, and remote sensing can enhance our ability to collect and analyze data on multitrophic pollination networks.
- **Synthetic biology:** Applying synthetic biology techniques to manipulate plant traits and enhance their attractiveness to specific pollinators can be a future direction for optimizing pollination services.
- **Ecosystem modeling:** Developing sophisticated models that simulate the dynamics of multitrophic interactions under different scenarios can provide valuable insights into network resilience and response to environmental changes.
- **Community engagement:** Involving local communities in conservation efforts and promoting sustainable land use practices can contribute to the preservation of multitrophic pollination networks.
- **Policy implementation:** Implementing and enforcing policies that prioritize biodiversity conservation and sustainable agriculture practices will be crucial for the long-term health of multitrophic systems.
- **Global collaboration:** Facilitating international collaboration among researchers, policymakers, and communities can lead to a more holistic approach to addressing the challenges and promoting the conservation of multitrophic pollination networks worldwide.

7.9 CONCLUSION

In conclusion, the chapter underscores the critical importance of understanding and preserving multitrophic pollination networks as a strategic approach in the face of declining biodiversity and escalating global food demand. By emphasizing the interconnectedness of plants, pollinators, and associated species, the chapter highlights the complexity of these interactions and their profound implications for ecosystem health and stability. It stresses the need for conservation efforts that promote biodiversity through the maintenance of multitrophic networks, contributing to ecosystem resilience. The chapter further accentuates the benefits of such networks in enhancing crop yields, regulating pests, and ensuring ecological stability. In the context of environmental challenges, it calls for an urgent appreciation of the intricate nature of multitrophic pollination networks, recognizing them as powerful tools for fostering biodiversity conservation and sustainable agriculture. Ultimately, the chapter advocates for a comprehensive and collaborative approach involving technological advancements, community engagement, and global cooperation to address the challenges and promote the long-term health of these complex ecological systems.

KEYWORDS

- **multitrophic interactions**
- **biodiversity enhancement**
- **crop pollination**
- **pest regulation**
- **ecological stability**

REFERENCES

1. Arthington, Á. H., Naiman, R. J., Mcclain, M. E., Nilsson, C. (2010). Preserving the biodiversity and ecological services of rivers: New challenges and research opportunities. *Freshwater Biology*, 55(1), 1–16.
2. Winfree, R., Reilly, J. R., Bartomeus, I., Cariveau, D. P., Williams, N. M., Gibbs, J. (2018). Species turnover promotes the importance of bee diversity for crop pollination at regional scales. *Science*, 359(6377), 791–793.

3. Ollerton, J. (2017). Pollinator diversity: Distribution, ecological function, and conservation. *Annual Review of Ecology, Evolution, and Systematics*, 48, 353–376.
4. Klein, R. J., Nicholls, R. J., Thomalla, F. (2003). Resilience to natural hazards: How useful is this concept? *Global Environmental Change Part B: Environmental Hazards*, 5(1–2), 35–45
5. Garibaldi, L. A., Steffan-Dewenter, I., Winfree, R., Aizen, M. A., Bommarco, R., Cunningham, S. A., Klein, A. M. (2013). Wild pollinators enhance fruit set of crops regardless of honey bee abundance. *Science*, 339(6127), 1608–1611.
6. Paschapur, A. U., Bhat, S., Subbanna, A. R. N. S., Hedau, N. K., Mishra, K. K., Kant, L. (2022). Insect pollinators of eggplant (*Solanum melongena* L.) in the Indian Himalayas and their role in enhancement of fruit quality and yield. *Arthropod-Plant Interactions*, 16(3–4), 349–360.
7. Paschapur, A. U., Bhat, S., Subbanna, A. R., Choudhary, G., Hedau, N. K., Mishra, K. K., Kant, L. (2022). Role of entomophily and artificial pollination in enhancing quality and yield of seed onion (L.) in Indian Himalayas. *Journal of Apicultural Science*, 66(2), 133–147.
8. Buzhdygan, O. Y., Petermann, J. S. (2023). Multitrophic biodiversity enhances ecosystem functions, services and ecological intensification in agriculture. *Journal of Plant Ecology*, 16, rtad019.
9. Aizen, M. A., Harder, L. D. (2009). The global stock of domesticated honey bees is growing slower than agricultural demand for pollination. *Current Biology*, 19(11), 915–918.
10. Chaplin-Kramer, R., Dombeck, E., Gerber, J., Knuth, K. A., Mueller, N. D., Mueller, M., Klein, A. M. (2014). Global malnutrition overlaps with pollinator-dependent micronutrient production. *Proceedings of the Royal Society B: Biological Sciences*, 281(1794), 20141799.
11. Ellis, A. M., Myers, S. S., Ricketts, T. H. (2015). Do pollinators contribute to nutritional health. *PLoS One*, 10(1), e114805.
12. Eilers, E. J., Kremen, C., Smith Greenleaf, S., Garber, A. K., Klein, A. M. (2011). Contribution of pollinator-mediated crops to nutrients in the human food supply. *PLoS One*, 6(6), e21363.
13. Biesmeijer, J. C., Roberts, S. P., Reemer, M., Ohlemuller, R., Edwards, M., Peeters, T., Kunin, W. E. (2006). Parallel declines in pollinators and insect-pollinated plants in Britain and the Netherlands. *Science*, 313(5785), 351–354.
14. Kremen, C., Chaplin-Kramer, R. (2007, June). Insects as providers of ecosystem services: Crop pollination and pest control. In: *Insect Conservation Biology: Proceedings of the Royal Entomological Society's 23rd Symposium* (pp. 349–382). CABI Publishing: Wallingford, UK.
15. Potts, S. G., Biesmeijer, J. C., Kremen, C., Neumann, P., Schweiger, O., Kunin, W. E. (2010). Global pollinator declines: Trends, impacts and drivers. *Trends in Ecology & Evolution*, 25(6), 345–353.
16. Mallinger, R. E., Gaines-Day, H. R., Gratton, C. (2017). Do managed bees have negative effects on wild bees?: A systematic review of the literature. *PLoS One*, 12(12), e0189268.
17. Thomson, D. M., Page, M. L. (2020). The importance of competition between insect pollinators in the Anthropocene. *Current Opinion in Insect Science*, 38, 55–62.

18. Inouye, D. W. (1978). Resource partitioning in bumblebees: Experimental studies of foraging behavior. *Ecology*, 59(4), 672–678.
19. Morse, D. H. (1977). Feeding behavior and predator avoidance in heterospecific groups. *BioScience*, 27(5), 332–339.
20. Thomson, J. D. (1988). Effects of variation in inflorescence size and floral rewards on the visitation rates of traplining pollinators of Aralia hispida. *Evolutionary Ecology*, 2, 65–76.
21. Briggs, D., Walters, S. M. (2016). *Plant Variation and Evolution*. Cambridge University Press.
22. Brosi, B. J., Briggs, H. M. (2013). Single pollinator species losses reduce floral fidelity and plant reproductive function. *Proceedings of the National Academy of Sciences*, 110(32), 13044–13048.
23. Cnaani, J., Thomson, J. D., Papaj, D. R. (2006). Flower choice and learning in foraging bumblebees: Effects of variation in nectar volume and concentration. *Ethology*, 112(3), 278–285.
24. Heinrich, B. (1976). The foraging specializations of individual bumblebees. *Ecological Monographs*, *46*(2), 105–128.
25. Westphal, C., Steffan-Dewenter, I., Tscharntke, T. (2009). Mass flowering oilseed rape improves early colony growth but not sexual reproduction of bumblebees. *Journal of Applied Ecology*, 46(1), 187–193.
26. Heinrich, B., Raven, P. H. (1972). Energetics and pollination ecology: The energetics of pollinators may have wide implications in floral biology and community ecology. *Science*, 176(4035), 597–602.
27. Classen, A., Eardley, C. D., Hemp, A., Peters, M. K., Peters, R. S., Ssymank, A., Steffan-Dewenter, I. (2020). Specialization of plant–pollinator interactions increases with temperature at Mt. Kilimanjaro. *Ecology and Evolution*, 10(4), 2182–2195.
28. Bardgett, R. D., Wardle, D. A., Yeates, G. W. (1998). Linking above-ground and below-ground interactions: How plant responses to foliar herbivory influence soil organisms. *Soil Biology and Biochemistry*, 30(14), 1867–1878.
29. Senapathi, D., Biesmeijer, J. C., Breeze, T. D., Kleijn, D., Potts, S. G., Carvalheiro, L. G. (2015). Pollinator conservation—The difference between managing for pollination services and preserving pollinator diversity. *Current Opinion in Insect Science*, 12, 93–101.
30. Hooper, E. R., Ashton, M. S. (2020). Fragmentation reduces community-wide taxonomic and functional diversity of dispersed tree seeds in the Central Amazon. *Ecological Applications*, 30(5), e02093.
31. Naeem, S., Duffy, J. E., Zavaleta, E. (2012). The functions of biological diversity in an age of extinction. *Science*, 336(6087), 1401–1406.
32. Cardinale, B. J. (2011). Biodiversity improves water quality through niche partitioning. *Nature*, 472(7341), 86–89.
33. Hooper, D. U., Adair, E. C., Cardinale, B. J., Byrnes, J. E., Hungate, B. A., Matulich, K. L., O'Connor, M. I. (2012). A global synthesis reveals biodiversity loss as a major driver of ecosystem change. *Nature*, 486(7401), 105–108.
34. Thébault, E., Loreau, M. (2003). Food-web constraints on biodiversity–ecosystem functioning relationships. *Proceedings of the National Academy of Sciences*, 100(25), 14949–14954.

35. Duffy, J. E., Cardinale, B. J., France, K. E., McIntyre, P. B., Thébault, E., Loreau, M. (2007). The functional role of biodiversity in ecosystems: Incorporating trophic complexity. *Ecology Letters*, 10(6), 522–538.
36. Hillebrand, H., Matthiessen, B. (2009). Biodiversity in a complex world: Consolidation and progress in functional biodiversity research. *Ecology Letters*, 12(12), 1405–1419.
37. Koh, L. P., Dunn, R. R., Sodhi, N. S., Colwell, R. K., Proctor, H. C., Smith, V. S. (2004). Species coextinctions and the biodiversity crisis. *Science*, 305(5690), 1632–1634.
38. Pimm, S. L. (1980). Properties of food webs. *Ecology*, 61(2), 219–225.
39. Bruno, J. F., Boyer, K. E., Duffy, J. E., Lee, S. C. (2008). Relative and interactive effects of plant and grazer richness in a benthic marine community. *Ecology*, 89(9), 2518–2528.
40. Allesina, S., Alonso, D., Pascual, M. (2008). A general model for food web structure. *Science*, 320(5876), 658–661.
41. Borrvall, C., Ebenman, B. (2006). Early onset of secondary extinctions in ecological communities following the loss of top predators. *Ecology Letters*, 9(4), 435–442.
42. Hegland, S. J., Nielsen, A., Lázaro, A., Bjerknes, A. L., Totland, Ø. (2009). How does climate warming affect plant-pollinator interactions. *Ecology Letters*, 12(2), 184–195.
43. Godfray, H. C. J., Beddington, J. R., Crute, I. R., Haddad, L., Lawrence, D., Muir, J. F., Toulmin, C. (2010). Food security: The challenge of feeding 9 billion people. *Science*, *327*(5967), 812–818.
44. Foley, J. A., DeFries, R., Asner, G. P., Barford, C., Bonan, G., Carpenter, S. R., Snyder, P. K. (2005). Global consequences of land use. *Science*, 309(5734), 570–574.
45. O'sullivan, C. A., Bonnett, G. D., McIntyre, C. L., Hochman, Z., Wasson, A. P. (2019). Strategies to improve the productivity, product diversity and profitability of urban agriculture. *Agricultural Systems*, 174, 133–144.
46. Scherr, S. J., McNeely, J. A. (2008). Biodiversity conservation and agricultural sustainability: Towards a new paradigm of 'ecoagriculture' landscapes. *Philosophical Transactions of the Royal Society B: Biological Sciences*, 363(1491), 477–494.
47. Wratten, S. D., Gillespie, M., Decourtye, A., Mader, E., Desneux, N. (2012). Pollinator habitat enhancement: Benefits to other ecosystem services. *Agriculture, Ecosystems & Environment*, 159, 112–122.
48. Wäckers, F. L., Romeis, J., van Rijn, P. (2007). Nectar and pollen feeding by insect herbivores and implications for multitrophic interactions. *Annual Review of Entomology*, 52, 301–323.
49. Classen, A., Peters, M. K., Ferger, S. W., Helbig-Bonitz, M., Schmack, J. M., Maassen, G., Steffan-Dewenter, I. (2014). Complementary ecosystem services provided by pest predators and pollinators increase quantity and quality of coffee yields. *Proceedings of the Royal Society B: Biological Sciences*, 281(1779), 20133148.
50. Thomas, M. B. (1999). Ecological approaches and the development of "truly integrated" pest management. *Proceedings of the National Academy of Sciences*, 96(11), 5944–5951.
51. Ekström, G., Ekbom, B. (2011). Pest control in agro-ecosystems: An ecological approach. *Critical Reviews in Plant Sciences*, 30(1–2), 74–94.

52. Chapman, M., Klassen, S., Kreitzman, M., Semmelink, A., Sharp, K., Singh, G., Chan, K. M. (2017). 5 key challenges and solutions for governing complex adaptive (food) systems. *Sustainability*, 9(9), 1594.
53. Flint, M. L., Dreistadt, S. H. (1998). *Natural Enemies Handbook: The Illustrated Guide to Biological Pest Control* (Vol. 3386). University of California Press.
54. Michaud, J. P. (2018). Problems inherent to augmentation of natural enemies in open agriculture. *Neotropical Entomology*, 47, 161–170.
55. Van Der Meijden, E., Klinkhamer, P. G. (2000). Conflicting interests of plants and the natural enemies of herbivores. *Oikos*, 89(1), 202–208.
56. Bhat, S., Kumar, S. (2024). Conventional and recent approaches of integrated pest management in greenhouse cultivation. In: *Protected Cultivation: Structural Design, Crop Management Modeling, and Automation* (p. 255).
57. Chau, N. N. B., Maeto, K. (2009). Temporary host paralysis and avoidance of self-superparasitism in the solitary endoparasitoid Meteorus pulchricornis. *Entomologia Experimentalis et Applicata*, 132(3), 250–255.
58. Powell, W., Pickett, J. A. (2003). Manipulation of parasitoids for aphid pest management: Progress and prospects. *Pest Management Science: Formerly Pesticide Science*, 59(2), 149–155.
59. Hayes, K. R., Inglis, G. J., Barry, S. C. (2019). The assessment and management of marine pest risks posed by shipping: The Australian and New Zealand experience. *Frontiers in Marine Science*, 6, 489.
60. Ferracini, C., Bueno, V. H., Dindo, M. L., Ingegno, B. L., Luna, M. G., Salas Gervassio, N. G., Tavella, L. (2019). Natural enemies of Tuta absoluta in the Mediterranean basin, Europe and South America. *Biocontrol Science and Technology*, *29*(6), 578–609.
61. Bodino, N., Cavalieri, V., Pegoraro, M., Altamura, G., Canuto, F., Zicca, S., Bosco, D. (2021). Temporal dynamics of the transmission of Xylella fastidiosa subsp. pauca by Philaenus spumarius to olive plants. *Entomologia Generalis*, 41(5), 1–17.
62. Deguine, J. P., Gloanec, C., Laurent, P., Ratnadass, A., Aubertot, J. N., Laurent, P. Aubertot, J. N. (2017). Agroecological transition keys. In: *Agroecological Crop Protection* (pp. 163–246). Springer Netherlands: Dordrecht.
63. Davis, J. K., Cohen, A. D., Getman-Pickering, Z. L., Grab, H. L., Hodgden, B., Maher, R. M., Thaler, J. S. (2023). Agricultural soil legacy influences multitrophic interactions between crops, their pathogens and pollinators. *Proceedings of the Royal Society B*, 290(2011), 20231453.
64. Bascompte, J., Jordano, P. (2013). *Mutualistic Networks*. Princeton University Press.
65. Del-Claro, K., Dirzo, R. (2021). Impacts of anthropocene defaunation on plant-animal interactions. In: *Plant-Animal Interactions: Source of Biodiversity* (pp. 333–345).
66. Zariman, N. A., Omar, N. A., Huda, A. N. (2022). Plant attractants and rewards for pollinators: Their significant to successful crop pollination. *International Journal of Life Sciences and Biotechnology*, 5(2), 270–293.
67. LaBar, T., Campbell, C., Yang, S., Albert, R., Shea, K. (2013). Global versus local extinction in a network model of plant–pollinator communities. *Theoretical Ecology*, 6, 495–503.
68. Bascompte, J., Scheffer, M. (2023). The resilience of plant–pollinator networks. *Annual Review of Entomology*, 68, 363–380.

69. Green, D. G., Sadedin, S. (2005). Interactions matter—Complexity in landscapes and ecosystems. *Ecological Complexity*, 2(2), 117–130.
70. Traveset, A., Richardson, D. M. (2014). Mutualistic interactions and biological invasions. *Annual Review of Ecology, Evolution, and Systematics*, 45, 89–113
71. Haines-Young, R., Potschin, M. (2010). The links between biodiversity, ecosystem services and human well-being. *Ecosystem Ecology: A new synthesis*, 1, 110–139.
72. Martinez, N. D. (2020). Allometric trophic networks from individuals to socio-ecosystems: Consumer–resource theory of the ecological elephant in the room. *Frontiers in Ecology and Evolution*, 8, 92.
73. Steffan-Dewenter, I., Tscharntke, T. (1999). Effects of habitat isolation on pollinator communities and seed set. *Oecologia*, 121, 432–440.
74. Ricketts, T. H. (2004). Tropical forest fragments enhance pollinator activity in nearby coffee crops. *Conservation Biology*, 18(5), 1262–1271.
75. Laverty, T. M. (1992). Plant interactions for pollinator visits: A test of the magnet species effect. *Oecologia*, 89, 502–508.
76. Isaacs, R., Kirk, A. K. (2010). Pollination services provided to small and large highbush blueberry fields by wild and managed bees. *Journal of Applied Ecology*, 47(4), 841–849.
77. Blaauw, B. R., Isaacs, R. (2014). Flower plantings increase wild bee abundance and the pollination services provided to a pollination-dependent crop. *Journal of Applied Ecology*, 51(4), 890–898.
78. Michener, W. K., Beach, J. H., Jones, M. B., Ludäscher, B., Pennington, D. D., Pereira, R. S., Schildhauer, M. (2007). A knowledge environment for the biodiversity and ecological sciences. *Journal of Intelligent Information Systems*, 29(1), 111–126.
79. Soucy, S. L. (2002). Nesting biology and socially polymorphic behavior of the sweat bee Halictus rubicundus (Hymenoptera: Halictidae). *Annals of the Entomological Society of America*, 95(1), 57–65.
80. Richards, S. A., Williams, N. M., Harder, L. D. (2009). Variation in pollination: Causes and consequences for plant reproduction. *The American Naturalist*, 174(3), 382–398.
81. Danforth, B. N., Minckley, R. L., Neff, J. L. (2019). *The Solitary Bees: Biology, Evolution, Conservation*. Princeton University Press.
82. Packer, L., Ali, E., Dumesh, S., Walker, K. THE IDENTIFICATION OF POLLINATORS. (2016). *Pollination Services to Agriculture: Sustaining and Enhancing a Key Ecosystem Service* (p. 57).
83. Jat, S. L., Parihar, C. M., Singh, A. K. (2020). *Conservation Agriculture for Sustainable Food Security in India*. Shri Vaishnav Vidyapeeth Vishwavidhyalaya.
84. Shakeel, Q., Ali, S., Raheel, M., Bajwa, R. T., Aqueel, M. A., Iftikhar, Y. Tariq, KHabitat (2022). Manipulation in crops for integrated insect pest and disease management. In: *Advances in Integrated Pest Management Technology* (pp. 83–119). Springer: Cham.

CHAPTER 8

Plant–Microbe Synergy in Agroforestry Microbial Partnerships

NUTAN KARKI, SUNAULLAH BHAT, and AMIT UMESH PASCHAPUR*

Crop Protection Division, ICAR-Vivekananda Parvatiya Krishi Anusandhan Sansthan, Almora, Uttarakhand, India

*Corresponding author

ABSTRACT

Agroforestry, a sustainable land-use system integrating trees and crops, offers a promising solution to agricultural and environmental challenges. Central to its success are the complex interactions between plants and soil microorganisms that influence productivity, resilience, and ecosystem health. This abstract highlights the crucial role of plant–microbe partnerships in enhancing soil fertility, nutrient cycling, and crop performance within agroforestry systems. Mycorrhizal fungi facilitate efficient uptake of nutrients, particularly phosphorus, while nitrogen-fixing rhizobia increase nitrogen availability to plants. Endophytic microbes further support plant growth by improving stress tolerance and providing resistance against pathogens. Together, these microbial partners not only strengthen individual plants but also enhance ecosystem functions. They contribute to soil structure by forming stable aggregates, improve water retention, and accelerate organic matter decomposition. This, in turn, boosts carbon sequestration and reduces greenhouse gas emissions, positioning agroforestry as an effective climate change mitigation strategy. Understanding these microbial relationships is essential for optimizing tree–crop combinations and improving land-use efficiency. Harnessing microbial diversity can advance sustainable farming, improve food security, and reduce

Agroforestry and Climate Change Mitigation: Harnessing Multitrophic Interactions for Sustainable Plant Protection and Ecosystem Resilience. Hanuman Singh Jatav, Vishnu D. Rajput, Ipsita Samal, Tanmaya Kumar Bhoi (Eds.)

DOI: 10.1201/9781779641397-8

environmental degradation. In summary, plant–microbe interactions in agroforestry present vast opportunities to build resilient agricultural systems and promote ecological sustainability.

8.1 INTRODUCTION

Agroforestry is an ecologically dynamic form of resource management that combines the fields of agriculture and forestry, with the potential to boost production sustainability and biodiversity. The term "Agroforestry" was coined in 1977 [1] in response to an early International campaign for research on integrated production systems with crops and trees [2]. During these early days, many meetings and debates took place on the subject of defining and labeling agroforestry. A scientific definition of agroforestry should highlight two traits that set it apart from other land-use practices and are shared by all varieties of agroforestry: the intentional planting of woody perennials in a spatial variety or sequence on the same area of land shared with livestock, agricultural crops, or both; and the ecosystem's woody components and nonwoody components must interact significantly, in a positive or negative way, in an ecological or economic sense [2]. The International Centre for Research in Agroforestry (ICRAF) defined agroforestry as "a collective name for land-use systems and technologies where woody perennials are deliberately used on the same land-management units as agricultural crops and/or animals, in some form of spatial arrangement or temporal sequence." Additionally, agroforestry is thought to be a successful way to maintain or even boost crop and tree productivity in accordance with promoting other ecosystem services and functions [3].

According to the World Agroforestry (ICRAF) estimates, 43% of farming lands around the world have 10% tree cover, whereas over 160 million hectares have more than 50% tree cover [4], indicating the potential for revegetating agricultural land with appropriate tree species [5]. Depending on the combination of the tree and the other elements, agroforestry systems (AFS) can be classified into different types: silvopastoral systems (trees plus pasture), agrisilviculture systems (trees plus crops), agrosilvopastoral systems (tree plus crops and livestock), and others (home gardens, riparian buffer strips, apiculture with trees, multipurpose woodlots, aquaculture with trees, etc. The trees in AFS facilitate the creation of many products, such as fodder, fuel, timber, fruits, fiber, gums, resin, craft items, gardening supplies, health care items, recreational activities, and ecological services. In relation to soil nutrient dynamics, it reduces nutrient loss, increases the internal cycling of nutrients, and makes the nutrients available to the crops. The AFS is a closed ecosystem that helps crops retain nutrients in the soil in the best possible way [3]. AFSs increase soil nutrient pools and lessen dependency on outside inputs by encouraging a closed system with internal nutrient recycling. The nutrients are accessed from lower soil horizons by roots and returned to the soil through leaf fall [6, 7]. AFS can improve and manage the fertility, productivity, and sustainability of soil for a long period. It can boost or improve the soil's physico-chemical as well as biological quality by adding and decomposing organic matter. It has been extensively documented that the presence of trees in AFSs increases microbiota in soil [7–9]. The biological component of the soil is crucial for the maintenance and functioning of ecosystems. Soil organisms are responsible for sustaining various soil cycles, including the fixation of nitrogen, cycling of nutrients, infiltration of water, aeration, capture and storage of carbon, and degradation of organic matter [10].

Nowadays, most people agree that agroforestry is used to achieve a number of goals. It is an interface between forestry and agriculture that includes mixed land-use practices, as shown in Figure 8.1. The main reason these practices were created is that advances in conventional agriculture and forestry have not adequately addressed the unique requirements and conditions of tropical developing nations [2]. Additionally, the agricultural soils around the world are in danger of degrading as a result of the management practices used [11]. Huge amounts of nitrogenous fertilizers are required for the cultivation of crops and forages. Rather than relying solely on fertilizers, AFS can provide several advantages. Numerous ecological advantages are provided by multifunctional AFSs, including carbon sequestration, mitigating the effects of climate change, improving soil fertility and water efficiency, soil erosion control, conserving biodiversity, enhancing pollination, controlling pests, promoting sustainable land use, providing shelter and windbreaks, improving microclimates, reducing poverty, and reducing food insecurity [12]. In addition to these, AFSs increase microbial diversity and organic matter in soil, which reduces greenhouse gas emissions. Consequently, by combining forestry and agricultural technologies, AFS can produce land-use systems that are more sustainable, productive, and profitable.

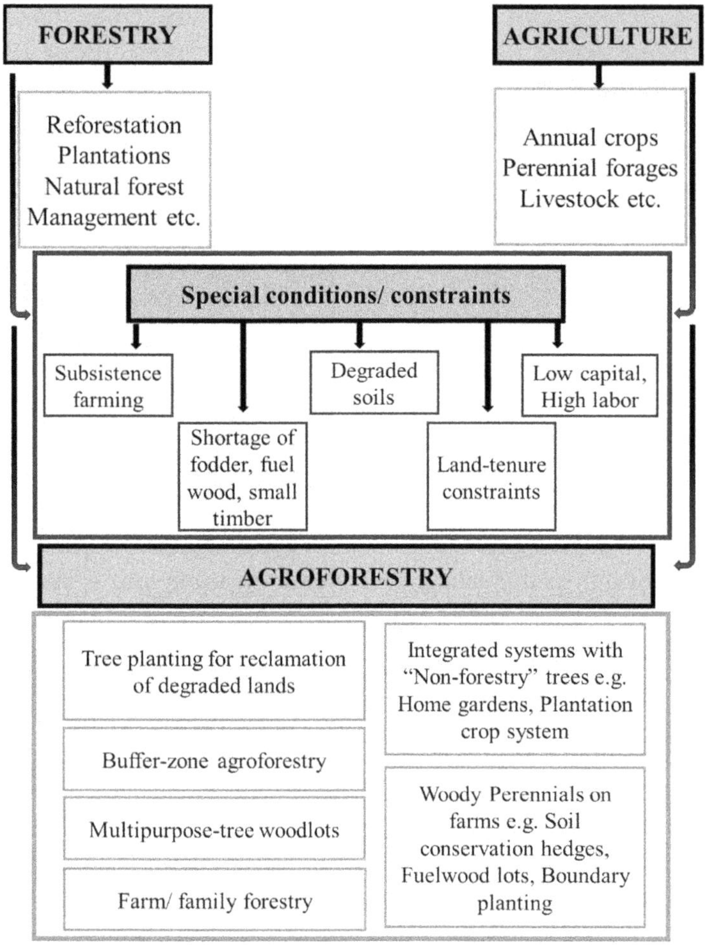

FIGURE 8.1 The unique demands and circumstances of tropical developing nations have led to the development of agroforestry as a bridge between agriculture and forestry.

Source: Adapted from Ref. [2].

Due to the expansion of contemporary agricultural practices that reduce the amount of arable land available for cultivation, the deforestation of virgin backwoods is occurring at a disturbing rate on a global scale. The biodiversity linked to forests is destroyed when virgin forests are destroyed. The loss of forest cover will primarily be responsible for visible diversity; however, invisible diversity loss (such as microbial diversity) will occur on a much larger scale and be incalculable [5]. The rhizosphere, canopy/

phyllosphere, and AFS's immediate surroundings are crucial to its development. The diversity of microbes in the soil and canopy is essential to maintaining and preserving the AFSs. Microbes and tree species can interact in ways that are advantageous, detrimental, or neutral [5, 13]. Since microbial diversity forms the foundation of the AFS, evaluating, managing, and analyzing it is crucial for drawing comparisons across a broad geographic region with other kinds of forest ecosystems and agroforests. The importance of plant–microbe synergy in agroforestry is projected in this chapter, with a focus on arbuscular mycorrhizal fungi (AMF), ectomycorrhizal fungi (EMF), endophytes, and nitrogen-fixing microbes.

8.2 MICROBIAL DIVERSITY IN AFS

Microorganisms are ubiquitous in nature and are thought to coexist with all living organisms [14]. From the extensive literature, it has been discovered that microorganisms exist in every part of a plant, including the rhizosphere, phyllosphere, and phylloplane areas [15]. The phyllosphere and phylloplane areas are primarily involved in activities on the plant's exterior and, in extreme situations, may also play a part in plant diseases [16]. On the contrary, the microbial community present in the rhizosphere can directly affect plant health, growth, and nutrition [15]. The scientific community has also paid huge attention to microbes residing in the rhizosphere during the past 10 years. Because high levels of chemical and biological activities are found in the rhizosphere [15]. In agroecosystems, soil microorganisms, particularly those that are beneficial, regulate a variety of processes. The addition of various crops and permanent vegetation cover to AFS typically improves the accumulation of organic matter in soils, which is predicted to raise soil microbial biomass. Increases in fine-root turnover and fallen litter can raise the concentration of organic matter in the soil. In addition to improving the microclimate, trees can also change the biological characteristics of the soil (both above and below ground around plant roots). Previous studies have revealed that AFS soil has a higher microbial biomass than monocrop or open grassland soils [17–22]. There are two main categories of microbial diversity in agroforestry: aboveground diversity and belowground diversity. The phyllosphere, abiotic factors, and nature of tree species affect the aboveground microbial diversity [23]. Similar to this, the rhizosphere, soil quality, abiotic factors, and the nature of AFS (such as crop system composition) affect the diversity of belowground microbes and their functions [24]. According to several authors [25–27], the AFS has higher soil microbial diversity and biomass due to the ameliorative impacts of trees and natural matter contributions, as well as contrasts in litter amount and quality and root exudates. The productivity of any agroecosystem depends on the existence of high microbial diversity in soil. The diverse microorganisms linked to AFS can be broadly classified into two groups:

1. Mutualistic fungi and
2. Nitrogen-fixing microbes

8.2.1 MUTUALISTIC FUNGI

Fungi have evolved mutualism with several organisms, including algae, cyanobacteria, plants, and animals. Fungus–plant mutualism (both organisms benefit from the relationship) plays an important role in AFS. Two general categories can be used to group the various mutualistic fungi associated with AFS:

1. Mycorrhizal fungi and
2. Nonmycorrhizal fungi

8.2.1.1 MYCORRHIZAL FUNGI

The mycorrhizal fungi are a significant class of soil microorganisms that form a symbiotic association with the roots of 97% terrestrial plant species [28]. The fungal mycelia in a mycorrhizal association channel water and minerals from the soil into the plant through their vast network of fungal hyphae [29]. In return, the fungus uses the plant's photosynthesis byproducts as fuel for its metabolism [30]. Two types of symbiotic fungi are known to form symbiotic associations with plant roots: (1) AMF (also known as endomycorrhizal fungi; penetrate deep into the cortical cells of plants), and (2) EMF (do not penetrate the cortical cells of plants).

8.2.1.1.1 Arbuscular Mycorrhizal Fungi

A symbiotic and endomycorrhizal relationship between fungi belonging to the phylum Glomeromycota and plant roots results in the formation of arbuscular mycorrhizae (AM). The phylum Glomeromycota includes the AMF as obligate/nonhost-specific symbionts [31–34]. It has 3 classes (Glomeromycetes, Archaeosporomycetes, and Paraglomeromycetes), 5 orders (Archaeosporales, Gigasporales, Diversisporales, Glomerales, and Paraglomerales), 15 families, and 31 genera [35]. The most common genera of AMF are *Funneliformis*, *Glomus*, *Claroideoglomus*, and *Acaulospora*. Their hyphae develop from a spore. On the epidermis of the root, they produce hyphopodium and form arbuscules inside cortical cells of the host plant roots (Figure 8.2). The AMF are found in a variety of ecosystems, including desert and aquatic habitats. Therefore, there are many opportunities for them to adapt to the AFS [36]. Ecosystem productivity and stability are said to be significantly impacted by AMF [37]. Most terrestrial plants, including most crops, have associations with AMF [30, 38, 39]. The hyphae of AMF enter the cells of plant roots. They give plants mineral nutrients [31, 40] in exchange for energy and carbon from the host plant [30, 41, 42]. Under stressed conditions, AMF enhances the uptake of several major nutrients [43, 44] and improves plant growth parameters [45, 46]. Additionally, under various agroecological settings, the AMF is essential for safeguarding their hosts from biotic [47, 48] and abiotic [31, 49] stresses. However, a variety of factors, including host plant species [50], environmental conditions such as soil pH [51], soil organic carbon [52], soil nitrogen [53], and land-use intensity [54], have the capacity to alter the diversity and composition of the AMF community.

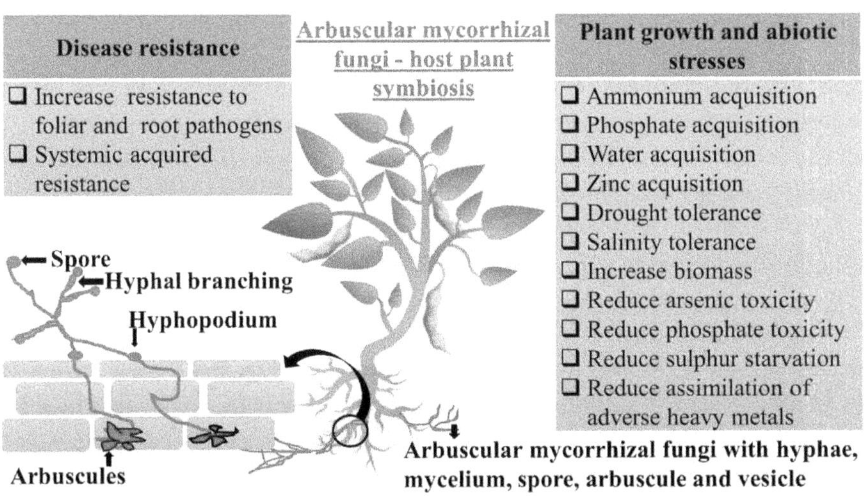

FIGURE 8.2 Benefits of AMF.

Many researchers have shown the significance of preserving AMF (active population) in the soil of AFS to increase the crop yield [22, 30]. There are numerous reports available regarding the impact of AFS on AMF abundance and diversity. According to the majority of these studies, AFS benefits the AMF communities [22, 30, 55]. Da Silva Maia et al. [56] evaluated the AMF colonization of oil palm (*Elaeis guineensis*) grown in monocultures and biodiverse AFSs in the Amazon. In AFSs, the percentage of mycorrhizal colonization of oil palm was three times higher (33.5%±1.2%) compared to monoculture (11.4%±1.2%). The cultivation of oil palm in AFSs will promote mycorrhizal colonization in the Amazon. Laurindo et al. [22] compared the benefits of planting native trees such as *Araucaria angustifolia*, *Mimosa scabrella*, and *Ilex paraguariensis* in a 5-year AFS to overcome the soil disturbance caused by pine tree invasion on the composition of the AMF community in Southern Brazil to the unassisted forest ecosystem. The diversity and dominance of AMF in the natural restoration and native tree species were similar. However, these conditions displayed greater values in comparison to the natural ecosystem. These variations in the AMF community structure and soil characteristics were linked to (1) disturbance of the ecosystem; (2) modifications to the soil, such as changes in pH; (3) rootability through fine root production; and (4) rhizo-deposition by soil microbial carbon. These findings provided new insights into the AMF communities in AFSs and offered new avenues for managing degraded lands caused by biological invasions involving native tree species and AMF species in Southern Brazil. Masebo et al. [34] analyzed the AMF diversity, spore density, and root colonization in various agroforestry practices in southern Ethiopia. Forty-three AMF morphotypes were identified in this study. These morphotypes belonged to eleven genera, that is, *Acaulospora, Ambispora, Claroideoglomus, Funneliformis, Gigaspora, Glomus, Paraglomus, Rhizophagus, Scutellospora, Septoglomus,* and one was unidentified. *Acaulospora* dominated with 32.56% of the total, followed by *Claroideoglomus* with 18.60% abundance. The greatest spore density (7641.5 spore 100 g^{-1} dry soil) was found in home gardens, whereas the lowest (683.6 spore 100 g^{-1} dry soil) was found in croplands. The highest percentage of roots colonized was found in woodlots (54.75%), followed by home gardens (48.25%). These agroforestry practices, such as parklands, home gardens, alley cropping, woodlots, trees planted on soil and water conservation structures, windbreaks, and border planting, support AMF and potentially increase their diversity and abundance [34]. Thus, agroforestry practices are more suited to support the diverse and abundant AMF community as compared to traditionally managed agricultural systems.

The AMF also protects AFS from a variety of pathogenic nematodes (*Pratylenchus*, *Rotylenchulus*, and *Meloidogyne*) and fungi (*Aphanomyces*, *Fusarium*, *Gaeumannomyces*, *Phytophthora*, *Pythium*, *Rhizoctonia*, *Sclerotium*, *Thielaviopsis*, and *Verticillium*) [5, 57].

8.2.1.1.2 Ectomycorrhizal Fungi

EMF are facultative saprotrophs. Their hyphae grow around the epidermal cells, forming a Hartig net, and also surround the outside of the plant root tip like a sheath (Figure 8.3). EMF can belong to the phyla *Ascomycota*, *Basidiomycota*, and *Zygomycota*. Approximately 7,750 fungal species worldwide are known to form ectomycorrhiza [58]. EMF is another significant and economically beneficial part of the AFS (Figure 8.3). Several studies on soil microbiomes reported that the tree rows in AFS increase EMF population size in soil [7, 21]. Additionally, the AFS has higher microbial biomass than mono-crop soils. Beule et al. [21] conducted a study that revealed that the soil of poplar tree rows in an alley-cropping system had up to 330 times more Basidiomycota in comparison to the nearby cropland monocultures. They also examined the soil fungal community using amplicon sequencing and discovered that the poplar trees specifically supported certain EMF (*Cortinarius*, *Geopora*, and *Inocybe*), which have the potential to significantly enhance nutrient acquisition in AFS. Vaupel et al. [7] also recorded that the tree rows changed the composition of the fungal community, leading to a higher percentage of EMF such as *Preussia* and *Laccaria* spp. in the mycobiome associated with the tree rows in AFS.

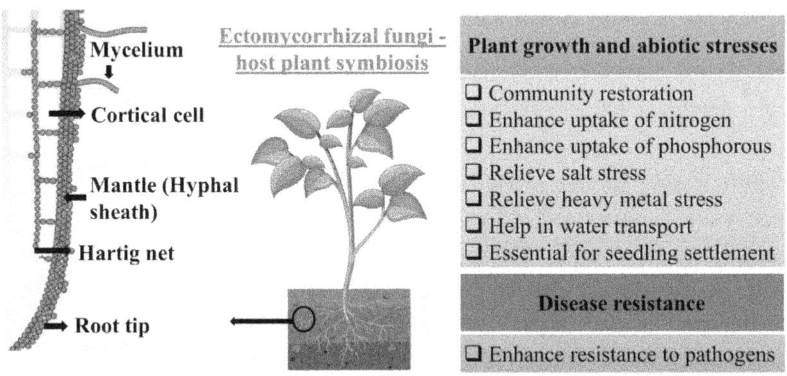

FIGURE 8.3 Benefits of EMF.

8.2.1.2 NONMYCORRHIZAL FUNGI (ENDOPHYTES)

The nonmycorrhizal fungi frequently form associations with the aboveground and underground healthy tissues of various plants as endophytes. Endophytes are microorganisms that are widely distributed within the tissues of their host plants without causing any noticeable symptoms of disease and produce a variety of bioactive compounds that are advantageous to plants [59, 60]. An assessment of their correlation with AFS is extremely valuable as it will add more context to the significance of these nonmycorrhizal fungi in terms of plant defense, promoting plant growth, and producing fungal metabolites. Endophytic fungal species belonging to the genus *Trichoderma* spp., including *Trichoderma viride* and *Trichoderma harzianum*, have shown their capacity to function as disease-control agents in a variety of cultivated plants [61]. Hatamzadeh et al. [62] isolated and identified endophytic fungi *Anthemis altissima*, *Achillea millefolium*, *Achillea filipendulina*, *Cichorium intybus,* and *Matricaria parthenium* with the ability to promote the growth of *Zea mays* in northeast Iran. *Alternaria consortiale*, *Alternaria embellisia*, *Bjerkandera adusta*, *Fusarium avenacearum*, *Preussia africana*, *Schizophyllum commune*, *Septoria malagutii*, *Trametes versicolor*, *Trichaptum biforme*, and *Verticillium dahliae* significantly improved seed germination. Among these endophytes, *Alternaria consortiale* produced the highest amount of growth-promoting compounds like indole-3-acetic acid. Agroecosystems may find it easier to use less chemical fertilizer if they can identify isolates from these endophytes that are effective and have the right ability to promote the growth of plants. Thus, AMF [30, 34, 63, 64], EMF [7, 21], and endophytes [60, 62] are important mutualistic fungi linked to plant species in AFS.

8.2.2 NITROGEN-FIXING MICROBES

Nitrogen is one of the most important macronutrients [15]. Nitrogen is a necessary component of adenosine triphosphate, urea, amino acids (proteins), nucleic acids (DNA and RNA), and nicotinamide adenine dinucleotide in all living cells [65]. It is essential for plant growth and productivity because it is a major component of chlorophyll, the most important pigment needed for photosynthesis [66]. Nitrogen is the most available gas and a primary component of the atmosphere, constituting up to 78.084%

of the dry air's volume. However, nitrogen present in gaseous form in the atmosphere is inaccessible to plants. Plants can only absorb reactive forms of nitrogen, such as oxidized or reduced (such as amines and NH_3) nitrogen types [65]. Given their substantial influence on crop productivity in agricultural systems, farmers use nitrogen fertilizers, despite their high cost, to boost crop yields. However, the extensive use of synthetic nitrogenous fertilizers to increase crop yield in modern agriculture has a negative impact on the environment. In order to maintain environmental health, nitrogen-fixing microbes are the natural bio-fertilizers that can be used to meet plants' needs for nitrogen and other nutrients (such as Potassium and Phosphorus) without endangering the environment's health or plant productivity. Nitrogen-fixing microorganisms live in association with plants and help the plants to assimilate the nitrogen present in the atmosphere. They convert nitrogen (N_2) present in the atmosphere to ammonia (NH_3; plant available form) by a nitrogenase enzymatic complex, and this process is called biological nitrogen fixation [67]. Biological nitrogen fixation is the second most significant biochemical reaction that sustains life on Earth, after photosynthesis. The microorganisms that produce the enzyme nitrogenase are either free-living organisms or form symbiotic relationships with specific plants, particularly legumes. Based on the potential of microbes to form mutualistic associations with eukaryotes or on their saprophytic life as part of ecological/environmental microflora, the nitrogen-fixing microbes are classified into two groups [68]:

1. Symbiotic nitrogen-fixing microbes.
2. Nonsymbiotic nitrogen-fixing microbes.

8.2.2.1 SYMBIOTIC NITROGEN-FIXING MICROBES

Symbiosis is defined as the acquisition of an organism or organisms by another organism that is unlike them, followed by a long-term integration that results in the emergence of new structures and metabolisms [69]. The symbiotic nitrogen-fixing microbes and plant species like legumes interact in a mutually beneficial symbiosis that benefits both plant species and microbes. In this symbiotic relationship, legumes host and supply carbon sources to symbiotic nitrogen-fixing microbes, and microbes provide ammonia to legumes in return (Figure 8.4). Plants receive approximately 60% of their nitrogen from biological fixation, 50% of which is supplied through symbiosis between plants and bacteria [57].

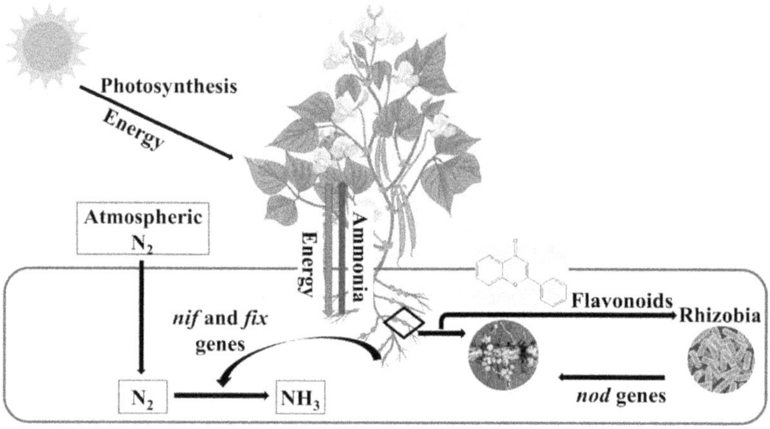

FIGURE 8.4 An overview of the rhizobia-mediated symbiotic nitrogen fixation process in legumes.

Source: Adapted from Ref. [70].

Several bacteria that fix nitrogen symbiotically are saprophytes. Few bacteria have evolved high symbiotic dependence or have become obligate symbionts [5]. The symbiotic relationship between nitrogen-fixing bacteria and plants primarily comprises three groups: (1) actinobacteria (belonging to family Frankiaceae), (2) alphaproteobacteria and betaproteobacteria, and (3) cyanobacteria. The alphaproteobacteria, betaproteobacteria, and actinobacteria are found in noduled plants, whereas cyanobacteria are found in mucilage cavities and intercellular spaces in cycads, bryophytes, and higher plants [5, 70].

The term "rhizobia" refers to a particular class of Gram-negative Alphaproteobacteria and Betaproteobacteria that fix dinitrogen in symbiosis with plant species belonging to the family Fabaceae (leguminous species) and Cannabiaceae (nonleguminous species belonging to genus *Parasponia*) as their host plants. These bacteria can form nodules on the roots, or occasionally on the stems, of their hosts [71, 72]. The name *Rhizobium* has been applied to the nitrogen-fixing bacteria of legumes a lot because it was the first genus of these bacteria to be described. At present, this group of symbiotic bacteria is dispersed across 20 genera in the families Bradyrhizobiaceae (*Bradyrhizobium, Blastobacter*), Brucellaceae (*Ochrobactrum*), Burkholderiaceae (*Burkholderia, Cupriavidus, Paraburkholderia, Ralstonia*), Hyphomicrobiaceae (*Devosia*), Methylobacteriaceae (*Methylobacterium, Microvirga*), Phyllobacteriaceae (*Mesorhizobium,*

Aminobacter, *Phyllobacterium*), Rhizobiaceae (*Allorhizobium*, *Ensifer/Sinorhizobium*, *Neorhizobium*, *Pararhizobium*, *Rhizobium*, *Shinella*), and Xanthobacteraceae (*Azorhizobium*). With 133 species, *Rhizobium* is the largest genus among rhizobia [73]. Recent research has revealed the existence of gamaproteobacteria in the nodules of legumes belonging to the genera *Cohnella*, *Enterobacter*, *Pantoea*, and *Pseudomonas* [11]. Therefore, it is suggested that bacteria that form structure knobs (nodules) in legumes to be referred to as *legume-nodulating bacteria* rather than rhizobia.

Many AFSs include nitrogen-fixing plant species that form symbiotic relationships with nitrogen-fixing bacteria in their root nodules [74]. The different nitrogen-fixing microbiota connected with legumes are useful in AFS. The bacteria linked with *Acacia* spp. are *Allorhizobium undicola*, *Mesorhizobium plurifarium*, *Ensifer mexicanus* [75], and *Sinorhizobium fredii* [76]. The bacteria linked with *Albizia* spp. are *Mesorhizobium albiziae* [77], and with *Calliandra* spp. are *Rhizobium tropici* [78]. The bacteria linked with *Erythrina* spp. are *Bradyrhizobium liaoningense* [76], and with *Faidherbia* spp. are *Bradyrhizobium elkanii* [76]. The bacteria linked with *Gliricidia* spp. (*Rhizobium etli* [79]), *Inga* spp. (*Bradyrhizobium japonicum* [80]), *Leucena* spp. (*Rhizobium gallicum* [79]), *Prosopis* spp. (*Sinorhizobium kostiense* [81]), *Robinia* spp. (*Rhizobium multihospitium* [82]), *Sesbania* spp. (*Azorhizobium caulinodans* [83]), and *Tephrosia* spp. (*Rhizobium hainanense* [84]) also plays an important role in AFS. Consequently, nitrogen-fixing plant not only sustains their own growth but also improve the availability of nitrogen for related crops. Synergistic relationships are formed by these interactions, which optimize the cycling of nutrients and promote a balanced supply of nutrients for the entire AFS [85]. In addition to this, they encourage more robust plant growth and lower the requirement for external nitrogen fertilizers. In Sub-Saharan Africa, alley cropping is a popular agroforestry technique that involves alternatively planting rows of shrubs or trees with crop alleys. According to a case study conducted in Nigeria, adding nitrogen-fixing trees such as *Gliricidia sepium* or *Leucaena leucocephala* to alley cropping systems significantly increased crop yields and soil fertility [86]. Because of the tree's natural ability to fix nitrogen, there was an increased supply of fertilizer available, which improved crop growth and productivity.

Besides rhizobia, prokaryotes belonging to Frankiaceae and cyano-bacteria can also fix dinitrogen symbiotically with plants. The bacteria belonging to Frankiaceae can form root nodules in nonleguminous plants, that is, actinorhizal plants belonging to angiosperms. However, cyanobacteria can form symbiotic relationships with a variety of plants [87]. More forest tree species are associated with Frankiaceae than Proteobacteria, and this family is highly significant for both AFS and forests. According to Nygren et al. [68], actinomycetes that belong to the genus *Frankia* are well known to form root knobs (nodules) in 280 nonleguminous plants belonging to 25 genera. Actinorhizal symbiosis is found in a vast array of plant species and geographical regions, ranging from semi-deserts to rainforests and the Arctic to the Tropics [88]. *Allocasuarina, Casuarina* (Casuarinaceae), *Alnus* (Betulaceae), and *Hippophae* (Elaeagnaceae) are the main tree species of significance in tropical AFS. The actinorhizal symbiosis with tree species is extremely valuable in agroforestry.

In farming systems, only the water ferns belonging to the *Azolla* genus hold economic significance [89], whereas none of the other plants symbiotically associated with cyanobacteria are significant in AFS. On the contrary, the AFSs understorey may contain a large number of bryophytes that undoubtedly contain cyanobacteria that enrich soils by fixing nitrogen. Comparably, several bacteria are well known to be endophytic in the xylem, leaf, bark, and root tissues of live trees, though it is unclear exactly how these bacteria contribute to AFS [5].

8.2.2.2 NONSYMBIOTIC NITROGEN-FIXING MICROBES

Nonsymbiotic microbes fix free nitrogen in the atmosphere without interacting with other organisms. Nonsymbiotic nitrogen-fixing bacteria supply ready-to-use nitrogen in soil. A number of nonsymbiotic nitrogen-fixing bacteria, specifically free-living plant growth-promoting rhizobacteria (PGPR) [90], are important elements of AFS. According to the published research, only 2–5% of rhizosphere bacteria are PGPR [15]. In AFS, the PGPR are advantageous as they have a positive impact on both yield and ecosystem functioning [91] (Figure 8.5). They promote plant growth [15, 92], fix nitrogen, solubilize insoluble phosphates [15], maintain systemic resistance in plants [93], control diseases in plants [94], and generate useful biochemicals [15]. Additionally, PGPR synthesizes growth-promoting metabolites such as auxins, abscisic acid, cytokinin, ethylene, gibberellins, and indoleacetic acid [95]. PGPR also supports the synthesis of chemical and physical barriers to abiotic stress; these materials include antibiotics, organic acids, siderophores, pigments, water-soluble vitamins, and various volatile organic compounds like alcohols containing monoterpenes [96]. Therefore, metabolites produced by the PGPR, including antibiotics (*Agrobacterium radiobacter, Bacillus* sp., *Pseudomonas* sp., *Burkholderia, Brevibacterium, Streptomyces*), amylase (*Stenotrophomonas rhizophila*), gibberellin (*Azotobacter chroococcum*), phytohormone (*Azotobacter, Bacillus* sp., *Pseudomonas* sp., *Rhizobium*), macronutrients (*Bacillus* spp., *Pseudomonas* spp., *Burkholderia* spp.), lytic enzyme (*Bacillus* spp., *Pseudomonas* spp.) and volatile metabolite (*Bacillus* spp., *Pseudomonas* spp., *Burkholderia* spp., *Agrobacterium* sp., *Xanthomonas*), etc. [15]. Two categories of PGPR have been identified as bioprotectants: (1) those that have the capacity to regulate plant growth through biocontrol, thereby indirectly promoting plant growth, and (2) those that have a direct impact on the seed emergence and growth of a plant, thereby increasing crop productivity [97]. The PGPR significantly contributes to the production of agricultural crops in a sustainable manner. In exchange, plants enable PGPR to become more competitive in their niche colonization, allowing them to interfere with the quorum detection signal [98]. In addition to this, plants enable PGPR to inhibit biofilm production by pathogenic bacteria [99]. Thus, the demand for PGPR in agroforestry management is high due to its multifunctional advantages.

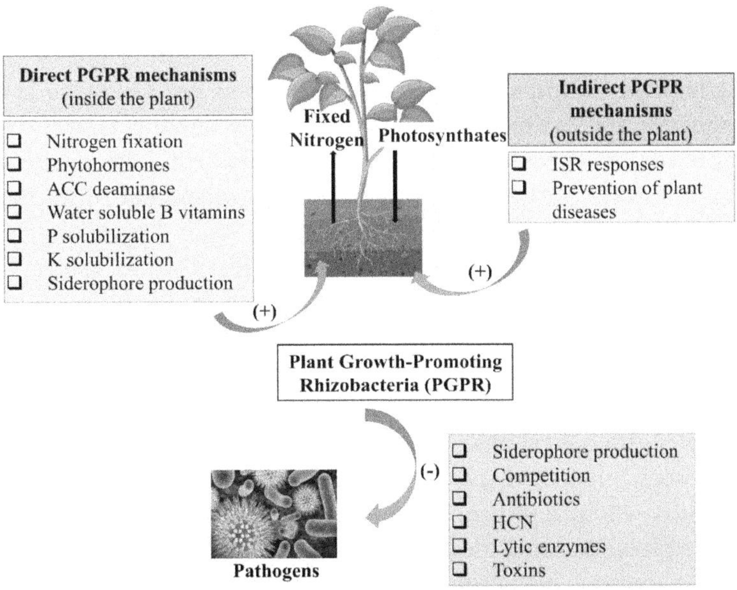

FIGURE 8.5 PGPR mechanism in the rhizosphere.

Source: Adapted from Ref. [15].

8.3 AGROFORESTRY FOR SOIL HEALTH IMPROVEMENT

Soil is one of the most important resources. According to the United States Department of Agriculture National Resources Conservation Service [100], soil health is defined as the soil's continued ability to function as an extremely important ecosystem that supports plants, animals, and humans. Moreover, soil health refers to the functioning of ecosystem services that soils provide and includes the capacity of soils to improve and sustain agricultural output, absorb carbon, decrease nutrient leaching, maintain biogeochemical cycling, and preserve biodiversity. Currently, there is significant degradation occurring worldwide. Approximately 47% of arable marginal lands (under rainfed conditions), 73% of rangelands (found in drylands), and significant portions of irrigated arable lands are being degraded [101]. According to the Food and Agriculture Organization (FAO) of the United Nations [102], 50% of the world's soil is moderately damaged, whereas 25% is severely deteriorated [103]. The destruction of soil fertility is caused by disruptions in the natural fallow cycle brought on by increasing population and a decrease in per capita available land. In temperate climates, conversion of natural land to farmland lost 60% of its soil organic carbon, whereas in tropical regions, the loss was approximately 75% [104]. Due to intensive cropping, heavy tillage, and inadequate C inputs, soil organic carbon has been reduced [105]. It is recognized that the main biophysical factors causing food insecurity are significant soil erosion and low soil fertility [106]. Thus, enhancing soil health is one of the key goals in the establishment of sustainable farming practices.

Farming techniques, such as zero tillage, residue return, cover cropping, minimum tillage, and crop rotation, have been created to reduce soil degradation and lift farmland productivity [107]. These techniques aim to improve soil health by raising the quantity of plant residue on the surface, reducing soil disturbance, and enhancing biodiversity in agroecosystems. Agroforestry is an agricultural practice that has been identified as a promising approach for increasing the sustainability of agriculture. In addition, agroforestry can influence the chemical, physical, and biological qualities of the soil and stimulate plant development by acting as a substantial source of soil organic matter. It can help restore soil health by stabilizing soils, storing and capturing carbon in soils, improving nutrient availability (Especially nitrogen, phosphorus, and potassium), promoting aggregate formation, and maintaining and protecting a healthy soil biota. In AFS, the primary source of organic carbon and nutrients is the buildup of litter from the shedding of twigs and leaves. This addition of organic matter into the soil plays a significant role in soil aggregation. It also reduces the density of soil, which helps in air circulation, improves the soil nutrient quality, helps in water distribution in the rhizosphere, and also recharges the groundwater [27]. According to a study conducted in Bangladesh by Riyadh et al. [108], the soils of various crops grown in jackfruit-based agroforestry showed 10–20% higher soil moisture, 3.37–9.25% lower soil temperature, and 9–19% higher total nitrogen content than those of open seasonal crop fields. Soil fertility may be increased by using jackfruit-based AFS, as they are thought to improve the physicochemical properties of the soil. The practice of agroforestry results in increased rates of soil C-sequestration, moisture contents, and available soil nitrogen, phosphorus, and potassium levels. These factors enable more sustainable farming practices in the upcoming seasons and minimize the need for chemical fertilizers, as the residues from these practices can be used for future crops [109]. The incorporation of trees alongside agricultural crops affects not only the physicochemical characteristics of the soil but also the soil microbiota. The soil microbiota indirectly encourages plant growth because it plays an important role in nutrient cycling and also improves plant health and productivity. Microbes receive the essential energy supply from root exudation and falling litter in the form of sugars, amino acids, organic acids, and other substances [110]. Agroforestry can enhance soil biodiversity in comparison to monocropping by offering specific habitats, a refuge for epigenic organisms, microclimate heterogeneity, buffering action, soil moisture, and humidity [27]. We have already covered the significance of different microorganisms in agroforestry in the chapter's previous section. In order to implement and support AFS, the management of microbial biodiversity is crucial. To achieve sustainable AFS, a variety of environmentally friendly techniques that are less intrusive than conventional methods could be used to support microbial biodiversity (Figure 8.6).

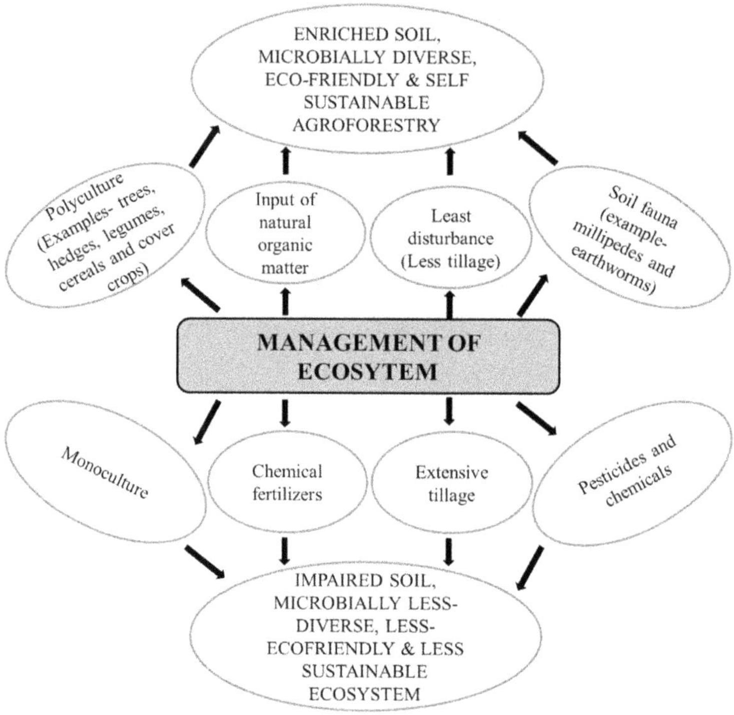

FIGURE 8.6 Ecosystem management plan to protect microbial biodiversity for sustainable agroforestry.

Source: Adapted from Ref. [5]

8.4 FUTURE DIRECTIONS AND RESEARCH NEEDS

Plant microbiome research has made substantial progress in the past 10 years because of its impact on overall plant development and productivity. Although plant–microbiome interactions hold immense potential for agroforestry, little is known about the microbial communities involved and how they interact. It may be possible to create strategies for sustainable agricultural practices by studying the interactions between plants and their microbiomes. To effectively use the plant microbiome's enormous potential in agriculture, it is necessary to unravel its intricate structural and functional diversity (Table 8.1). AFS can be supported by a wide range of trees that are suited to a particular geographic area in order to increase economic gains. The microbes use organic matter as a source of energy in the soil, which increases microbial diversity. They encourage the

TABLE 8.1 Overview of Microbial Diversity in Agroforestry

Diverse Microorganisms Linked to AFS			Important Genera
Mutualistic fungi	Mycorrhizal fungi	Arbuscular mycorrhizal fungi	*Acaulospora, Claroideoglomus, Funneliformis*, and *Glomus*
		Ectomycorrhizal fungi	*Cortinarius, Geopora, Inocybe, Laccaria,* and *Preussia,*
	Nonmycorrhizal fungi	Endophytes	*Achillea, Alternaria, Anthemis, Bjerkandera, Cichorium, Fusarium, Matricaria, Preussia, Schizophyllum, Septoria, Trametes, Trichaptum,*
			Trichoderma, and *Verticillium*
Nitrogen-fixing microbes	Symbiotic nitrogen-fixing microbes	Alphaproteobacteria and Betaproteobacteria	*Allorhizobium, Aminobacter, Azorhizobium, Blastobacter, Bradyrhizobium, Burkholderia, Cupriavidus, Devosia, Mesorhizobium, Methylobacterium, Microvirga, Neorhizobium, Ochrobactrum, Paraburkholderia, Pararhizobium, Phyllobacterium, Ralstonia, Rhizobium, Shinella,* and *Sinorhizobium,*
		Actinobacteria	*Frankia*
		Cyanobacteria	*Anabaena*
	Nonsymbiotic nitrogen-fixing microbes	Plant growth-promoting rhizobacteria	*Acetobacter, Azotobacter, Arthrobacter, Acinetobacter, Azoarcus, Alcaligenes, Arthrobacter, Azospirillum, Bacillus, Burkholderia, Pantoae, Beijerinckia, Enterobacter, Herbaspirillum, Derxia, Gluconacetobacter, Ochrobactrum, Klebsiella, Serratia, Lactobacillus, Paenobacillus, Stenotrophomonas, Pseudomonas, Rhodococcus*, and *Zoogloea*

absorption of important minerals, breakdown of organic matter, nitrogen fixation, and also provide disease resistance. These processes take a long time to stabilize the AFS in favor of production. The production and demand for microbial inoculants differ significantly, but this could be addressed by screening and picking viable/ effective microbial consortia (comprised of PGPR and AMF) for the crops and trees used in AFS. The chosen effective microbial consortia could then be used for inoculation in AFS. Besides this, it is essential to use multiple strain combinations to fully use the potential of the microbes in AFS. Additionally, coordinated global methods for assessing the microbial diversity at the area level and AFS level will produce important findings that will help solve issues pertaining to microbial diversity under AFS on a larger geographic scale.

8.5 CONCLUSION

In conclusion, AFS offers a promising way to boost crop productivity while encouraging sustainable land management techniques. The integration of trees with agricultural crops, along with appropriate species selection and management techniques, helps to improve the degraded soil's biological as well as physico-chemical properties. This agricultural practice improves the soil quality, which in turn increases soil microbiota. Several soil microorganisms are known to form symbiotic associations with plants. The plant–microbe synergy in AFS has several advantages, ranging from enhanced nutrient pumping and cycling and soil fertility to increased climate change adaptation and resilience of AFS. Researchers, policymakers, and farmers can use these insights to further develop and adopt agroforestry practices in order to address the challenges of sustainable agriculture and food security in a changing global context.

KEYWORDS

- **agroforestry**
- **microbial partnerships**
- **plant-microbe synergy**
- **soil health**
- **sustainable agriculture**

REFERENCES

1. Bene, J. G., Beall, H. W., Côté, A. (1977). *Trees, Food and People: Land Management in the Tropics*. IDRC: Ottawa, Canada.
2. Nair, P. R., Kumar, B. M., Nair, V. D., Nair, P. R., Kumar, B. M., Nair, V. D. (2021). Definition and concepts of agroforestry. In: *An Introduction to Agroforestry: Four Decades of Scientific Developments* (pp. 21–28).
3. Widyati, E., Siarudin, M., Indrajaya, Y. (2022). The dynamic of functional microbes community under auri (*Acacia auriculiformis* Cunn. Ex Benth) agroforestry system. *Jurnal Manajemen Hutan Tropika*, 28(2), 119.
4. Zomer, R. J., Trabucco, A., Coe, R., Place, F. (2009). *Trees on Farm: Analysis of Global Extent and Geographical Patterns of Agroforestry* (p. 72). ICRAF Working Paper-World Agroforestry Centre.
5. Sridhar, K. R., Bagyaraj, D. J. (2017). Microbial biodiversity in agroforestry systems. In: *Agroforestry: Anecdotal to Modern Science* (pp. 645–667), Springer Nature.
6. Smith, J., Pearce, B. D., Wolfe, M. S. (2013). Reconciling productivity with protection of the environment: Is temperate agroforestry the answer. *Renewable Agriculture and Food Systems*, 28(1), 80–92.
7. Vaupel, A., Bednar, Z., Herwig, N., Hommel, B., Moran-Rodas, V. E., Beule, L. (2023). Tree-distance and tree-species effects on soil biota in a temperate agroforestry system. *Plant and Soil*, 17, 1–8.
8. Guillot, E., Bertrand, I., Rumpel, C., Gomez, C., Arnal, D., Abadie, J., Hinsinger, P. (2021). Spatial heterogeneity of soil quality within a Mediterranean alley cropping agroforestry system: Comparison with a monocropping system. *European Journal of Soil Biology*, 105, 103330.
9. Luo, J., Beule, L., Shao, G., Veldkamp, E., Corre, M. D. (2022). Reduced soil gross N_2O emission driven by substrates rather than denitrification gene abundance in cropland agroforestry and monoculture. *Journal of Geophysical Research: Biogeosciences*, 127(3), e2021JG006629.
10. Vega, H. B., Espinosa-Victoria, D., Salaya-Domínguez, J. M., Gómez-Méndez, E. (2013). The soil biota: Importance in agroforestry and agricultural systems. *Tropical and Subtropical Agroecosystems*, 16(3), 445–453.
11. Araujo, A. S., Leite, L. F., Iwata, B. D., Lira, M. D., Xavier, G. R., Figueiredo, M. D. (2012). Microbiological process in agroforestry systems. A review. *Agronomy for Sustainable Development*, 32, 215–226.
12. Fagerholm, N., Torralba, M., Burgess, P. J., Plieninger, T. (2016). A systematic map of ecosystem services assessments around European agroforestry. *Ecological Indicators*, 62, 47–65.
13. Desai, S., Kumar, G. P., Amalraj, L. D., Bagyaraj, D. J., Ashwin, R. (2016). Exploiting PGPR and AMF biodiversity for plant health management. *Microbial Inoculants in Sustainable Agricultural Productivity: Research Perspectives*, 1, 145–160.
14. Kirchman, D. L. (2018). The physical-chemical environment of microbes. In: *Processes in Microbial Ecology* (pp. 35–52), Oxford.
15. Singh, A., Yadav, V. K., Chundawat, R. S., Soltane, R., Awwad, N. S., Ibrahium, H. A., Yadav, K. K., Vicas, S. I. (2023). Enhancing plant growth promoting rhizobacterial

activities through consortium exposure: A review. *Frontiers in Bioengineering and Biotechnology*, 11, 1099999.

16. Saeed, Q., Xiukang, W., Haider, F. U., Kučerik, J., Mumtaz, M. Z., Holatko, J., Naseem, M., Kintl, A., Ejaz, M., Naveed, M., Brtnicky, M. (2021). Rhizosphere bacteria in plant growth promotion, biocontrol, and bioremediation of contaminated sites: A comprehensive review of effects and mechanisms. *International Journal of Molecular Sciences*, 22(19), 10529.
17. Almeida, E. F., Polizel, R. P., Gomes, L. C., Xavier, F. A., Mendonça, E. S. (2007). Biomassa microbiana em sistemas agroflorestais na zona da mata mineira. *Revista Brasileira de Agroecologia*, 2(2), 739–742.
18. Kaur, B., Gupta, S. R., Singh, G. (2000). Soil carbon, microbial activity and nitrogen availability in agroforestry systems on moderately alkaline soils in northern India. *Applied Soil Ecology*, 15(3), 283–294.
19. Cardoso, I. M., Kuyper, T. W. (2006). Mycorrhizas and tropical soil fertility. *Agriculture, Ecosystems & Environment*, 116(1–2), 72–84.
20. Freitas, A. D., Sampaio, E. V., Santos, C. E., Fernandes, A. R. (2010). Biological nitrogen fixation in tree legumes of the Brazilian semi-arid caatinga. *Journal of Arid Environments*, 74(3), 344–349.
21. Beule, L., Arndt, M., Karlovsky, P. (2021). Relative abundances of species or sequence variants can be misleading: Soil fungal communities as an example. *Microorganisms*, 9(3), 589.
22. Laurindo, L. K., de Souza, T. A., da Silva, L. J., Casal, T. B., de Jesus Conceição Pires, K., Kormann, S., Schmitt, D. E., Siminski, A. (2021). Arbuscular mycorrhizal fungal community assembly in agroforestry systems from the Southern Brazil. *Biologia*, 76(4), 1099–1107.
23. Sridhar, K. R. (2009). Fungi in the tree canopy: An appraisal. In: *Applied Mycology* (pp. 73–91), CABI: Wallingford, UK.
24. Radhakrishnan, S., Varadharajan, M. (2016). Status of microbial diversity in agroforestry systems in Tamil Nadu, India. *Journal of Basic Microbiology*, 56(6), 662–669.
25. Mungai, N. W., Motavalli, P. P., Kremer, R. J., Nelson, K. A. (2005). Spatial variation of soil enzyme activities and microbial functional diversity in temperate alley cropping systems. *Biology and Fertility of Soils*, 42, 129–136.
26. Sørensen, J., Sessitsch, A. (2007). Plant-associated bacteria lifestyle and molecular interactions. *Modern Soil Microbiology*, 2, 211–236.
27. Fahad, S., Chavan, S. B., Chichaghare, A. R., Uthappa, A. R., Kumar, M., Kakade, V., Pradhan, A., Jinger, D., Rawale, G., Yadav, D. K. (2022). Agroforestry systems for soil health improvement and maintenance. *Sustainability*, 14, 14877.
28. Guo, L. (2012). Progress of microbial species diversity research in China. *Biodiversity Science*, 20(5), 572.
29. Polo-Marcial, M. H., Solís-Ramos, L. Y., Murillo-Cruz, R., Ávila-Arias, C., Andrade-Torres, A. (2023). Mycorrhizal and endophytic richness and colonization in *Cedrela odorata* L., in agroforestry systems and secondary forest from southeastern Costa Rica. *Agroforestry Systems*, 97(4), 647–658.
30. Qiao, X., Sun, T., Lei, J., Xiao, L., Xue, L., Zhang, H., Jia, J., Bei, S. (2022). Arbuscular mycorrhizal fungi contribute to wheat yield in an agroforestry system with different tree ages. *Frontiers in Microbiology*, 13, 1024128.

31. Smith, S. E., Read, D. J. (2010). *Mycorrhizal Symbiosis*. Academic Press.
32. Tedersoo, L., Sánchez-Ramírez, S., Koljalg, U., Bahram, M., Döring, M., Schigel, D., May, T., Ryberg, M., Abarenkov, K. (2018). High-level classification of the Fungi and a tool for evolutionary ecological analyses. *Fungal Diversity*, 90, 135–159.
33. van Tuinen, D., Tranchand, E., Hirissou, F., Wipf, D., Courty, P. E. (2020). Carbon partitioning in a walnut-maize agroforestry system through arbuscular mycorrhizal fungi. *Rhizosphere*, 15, 100230.
34. Masebo, N., Birhane, E., Takele, S., Belay, Z., Lucena, J. J., Pérez-Sanz, A., Anjul, O. A. (2023). Diversity of Arbuscular Mycorrhizal fungi under different agroforestry practices in the drylands of Southern Ethiopia. *BMC Plant Biology*, 23(1), 1–4.
35. Goto, B. T., Silva, G. A., Assis, D., Silva, D. K., Souza, R. G., Ferreira, A. C., Jobim, K., Mello, C., Vieira, H. E., Maia, L. C., Oehl, F. (2012). Intraornatosporaceae (Gigasporales), a new family with two new genera and two new species. *Mycotaxon*, 119(1), 117–132.
36. Bagyaraj, D. J. (1992). Vesicular-arbuscular mycorrhiza: Application in agriculture. In: *Methods in Microbiology* (Vol. 24, pp. 359–373). Academic Press: London.
37. Van Der Heijden, M. G., Bardgett, R. D., Van Straalen, N. M. (2008). The unseen majority: Soil microbes as drivers of plant diversity and productivity in terrestrial ecosystems. *Ecology Letters*, 11(3), 296–310.
38. Barrios, E. (2007). Soil biota, ecosystem services and land productivity. *Ecological Economics*, 64(2), 269–285.
39. Turrini, A., Avio, L., Giovannetti, M., Agnolucci, M. (2018). Functional complementarity of arbuscular mycorrhizal fungi and associated microbiota: The challenge of translational research. *Frontiers in Plant Science*, 9, 1407.
40. Wipf, D., Krajinski, F., van Tuinen, D., Recorbet, G., Courty, P. E. (2019). Trading on the arbuscular mycorrhiza market: From arbuscules to common mycorrhizal networks. *New Phytologist*, 223(3), 1127–1142.
41. Zhu, Y. G., Miller, R. M. (2003). Carbon cycling by arbuscular mycorrhizal fungi in soil-plant systems. *Trends Plant Science*, 8, 407–409.
42. Bryla, D. R., Eissenstat, D. M. (2005). Respiratory costs of mycorrhizal associations. *Plant Respiration*, 18, 207–224.
43. Jansa, J., Forczek, S. T., Rozmoš, M., Püschel, D., Bukovská, P., Hršelová, H. (2019). Arbuscular mycorrhiza and soil organic nitrogen: Network of players and interactions. *Chemical and Biological Technologies in Agriculture*, 6(1), 1–10.
44. Song, Z., Bi, Y., Zhang, J., Gong, Y., Yang, H. (2020). Arbuscular mycorrhizal fungi promote the growth of plants in the mining associated clay. *Scientific Reports*, 10(1), 2663.
45. Nakmee, P. S., Techapinyawat, S., Ngamprasit, S. (2016). Comparative potentials of native arbuscular mycorrhizal fungi to improve nutrient uptake and biomass of *Sorghum bicolor* Linn. *Agriculture and Natural Resources*, 50(3), 173–178.
46. Parihar, P., Bora, M. (2018). Effect of mycorrhiza (*Glomus mosseae*) on morphological and biochemical properties of Ashwagandha (*Withania somnifera*) (L.) Dunal. *Journal of Applied and Natural Science*, 10(4), 1115–1123.
47. Vos, C. M., Tesfahun, A. N., Panis, B., De Waele, D., Elsen, A. (2012). Arbuscular mycorrhizal fungi induce systemic resistance in tomato against the sedentary nematode

Meloidogyne incognita and the migratory nematode *Pratylenchus penetrans*. *Applied Soil Ecology*, 61, 1–6.

48. Kumari, S. M., Prabina, B. J. (2019). Protection of tomato, *Lycopersicon esculentum* from wilt pathogen, *Fusarium oxysporum* f. sp. lycopersici by arbuscular mycorrhizal fungi, *Glomus* sp. *International Journal of Current Microbiology and Applied Sciences*, 8, 1368–1378.
49. Hodge, A., Storer, K. (2015). Arbuscular mycorrhiza and nitrogen: Implications for individual plants through to ecosystems. *Plant and Soil*, 386, 1–9.
50. Liu, Y., Shi, G., Mao, L., Cheng, G., Jiang, S., Ma, X., An, L., Du, G., Collins Johnson, N., Feng, H. (2012). Direct and indirect influences of 8 yr of nitrogen and phosphorus fertilization on Glomeromycota in an alpine meadow ecosystem. *New Phytologist*, 194(2), 523–535.
51. Dumbrell, A. J., Nelson, M., Helgason, T., Dytham, C., Fitter, A. H. (2010). Relative roles of niche and neutral processes in structuring a soil microbial community. *The ISME Journal*, 4(3), 337–345.
52. Bai, C., He, X., Tang, H., Shan, B., Zhao, L. (2009). Spatial distribution of arbuscular mycorrhizal fungi, glomalin and soil enzymes under the canopy of *Astragalus adsurgens* Pall. in the Mu Us sandland, China. *Soil Biology and Biochemistry*, 41(5), 941–947.
53. Fitzsimons, M. S., Miller, R. M., Jastrow, J. D. (2008). Scale-dependent niche axes of arbuscular mycorrhizal fungi. *Oecologia*, 158, 117–127.
54. Oehl, F., Laczko, E., Bogenrieder, A., Stahr, K., Bösch, R., van der Heijden, M., Sieverding, E. (2010). Soil type and land use intensity determine the composition of arbuscular mycorrhizal fungal communities. *Soil Biology and Biochemistry*, 42(5), 724–738.
55. Araújo, T. M., Silva, K. D., Pereira, G. M., Curcino, A., Stürmer, S. L., Gomide, P. H. (2019). Diversity of arbuscular mycorrhizal fungi in agroforestry, conventional plantations and native forests in Roraima State, Northern Brazil. *Journal of Agricultural Science*, 11(14), 282.
56. da Silva Maia, R., Vasconcelos, S. S., Viana-Junior, A. B., Castellani, D. C., Kato, O. R. (2021). Oil palm (*Elaeis guineensis*) shows higher mycorrhizal colonization when planted in agroforestry than in monoculture. *Agroforestry Systems*, 95, 731–740.
57. Barea, J. M., Werner, D., Azcón-Guilar, C., Azcón, R. (2005). Interactions of arbuscular mycorrhiza and nitrogen-fixing symbiosis in sustainable agriculture. In: *Nitrogen Fixation in Agriculture, Forestry, Ecology, and the Environment* (pp. 199–222).
58. Liu, Y., Li, X., Kou, Y. (2020). Ectomycorrhizal fungi: Participation in nutrient turnover and community assembly pattern in forest ecosystems. *Forests*, 11(4), 453.
59. Strobel, G., Daisy, B., Castillo, U., Harper, J. (2004). Natural products from endophytic microorganisms. *Journal of Natural Products*, 67(2), 257–268.
60. Fanuel, K., David, M., Asenath, A. (2022). Diversity of culture dependent endophytic bacteria isolated from leguminous agroforestry trees in western Kenya.
61. Florindo, R. H. Bioprospecção de metabólitos secundários bioativos produzidos por fungos endofíticos associados à Piper sp. coletada no Parque Estadual do Rio Doce, Minas Gerais.

62. Hatamzadeh, S., Rahnama, K., White, J. F., Oghaz, N. A., Nasrollahnejad, S., Hemmati, K. (2023). Investigation of some endophytic fungi from five medicinal plants with growth promoting ability on maize (*Zea mays* L.). *Journal of Applied Microbiology*, 134(1), lxac015.
63. Jeffries, P., Gianinazzi, S., Perotto, S., Turnau, K., Barea, J. M. (2003). The contribution of arbuscular mycorrhizal fungi in sustainable maintenance of plant health and soil fertility. *Biology and Fertility of Soils*, 37, 1–6.
64. Guzman, A., Montes, M., Hutchins, L., DeLaCerda, G., Yang, P., Kakouridis, A., Dahlquist-Willard, R. M., Firestone, M. K., Bowles, T., Kremen, C. (2021). Crop diversity enriches arbuscular mycorrhizal fungal communities in an intensive agricultural landscape. *New Phytologist*, 231(1), 447–459.
65. Buren, S., Rubio, L. M. (2018). State of the art in eukaryotic nitrogenase engineering. *FEMS Microbiology Letters*, 365(2), fnx274.
66. Wagner, S. C. (2011). Biological nitrogen fixation. *Nature Education Knowledge*, 3(10), 15.
67. Li, W., Li, F., Zeng, H., Ma, L., Qi, L., Wang, X., Wang, W., Peng, Z., Degen, A. A., Bai, Y., Zhang, T. (2021). Diversity and variation of asymbiotic nitrogen-fixing microorganisms in alpine grasslands on the Tibetan Plateau. *Frontiers in Ecology and Evolution*, 9, 702848.
68. Nygren, P., Fernandez, M. P., Harmand, J. M., Leblanc, H. A. (2012). Symbiotic dinitrogen fixation by trees: An underestimated resource in agroforestry systems. *Nutrient Cycling in Agroecosystems*, 94, 123–160.
69. Zook, D. (2015). *Symbiosis-evolution's Co-author* (pp. 41–80). Springer International Publishing.
70. Lindström, K., Mousavi, S. A. (2020). Effectiveness of nitrogen fixation in rhizobia. *Microbial Biotechnology*, 13(5), 1314–1335.
71. Garrity, G. M., Bell, J. A., Lilburn, T. Class, I. (2005). Alphaproteobacteria class. nov. In: *Bergey's manual® of Systematic Bacteriology* (pp. 1–574), Springer: Boston, MA.
72. Sprent, J. I. (2008). 60Ma of legume nodulation. What's new? What's changing. *Journal of Experimental Botany*, 59(5), 1081–1084.
73. Frank, B. (1889). Über die Pilzsymbiose der Leguminosen. *Berichte Dtsch. Bot. Ges*, 7, 332–346.
74. Nair, P. K. R., Mohan Kumar, B., Nair, V. D. (2009). Agroforestry as a strategy for carbon sequestration. *Journal of Plant Nutrition and Soil Science*, 172(1), 10–23.
75. Lloret, L., Ormeno-Orrillo, E., Rincon, R., Martínez-Romero, J., Rogel-Hernández, M. A., Martínez-Romero, E. (2007). *Ensifer mexicanus* sp. nov. a new species nodulating *Acacia angustissima* (Mill.) Kuntze in Mexico. *Systematic and Applied Microbiology*, 30(4), 280–290.
76. Wolde-Meskel, E., Terefework, Z., Frostegård, Å., Lindström, K. (2005). Genetic diversity and phylogeny of rhizobia isolated from agroforestry legume species in southern Ethiopia. *International Journal of Systematic and Evolutionary Microbiology*, 55(4), 1439–1452.
77. Wang, F. Q., Wang, E. T., Liu, J., Chen, Q., Sui, X. H., Chen, W. F., Chen, W. X. (2007). *Mesorhizobium albiziae* sp. nov., a novel bacterium that nodulates *Albizia*

kalkora in a subtropical region of China. *International Journal of Systematic and Evolutionary Microbiology*, 57(6), 1192–1199.
78. Zhang, X., Harper, R., Karsisto, M., Lindström, K. (1991). Diversity of Rhizobium bacteria isolated from the root nodules of leguminous trees. *International Journal of Systematic and Evolutionary Microbiology*, 41(1), 104–113.
79. Hernandez-Lucas, I., Segovia, L., Martinez-Romero, E., Pueppke, S. G. (1995). Phylogenetic relationships and host range of *Rhizobium* spp. that nodulate *Phaseolus vulgaris* L. *Applied and Environmental Microbiology*, 61(7), 2775–2779.
80. Leblanc, H. A., McGraw, R. L., Nygren, P., Roux, C. L. (2005). Neotropical legume tree *Inga edulis* forms N_2-fixing symbiosis with fast-growing *Bradyrhizobium* strains. *Plant and Soil*, 275, 123–133.
81. Sprent, J. I. (2009). *Legume Nodulation: A Global Perspective* (p. 181). Wiley-Blackwell: Oxford.
82. Han, T. X., Wang, E. T., Wu, L. J., Chen, W. F., Gu, J. G., Gu, C. T., Tian, C. F., Chen, W. X. (2008). *Rhizobium multihospitium* sp. nov., isolated from multiple legume species native of Xinjiang, China. *International Journal of Systematic and Evolutionary Microbiology*, 58(7), 1693–1699.
83. Dreyfus, B., Garcia, J. L., Gillis, M. (1988). Characterization of *Azorhizobium caulinodans* gen. nov., sp. nov., a stem-nodulating nitrogen-fixing bacterium isolated from *Sesbania rostrata*. *International Journal of Systematic and Evolutionary Microbiology*, 38(1), 89–98.
84. Chen, W. X., Tan, Z. Y., Gao, J. L., Li, Y., Wang, E. T. (1998). *Rhizobium hainanense* sp. nov., isolated from tropical legumes. *International Journal of Systematic and Evolutionary Microbiology*, 47(3), 870–873.
85. Kaur, A., Paruchuri, R. G., Nayak, P., Devi, K. B., Upadhyay, L., Kumar, A., Pancholi, R., Yousuf, M. (2023). The role of agroforestry in soil conservation and sustainable crop production: A comprehensive review. *International Journal of Environment and Climate Change*, 13(11), 3089–3095.
86. Franzel, S., Coe, R., Cooper, P., Place, F. (2001). Assessing the adoption potential of agroforestry practices in sub-Saharan Africa. *Agricultural Systems*, 69(1–2), 37–62.
87. Mousavi, S. A. (2016). Revised taxonomy of the family Rhizobiaceae, and phylogeny of mesorhizobia nodulating *Glycyrrhiza* spp. *Dissertationes Schola Doctoralis Scientiae Circumiectalis, Alimentariae, Biologicae*.
88. Russo, R. O. (2005). Nitrogen-fixing trees with actinorhiza in forestry and agroforestry. In: *Nitrogen Fixation in Agriculture, Forestry, Ecology, and the Environment* (pp. 143–171). Springer Netherlands: Dordrecht.
89. Herridge, D. F., Peoples, M. B., Boddey, R. M. (2008). Global inputs of biological nitrogen fixation in agricultural systems. *Plant and Soil*, 311, 1–8.
90. Vega-Celedon, P., Bravo, G., Velasquez, A., Cid, F. P., Valenzuela, M., Ramirez, I., Vasconez, I. N., Alvarez, I., Jorquera, M. A., Seeger. M. (2021). Microbial diversity of psychrotolerant bacteria isolated from wild flora of andes mountains and patagonia of chile towards the selection of plant growth-promoting bacterial consortia to alleviate cold stress in plants. *Microorganisms*, 9(3), 538.
91. Barbosa, J. Z., de Almeida Roberto, L., Hungria, M., Corrêa, R. S., Magri, E., Correia, T. D. (2022). Meta-analysis of maize responses to *Azospirillum brasilense*

inoculation in Brazil: Benefits and lessons to improve inoculation efficiency. *Applied Soil Ecology*, 170, 104276.
92. Rosa, P. A., Galindo, F. S., Oliveira, C. E., Jalal, A., Mortinho, E. S., Fernandes, G. C., Marega, E. M., Buzetti, S., Teixeira Filho, M. C. (2022). Inoculation with plant growth-promoting bacteria to reduce phosphate fertilization requirement and enhance technological quality and yield of sugarcane. *Microorganisms*, 10(1), 192.
93. Renoud, S., Abrouk, D., Prigent-Combaret, C., Wisniewski-Dyé, F., Legendre, L., Moënne-Loccoz, Y., Muller, D. (2022). Effect of inoculation level on the impact of the PGPR *Azospirillum lipoferum* CRT1 on selected microbial functional groups in the rhizosphere of field maize. *Microorganisms*, 10(2), 325.
94. Kong, W. L., Li, P. S., Wu, X. Q., Wu, T. Y., Sun, X. R. (2020). Forest tree associated bacterial diffusible and volatile organic compounds against various phytopathogenic fungi. *Microorganisms*, 8(4), 590.
95. Patel, T., Saraf, M. (2017). Biosynthesis of phytohormones from novel rhizobacterial isolates and their in vitro plant growth-promoting efficacy. *Journal of Plant Interactions*, 12(1), 480–487.
96. Chandran, H., Meena, M., Swapnil, P. (2021). Plant growth-promoting rhizobacteria as a green alternative for sustainable agriculture. *Sustainability*, 13(19), 10986.
97. Chauhan, H., Bagyaraj, D. J., Selvakumar, G., Sundaram, S. P. (2015). Novel plant growth promoting rhizobacteria—Prospects and potential. *Applied Soil Ecology*, 95, 38–53.
98. Hartmann, A. (2020). Quorum sensing N-acyl-homoserine lactone signal molecules of plant beneficial Gram-negative rhizobacteria support plant growth and resistance to pathogens. *Rhizosphere*, 16, 100258.
99. Uruén, C., Chopo-Escuin, G., Tommassen, J., Mainar-Jaime, R. C., Arenas, J. (2020). Biofilms as promoters of bacterial antibiotic resistance and tolerance. *Antibiotics*, 10(1), 3.
100. United States Department of Agriculture National Resources Conservation Service, Soil Health. (2020). Available online: https: //www.nrcs.usda.gov/wps/portal/nrcs/main/soils/health.
101. Gisladottir, G., Stocking, M. (2005). Land degradation control and its global environmental benefits. *Land Degradation & Development*, 16(2), 99–112.
102. Food and Agriculture Organization of the United Nations - FAO & Global Mechanism of the UNCCD. (2015). *Sustainable Financing for Forest and Landscape Restoration: Opportunities, Challenges and the Way Forward*. FAO: Rome, Italy.
103. Srivastava, P., Giri, N., Mandal, D. (2019). 137 Cs technology for soil erosion and soil carbon redistribution. *Current Science*, 116(6), 888–889.
104. Lal, R. (2004). Soil carbon sequestration to mitigate climate change. *Geoderma*, 123(1–2), 1–22.
105. Salceda, M., Udawatta, R. P., Nelson, K. A., Mendis, S. S., Bardhan, S. (2022). Spatial and temporal variability of soil organic carbon on a corn–soybean watershed with 23 years of agroforestry. *Agronomy Journal*, 114(1), 440–451.
106. Mesfin, S., Haileselassie, H. (2022). Evaluation of soil physico-chemical properties as affected by canopies of scattered agroforestry trees on croplands. *South African Journal of Plant and Soil*, 39(2), 153–162.

107. Du, X., Jian, J., Du, C., Stewart, R. D. (2022). Conservation management decreases surface runoff and soil erosion. *International Soil and Water Conservation Research*, 10(2), 188–196.
108. Riyadh, Z. A., Rahman, M. A., Saha, S. R., Hossain, M. I. (2018). Soil properties under jackfruit based agroforestry systems in Madhupur tract of Narsingdi district. *Journal of the Sylhet Agricultural University*, 5, 173–179.
109. Surki, A. A., Nazari, M., Fallah, S., Iranipour, R., Mousavi, A. (2020). The competitive effect of almond trees on light and nutrients absorption, crop growth rate, and the yield in almond–cereal agroforestry systems in semi-arid regions. *Agroforestry Systems*, 94, 1111–1122.
110. Beule, L., Vaupel, A., Moran-Rodas, V. E. (2022). Abundance, diversity, and function of soil microorganisms in temperate alley-cropping agroforestry systems: A review. *Microorganisms*, 10(3), 616.

CHAPTER 9

Root-Mediated Interactions in the Agroforestry System as Seen Through the Window of Rhizosphere

SHARAT KOTHARI,[1*] ANN THERESA JOSE,[2] ALAPATI NYMISHA,[2] TANMAYA KUMAR BHOI,[1] ANSHUMAN PATEL,[3] and NAVEEN KUMAR[2]

[1]*ICFRE-Arid Forest Research Institute, Jodhpur, Rajasthan, India*

[2]*Division of Soil Science and Agricultural Chemistry, ICAR-IARI, New Delhi, India*

[3]*ICFRE-Rain Forest Research Institute, Jorhat, Assam, India*

*Corresponding author

ABSTRACT

Agroforestry, a sustainable land management approach, emphasizes the ecological and economic interactions among trees, crops, and sometimes animals within the same land unit. A critical component of these systems is the belowground interaction in the soil matrix, where plant roots and soil microorganisms dynamically influence growth and productivity. The rhizosphere, a highly active microecosystem surrounding plant roots, differs significantly from bulk soil in its physical, chemical, and biological properties. Through rhizodeposition, plants release root exudates that shape the rhizosphere microbiome, which in turn supports plant growth by enhancing nutrient solubilization, nitrogen fixation, siderophore production, and phytohormone regulation. When trees and crops are cultivated together, their root exudates vary in quality and quantity, influencing

Agroforestry and Climate Change Mitigation: Harnessing Multitrophic Interactions for Sustainable Plant Protection and Ecosystem Resilience. Hanuman Singh Jatav, Vishnu D. Rajput, Ipsita Samal, Tanmaya Kumar Bhoi (Eds.)

DOI: 10.1201/9781779641397-9

microbial activity, nutrient acquisition, and overall soil health. The tree rhizosphere generally exerts a greater influence than that of crops, leading to both positive and negative interactions. Competition for water, nutrients, or allelopathic effects may result in negative outcomes, while facilitation of nutrient and water availability supports positive interactions. Arbuscular mycorrhizal fungi, forming associations with both trees and crops, play a vital role in mediating these interspecific interactions. Therefore, understanding root and rhizosphere interactions is essential for selecting compatible components in agroforestry systems.

9.1 INTRODUCTION

In 2050, an aspirational year for Science and Technology, the biggest breakthroughs can be expected in the fields of computer science, artificial intelligence, machine learning, biotechnology, medicine, nanotechnology, and space exploration. However, among all the fields, the agriculture sector will face the greatest challenge [1]. The agriculture sector must feed around 9 billion people by 2050, for which our grain basket needs to be expanded. It needs around 50–60% more food and 20–30% more water than the 2019 level [2, 3]. The additional food production has to come from the shrinking and degrading land resources, as the global per capita land availability is going to reach 0.17 ha by 2050. This will further stretch the capacity of soil resources and catalyze the land degradation process. Along with this, the grain basket has to be enlarged when the world average temperature is rising at an alarming rate of 0.2?C /decade, which will further jeopardize the situation. On the contrary, urbanization is eating up agricultural lands at a faster rate [4].

In the era of artificial intelligence, intelligent management of natural resources is required to cope with the above challenges. Among the various options for addressing these issues, agroforestry remains a viable gateway to sustainability. Agroforestry, a sustainable land management option, mainly focuses on the ecological and economic interaction between tree, crop, and animal components in the same piece of land. In these systems, integrating trees and shrubs with traditional crops or livestock creates a unique ecological symphony that influences microclimates, biodiversity, and socio-economic dimensions. Apart from improving the physical, chemical, and biological properties of the soil, integrating trees into cultivated areas can sustain the nutrient reservoir and improve the overall soil fertility [5]. It can enhance water quality by mitigating pollution levels and reducing soil erosion, contributing to landscape amelioration. In addition, enhanced water-use efficiency and greater environmental sustainability can be achieved [6]. Agroforestry has emerged as a mitigating force against biodiversity decline by offering a protective tree cover alongside agricultural fields. The inclusion of trees not only reduces the loss of biodiversity but also fosters an environment that promotes diversity. Trees serve as shelters and habitats for various flora and fauna, contributing to an enriched ecological landscape [7]. Agroforestry systems (AFSs) present a significant opportunity for sequestering both aboveground and belowground (soil) carbon, playing a vital role in mitigating the greenhouse effect by curbing carbon emissions. The inclusion of trees under AFS, especially in tropical regions, can alleviate environmental degradation resulting from deforestation and ease the pressure on natural forests [8]. AFS serve as valuable sources of many marketable products, such as household food, medicinal plants, small timber, domestic wood supply, fiber, and fuel, and can enhance the livelihoods of small and marginal farmers by providing economic security. Livestock benefit from shaded areas provided by trees, contributing to animal welfare and potentially improving productivity [9].

AFSs intricately blend the elements of forestry and agriculture, unfolding a tapestry of aboveground and belowground interactions that shape the dynamics of sustainable land use. The study of plant–soil dynamics and interactions within agroforestry has emerged not only as a scientific pursuit but also as a beacon illuminating sustainable pathways toward food security and ecological balance. In particular, the interaction between different components of AFSs plays a pivotal role in shaping microclimates, influencing temperature, humidity, and wind patterns. Therefore, the interaction should be harnessed effectively for the success of the AFS. The aboveground interactions under AFSs have received much attention and have been more widely studied than the belowground interactions [10].

The belowground interaction under the AFS in the soil matrix, between the roots of trees and crops and the roots and soil microorganisms, plays a crucial role in the resilience of the AFS and requires proper attention. In AFSs, which involve the intentional integration of trees and crops on the same piece of land, interactions within the rhizosphere can play a crucial role in shaping the overall ecosystem. As we strive to develop resilient and sustainable agricultural practices, understanding and harnessing the potential of the rhizosphere in agroforestry is essential for the future of our food systems and the health of our planet. Therefore, in this chapter, we focus on root-related interactions in AFSs with special emphasis on the rhizosphere. First, we have discussed the rhizosphere and soil modification under its influence. The uniqueness of the rhizosphere under agroforestry and the modification of the root microbiome are discussed subsequently. The core of the chapter lies in the section on the interaction between tree roots and crop roots in the soil, where both positive and negative interactions are discussed, followed by their significance.

9.1.1 THE RHIZOSPHERE NEXUS OF LIFE

Rhizosphere is a narrow region of soil that is directly influenced by root activity and secretions. The rhizosphere is a complex dynamic microecosystem where diverse biotic and abiotic interactions occur. The rhizosphere can harbor as many as 100 billion (10^{11}) microbial cells per gram of root and more than 30,000 prokaryotic species [11, 12]. Recognized as one of the most intricate ecosystems on Earth, the rhizosphere is vital for plant–microbe interactions. Plant roots release up to 21% of their photosynthetic compounds, serving not only as nutrients for soil microbes but also as signaling molecules in plant–microbe interactions. Root exudates, comprising organic acids such as citric and malic acids, sugars, amino acids, flavonoids, phenolic compounds, lipids, fatty acids, enzymes, and growth regulators, play pivotal roles in the rhizosphere. Organic acids enhance nutrient availability by solubilizing minerals, sugars fuel microbial activity for nutrient cycling, amino acids and proteins act as nitrogen sources for beneficial microbes, flavonoids attract symbiotic mycorrhizal fungi, phenolic compounds serve in plant defense against pathogens, lipids contribute to soil structure, and enzymes break down organic matter for nutrient release. In addition, growth regulators influence neighboring plant and microbial growth. These diverse exudates collectively shape the rhizosphere environment, influencing nutrient cycling, microbial communities, and overall soil health, thereby affecting plant growth and productivity. This addition of carbon through root exudation to small volumes of soil makes it a biological hotspot. In addition, roots play a role in attracting beneficial microorganisms and deterring soil pests in the immediate vicinity of the plant [13]. Interactions between plants and microorganisms are crucial for ecosystems. Plants use the rhizosphere as a battleground for competitive interactions, releasing allelochemicals that are detrimental to other plant species, as well as harmful soil nematodes and insects [14].

9.2 PHYSICAL, CHEMICAL, AND BIOLOGICAL MODIFICATION OF SOIL IN THE RHIZOSPHERE

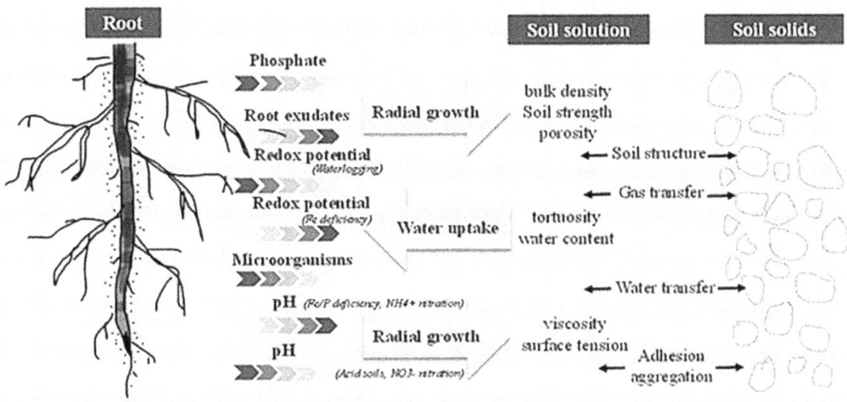

FIGURE 9.1 Physical–chemical and biological modification of soil in the rhizosphere.

The rhizosphere is a distinctive environment because of its diverse array of root activities. Distinct alterations in the biochemical, chemical, and physical characteristics of the soil surrounding the roots, as opposed to the surrounding bulk soil, result from processes directly influenced by the roots (Figure 9.1). In addition, microorganisms in the proximity of the roots become activated because of the release of rhizo-deposits by the roots, further contributing to these changes. This phenomenon is commonly referred to as the rhizosphere effect, which involves the activation of microorganisms that can exhibit either positive effects, such as being beneficial, or negative effects, such as the presence of pathogenic microorganisms [11].

Among the numerous activities occurring in the rhizosphere, significant focus has been on the process of rhizodeposition. The roots of higher plants release a diverse array of organic compounds, with sugars, polysaccharides, organic acids, amino acids, peptides, and proteins comprising the major components of the rhizodeposits. The release of these diverse compounds serves as the primary carbon source for rhizosphere microorganisms, leading to the observed rhizosphere effect. These compounds play a crucial role in establishing various trophic relationships and significantly enhancing the turnover of microbial biomass in the rhizosphere compared with bulk soil. Investigated extensively for over a century, the trophic and other biological interactions in the rhizosphere have seen rapid progress in understanding, thanks to recent advancements in molecular tools and approaches. In addition to these well-documented biological processes, roots can induce notable changes in the physical and chemical properties of the rhizosphere, ultimately affecting the activities of soil microorganisms and potentially influencing the roots themselves [15].

TABLE 9.1 Modification of Soil Properties in the Rhizosphere Zone

Physical Properties	Chemical Properties	Biological Properties
Bulk density	Organic matter content	Root growth
Pore size distribution	Redox potential	Rhizosphere effect
Infiltration rate	pH	Substrate availability
Water holding capacity	Nutrient solubility and availability	Microbial activity and diversity
Soil structure	Rhizodeposits	Pathogen suppression

The physical characteristics of the rhizosphere undergo direct alterations because of certain root activities, with the primary functions of plant roots, such as growth and water uptake, playing a central role (Table 9.1). Root growth exerts significant forces that ultimately change physical properties, including bulk density, porosity, and soil strength. Evidence of compaction around roots has been verified through micromorphological approaches using *in situ*-collected samples [16]. Moreover, changes in physical properties influencing soil structure within the rhizosphere highlight the substantial role played by polysaccharides, a major component of rhizodeposits [17]. The development of "rhizosheaths," distinctive structural elements in the rhizosphere, primarily depends on the presence of root hairs and the production of mucilage by roots. Although mycorrhizal hyphae may contribute to the attachment of soil particles to roots, studies indicate that exopolysaccharides generated by microorganisms in the rhizosphere play a crucial role in fostering the aggregation of soil around the roots. These processes can have feedback effects on the functioning of roots, influencing the transfer of water, nutrients, and toxic elements in either a facilitating or retarding manner. Additionally, these processes impact the habitat of rhizosphere microorganisms, leading to outcomes that can be either beneficial or detrimental to plant nutrition and health. For example, parameters such as soil structure and water potential have been identified as significant factors affecting the spatial exploration of the rhizosphere by root pathogens such as *Rhizoctonia solani* and *Gaeumannomyces graminis var. tritici* [18, 19].

Another crucial role of roots, with a pronounced impact on key physical properties of the rhizosphere, is their function in water uptake. These changes in water uptake can induce significant and rapid alterations in the water potential around roots. This not only influences the radial transfer of water into the plant but also impacts microbial activities within the rhizosphere. In addition, the production of mucilage by roots has been demonstrated to influence the viscosity and surface tension properties of the soil, thereby affecting the soil/water relationships in the rhizosphere. Apart from their effects on the physical dynamics of the rhizosphere, both root and microbial exopolysaccharides play roles in chemical processes, such as shielding the root apex from aluminum toxicity and aiding in the acquisition of nutrients such as phosphate.

The fundamental biological functions of plant roots, including absorption, respiration, and exudation, can significantly modify various chemical properties in the rhizosphere. These alterations encompass changes in nutrient concentrations, levels of toxic elements (such as aluminum) and pollutants, concentrations of complexing/chelating compounds, pH, redox potential , and partial pressures of O_2 and CO_2. The depletion of nutrients, which arises from the primary function of roots in nutrient uptake, has been traditionally regarded as the predominant, if not the exclusive, driving force in models explaining how plants acquire nutrients. The processes induced by roots in the rhizosphere, such as pH changes, have a profound impact on the acquisition of phosphorus [20].

Biochemical processes occurring in the rhizosphere exert a significant influence on soil functions. Notably, extracellular enzymes such as phosphatases, proteases, and arylsulfatases demonstrate heightened activity in the rhizosphere compared with that in the bulk soil. This increased enzymatic activity can have a profound impact on the cycling of nutrients, including phosphorus, nitrogen, and sulfur. Among these enzymes, only some are directly released by plant roots. Root symbionts, such as ectomycorrhizal fungi, along with various other microorganisms in the rhizosphere, play a crucial role in the secretion of these enzymes. Roots of genetically modified plants exposed to phosphorus deficiency exhibit increased release of phytase, particularly when grown in a low-phosphorus soil. Such feedback loops are likely to play a crucial role in the dynamics of the rhizosphere and offer additional avenues for manipulation. For instance, the heightened release of various phosphorus-solubilizing compounds produced by roots, including protons and organic anions, has been documented in response to phosphorus deficiency. The complexity of these interacting processes poses a significant challenge, limiting the prospects for a comprehensive understanding of the biochemistry and chemistry of the rhizosphere. Furthermore, manipulating the rhizosphere for management becomes an even more intricate task [21].

9.2.1 HOW THE RHIZOSPHERE STRUCTURES THE SOIL MICROBIOME

Rhizosphere refers to the limited soil area directly impacted by root secretions and the accompanying soil microorganisms, collectively recognized as the root microbiome. The rhizosphere induces a significant increase in the population of bacteria and other soil organisms compared with the adjacent bulk soil, often resulting in a tenfold increase in the bacterial population. Specifically, there is a preferential expansion of copiotrophic bacteria, which swiftly react to the availability of carbon [22]. Root colonization processes are initiated by root exudates in the rhizosphere, which attract and selectively proliferate specific microbial groups for coordinated root surface colonization (rhizoplane). Attachment involves cellular recognition and adhesion, potentially forming bacterial biofilms via quorum sensing. Root colonization within roots is specific and influences plant hormonal and systemic-induced resistance pathways via molecular cascades [23]. Intercellular endosphere colonization occurs through the apoplastic space, often entering roots via lateral root branch points or noncontinuous regions of the epi- and endodermal layers [24]. Together, these investigations highlight a robust correlation between the evolution of the rhizosphere and community composition, encompassing population structure and the existence of crucial functional genes. Although distinctly root-specific and compartmentalized within and along roots, this association is predominantly influenced by the makeup of root exudates [25].

The microbial community composition in the rhizosphere is significantly influenced by the strong rhizosphere effect. The primary determinant is soil type, which serves as the foundation for microbial diversity from which the rhizosphere community forms. Despite notable variations in soil diversity, several studies have shown the development of common patterns known as core communities. For instance, bacterial phyla, such as Proteobacteria, Firmicutes, and Bacteroidetes, thrive in the rhizosphere, whereas dominant soil phyla, such as Acidobacteria and Actinobacteria, are reduced. Another major factor shaping the rhizosphere microbiome composition is the plant species, with host differences attributed to distinct root system structures and functions [26]. Distinct types of roots, such as seminal and nodal roots, host different bacterial communities, as observed in research on the purple false broom. Seminal roots are characterized by Oxalobacteraceae, whereas nodal roots have Comamonadaceae. Microbial gene functions also vary among these root types. Root type influences arbuscular mycorrhizal fungi (AMF) communities, although soil type has a greater impact. Microbial communities differ between axial and lateral roots within the same system. Microbes at the root tip differ from those at the base, with more pronounced variations in community structure and gene content at the tip compared with those at the base [27].

9.3 UNIQUENESS OF THE RHIZOSPHERE UNDER THE AGROFORESTRY

In multicomponent systems, such as agroforestry, which hosts two or more root systems, a specific spatiotemporal behavior of the rhizosphere can be expected. The interaction between the two-root system further extends the complexity of the rhizosphere. Although the basic role of the root system remains the same for both the tree and crop components, spatial and temporal variability is observed between the rhizospheres of these root systems. The magnitude of the rhizosphere of the tree is much higher than that of the crop rhizosphere. Tree roots also possess higher plasticity than crop roots. Higher spatial variation in the soil volume occupied by roots can be observed in trees than in crops. Tree roots are slow to establish compared to the crop roots by which croproots can quickly explore the top soil volume. As the crop component is grown and harvested every year, it lacks permanency. The tree rhizosphere will be in place for several years.

Annual crops exhibit seasonal variability in root growth and development. Soil disturbance under AFSs is minimal. The nutrient and water uptake patterns also differ between the tree and crop components. The nutrient uptake in trees is gradual compared with that in annual crops [28].

Trees enhance soil fertility by engaging in biological nitrogen fixation and breaking down organic matter within their rhizosphere. These rhizosphere activities play a crucial role in supporting plant nutrition and preserving soil fertility. In addition, the presence of AMF shows potential for elevating crop yield and nutrient dynamics, serving as a pathway for nutrient cycling and a site for the accumulation of nutrients in the agroecosystem. Several investigations have indicated that root infections by AMF lead to heightened plant productivity, improved drought resistance, enhanced phosphorus nutrition [29], increased soil water extraction, and regulation of stomatal activity [30]. The common mycorrhizal network between the tree and crop components has special significance in terms of nutrient and water availability, which has been discussed further.

Root exudates, comprising H^+, OH^-, organic acids, and enzymes, represent a crucial carbon and nutrient source in soils and exhibit considerable variability within the root system. These exudates significantly contribute to the advantages observed at the root–soil interface, with such contributions in AFSs frequently originating from the tree component. Although niche partitioning in AFSs can enhance the overall efficiency of resource acquisition, there is often significant spatial overlap between the root systems of trees and crops. However, in such instances, spatial and temporal overlap in roots may not necessarily indicate straightforward competition between plants. The close proximity of active roots can offer advantages related to the enhancement of nutrient availability in the soil. Moreover, there is a growing body of evidence pointing to intricate multitrophic interactions within the rhizosphere that can influence interactions between roots [31].

9.4 MODIFICATION OF ROOT MICROBIOME UNDER AFSS

Land-use changes significantly influence the root microbiome, which governs the biogeochemical and ecological processes in the soil. Both the quantity and quality of organic matter input to the soil through litter deposits or root exudates vary according to the particular plant species.

These inputs to the soil act as substrates for microorganisms and fuel the food web in the soil. Therefore, modification of the root microbiome can be expected under AFSs compared with monoculture of tree or crop components. A stronger influence of land use patterns was observed on bacterial communities in the soil, with tree-containing plots promoting greater bacterial abundance [32]. The major determinants included soil pH and carbon contents, whereas specific bacterial taxa acted as keystone species in the AFSs. Diverse organic matter input in various agroforestry soils results in robust microbial diversity. In general, the bacterial population predominates, followed by actinomycetes and fungi under AFSs. A positive correlation can be observed between microbial populations and soil physicochemical properties, indicating that soil organic matter, vegetation, and nutrients influence the microbial community under AFSs [33].

In AFSs, tree species play an important role in shaping the rhizosphere fungal community structure of crop plants [34]. Root metabolic activity depends on the availability of resources. Tree components with better availability of resources can modify the linkage between crop and belowground communities by influencing the quality, quantity, timing, and spatial structure of plant-derived inputs [35]. In AFSs, including olive, the diversity of the microbial profile under the olive rhizosphere increased with the presence of cover crops, which was attributed to the decomposition of complex polymers of organic matter. In addition, olive trees affect the microbial profile of the annual crop rhizosphere. In olive AFSs, the crop component is the primary driver of fungal diversity. Olive trees can indirectly impact the soil beta diversity of annual crops through carbon input. The tree rhizosphere acts as a seed bank for the rhizosphere microbiome of the crop species. Bacterial phyla, such as Actinobacteria, Proteobacteria, and Firmicutes, and fungal phyla, such as Basidiomycota, Ascomycota, Mucoromycota, and Chytridiomycota, were increased under the crop rhizosphere under the olive-based AFS. This species was the signature microorganism under the olive tree, showing the transfer of abundance of these species from the tree to the crop. *Bradyrhizobium* and *Mesorhizobium* species increased in the rhizospheres of barley and maize under poplar-based AFSs [36]. Fungal species present in the rhizosphere zone profoundly influence plant growth. The relative abundance of certain rhizosphere fungal populations will increase under agroforestry. The relative abundance of fungal species belonging to the genus Chaetomium, which includes some biocontrol agents, was found to be increased under *Populus euramevicana*-barley *and Taxodium distichum*-barley AFSs [34].

9.5 INTERACTION BETWEEN TREE ROOTS AND CROP ROOTS IN THE SOIL

Plants can shape the microbial community near the root zone through root exudates and litter input. Distinct microbial communities with respect to their structure and functions are formed under different plant species because of the varying quality and quantity of organic matter input, which acts as a substrate for microbial growth. Thus, the interaction between the roots of trees and crop components can be mediated by soil microorganisms. Apart from root–microbe-mediated interactions, multicomponent systems such as agroforestry have direct root-to-root interactions in the rhizosphere zones. Belowground plant-to-plant communication occurs through the rhizosphere through chemical signaling [37]. Therefore, to harness the positive effect of agroforestry, the belowground interactions between tree roots and crop roots should be properly understood.

The rooting habit of the tree and crop species involved in the interaction, the age of the AFS, and the time point at which the interaction is considered play an important role in root–root interactions in AFSs. Tree roots require more time for initial establishment. Comparatively, the crop component is quick to establish; therefore, the interaction depends on the time point at which it is considered. The age at which belowground interactions begin depends on the initial spacing and rate of lateral root development. The root interaction between the components increases with the age of the system. Temporal variability in the extent of the tree rhizosphere is limited.

TABLE 9.2 Different Types of Root-to-Root Interactions in Soil

Positive Interaction	Negative Interaction
Hydraulic lift [28]	Competition for water [28, 38]
Safety net [28]	Competition for nutrients [28, 38, 39]
Nutrient pump [47]	Allelopathic interaction [37, 41–44]
Nutrient acquisition [39, 48, 49]	
VAM network [10, 37, 49, 50, 51]	

However, the crop component exhibits seasonal variation in the rhizosphere volume. The magnitude of the rhizosphere zone is much higher under the tree component than under the crop component [28] (Table 9.2).

Neutral, negative, and positive interactions were observed between the roots of the tree and the crop component. In this section, we focus on positive and negative interactions. Negative interactions occur because of the competition between roots for water and nutrients or because of the release of allelopathic compounds. The positive interaction occurs because of the facilitation of nutrient and water availability between the two components.

9.5.1 NEGATIVE INTERACTIONS

9.5.1.1 COMPETITION FOR NUTRIENTS AND WATER

Although the different rooting habits of tree and crop components in AFSs are idealized, during the initial few years of establishment or in case of fast-growing tree species, exploitation of the same soil depth can be observed, which results in competition between the roots of the tree and crop [28, 38]. The selection of two highly competing species reduces the growth of both components, particularly under resource-scarce conditions. Competition in soil mainly occurs for nutrients and water. Site-specific competition for nutrients is expected in zones with interspecific interactions. In the competitive contact zone of the two root systems, plants attempt to avoid competition by avoiding further lateral root growth by favoring deeper vertical root growth. The rhizosphere effect prevents the two roots from coming too close to each other. The overlapping of nutrient depletion zones between the two roots results in increased competition. However, roots can sense zones of nutrient depletion by neighboring roots through the root detectable compounds. It discourages the growth of the roots of other species in that area [39]. Root systems with higher fine root length density are more competitive than those with lower plant root length density. In addition, the ability to rapidly occupy fertile soil microsites makes the root system more competitive. Proper selection of tree and crop components should be performed to avoid excessive competition between species. Control of belowground competition by supplying nutrients may increase aboveground competition by the shading effect. Temporary root competition can be avoided if the peak growth period of the roots of the component plants does not merge, that is, they occur during different seasons of the year.

Spatial segregation of the root system allows the plant to avoid root competition. Root functions can be improved by root management options such as proper spacing, nutrient placement, and shoot pruning, which will improve the complementary and stimulative interactions in the root zone [38].

Competition between roots for nutrients and water results in a reduction in overall root growth and yield. For instance, under the jujube-wheat AFS, the root length density of both crops was negatively affected compared with that of the monocropping system. Preferential growth of jujube roots near the root zone of wheat resulted in a shallower distribution of roots in the profile than in the monoculture. With the increase in the age of the system, the wheat yield reduction also increased. However, the reduction in the productivity of jujube showed the opposite trend, where productivity increased with the age of trees. Even though individual productivity decreased, the land equivalent ratio increased under this AFS. This phenomenon has also been reported in the maize–walnut system. The reduction in yield was mainly due to root competition. The spacing between the tree and crop components and management practices determines the magnitude of competition for resources. The higher the root length density, the higher will be the competing ability of the plant. Perennial trees occupy a greater soil volume than the crop component. A reduction in root volume limits the volume of soil that can be grazed by the root system, which ultimately reduces growth and yield [40].

9.5.1.2 ALLELOPATHY

Allelopathy is another form of negative interaction between the roots of trees and crops in AFSs. The negative impact of one plant species on the growth and development of another through the release of certain chemicals into the environment is known as allelopathy. Release may take the form of root exudation, leaching, volatilization, and residue decomposition. Allelopathy results in severe crop losses, especially when there is an unsuitable combination of trees and crops. Many allelochemicals have their origins in the root system. Several factors contribute to the presence of these allelochemicals in soil, including the density of fallen leaves, the decomposition rate of this organic material, proximity to other plants, and the quantity and distribution of annual rainfall. Under natural conditions, only minute quantities of allelochemicals are typically leached or directly volatilized, subsequently interacting with the roots of receptor plants.

Spacing between tree and crop components determines the fate of allelopathy in AFSs. Moreover, once released into the soil, allelochemicals undergo modifications influenced by soil characteristics, such as absorption, migration, and decomposition facilitated by microorganisms and enzymes. In addition, the nutrient content of the soil may affect the outcomes of allelopathy [41].

Allelopathic chemicals alter the interaction between native plants and soil microbial communities. The crop belonging to the Brassicaceae family releases glucosinolates, which can change the soil microbial communities. These conditions can disrupt the mutualistic association between tree roots and AM fungi. *Eucalyptus* spp. can modify the bacterial community in the soil in addition to soil nutrient depletion, soil acidification, and water table depletion. It affects the microbial community structure and functions and reduces mycorrhizal spores and mycelial networks in the soil. Therefore, root-mediated allelopathy requires appropriate attention [37]. In AFSs involving *Moringa oleifera,* a reduction in the germination rate of mung beans has been reported [42]. Leaf extracts of mulberry, plum, and pomegranate negatively affect seed germination, root length, shoot length, and dry matter production in soybeans [43]. In certain tree species, bark can also negatively impact the growth of crop species during decomposition. *Populus nigra* bark has allelopathic effects on the germination, length of the plumule and radical, fresh weight, and dry weight of *Zea mays* [44]. Soil microbial activity influences the allelopathic effect either positively or negatively. Soil microbes can decompose phenolic allelochemicals, preventing them from reaching active concentrations in the soil. Allelopathic effects on seedlings can be reduced by AM fungi [37]. However, in a few cases, the microbes convert the ineffective form of the allelochemical to active forms and render the crop barren. For instance, ceratolin, a nonallelopathic active substance produced by *Ceratiola ericoides*, can be transformed into activated phenylpropionic acid by specific soil microorganisms, which is further converted into the more toxic chemical hypnone [45].

9.5.2 POSITIVE INTERACTIONS

9.5.2.1 HYDRAULIC LIFT AND SAFETY NET

The interaction between the tree root and crop root also plays some facilitative roles that are complementary to each other. Tree components with greater root depth can act as hydraulic lifters of subsurface water and bioirrigators of adjacent plants. Trees with deep taproots can access groundwater, reducing competition with shallow-rooted crops for surface water. The tree component can redistribute water to the upper horizons. The redistributed water may play a crucial role not only in regulating the water status of plants, such as by boosting transpiration, but also in promoting the survival and growth of crops that coexist in mixed systems. Both upward and downward movement of water is possible through a hydraulic lift, depending on the difference in the water potential. This phenomenon can be observed in trees with a dimorphic root system. Root length density and root surface area strongly influence root water dynamics. Under drought conditions, this hydraulic lift and subsequent redistribution helps in the survival of the crop component and is a crucial mechanism ensuring the productivity of the AFS. Under moisture stress conditions, this will increase the resilience of the AFS. *Grevillea* spp. extract over 80% of their water requirement from the soil layer below the crop root zone. In addition, trees can influence the microclimate, reducing evaporation and enhancing water-use efficiency, further contributing to drought resilience [46].

Deep-rooted crops can bring back the leached nutrients to the soil surface of shallow-rooted crops through litter fall, which is known as the safety net hypothesis. The safety net hypothesis complements the nutrient pump hypothesis. The nutrient pump hypothesis entails the uptake of nutrients, typically less mobile, from deep soil layers. These nutrients are then incorporated into aboveground plant tissues, such as leaves, and transferred to the upper soil layers through processes such as leaf shedding and decomposition. The nutrient pumping effect is usually observed for Ca and K [47]. The nutrient uptake in the tree component was gradual, whereas that in the crop component had comparatively faster nutrient uptake from the concentrated zone. Therefore, after establishment, the tree component suffers less from root competition than the crop component [28].

9.5.2.2 NUTRIENT ACQUISITION

The rhizosphere-mediated nutrient acquisition strategy of one crop component also influences the other component, particularly when both are closely associated. For example, facilitation of P uptake occurs when one species increases P availability to the other. This can occur when the roots are in proximity or intermingled. P acquisition strategies in the rhizosphere, such as carboxylate release, root-induced alkalization or acidification, and phosphatase enzyme release, will also have a positive impact on the component crop. Mobilized P can reach other components [48]. N fixed by leguminous trees becomes available for component crops through litter decomposition or mycorrhizal networks [49]. Root distribution factors dictate the nutrient acquisition strategy in an AFS. The tree component can add more nitrogen through biological nitrogen fixation or uplift from the deeper soil layers. Management practices, such as trenching, can alter the nutrient acquisition pattern of component trees under an AFS [39].

9.5.2.3 ROLE OF AM FUNGI IN ROOT–ROOT INTERACTIONS

Mycorrhizal fungi engage in a mutually beneficial association with the roots of most plants, with certain plant families, newly established plants, and those cultivated in mycorrhizal-free soils or soilless potting mixtures being exceptions. Among the two primary mycorrhizal types—ectomycorrhiza and endomycorrhiza—ectomycorrhiza creates a noticeable covering of fungal hyphae around root tips, leading to significant morphological changes in the roots. Endomycorrhiza, particularly the vesicular–arbuscular mycorrhizae (AM), are prevalent in many agricultural crop plants, inducing minimal morphological alterations in the roots. Both types of mycorrhizae possess an external hyphal phase that extends into the soil, exerting profound effects on the physiological processes of their host plants. Plants associated with mycorrhizae exhibit a modified nutritional status, elevated rates of photosynthesis, changes in the levels of growth-regulating substances, and adjustments in root exudation patterns attributed to shifts in membrane permeability. These physiological transformations, coupled with the physical and chemical influence of external hyphae from mycorrhizal fungi, have a substantial impact on the chemical, physical, and microbiological composition of the soil in the rhizosphere [49].

AM fungi are key components of soil microorganisms that affect plant growth and nutrient acquisition. They act as an interlink between the root and soil environment and play a crucial role in root-to-root interactions. AM fungi form associations with both tree and crop components and are strongly involved in interspecific interactions between the two components.

Rhizodeposition of the root system increases in the presence of AM fungi. AM association can change the microbial composition of the entire rhizosphere. The region of soil that is under the influence of mycorrhizal activity is known as the mycorrhizosphere. The mycorrhizosphere, similar to the rhizosphere, is the zone of carbon-containing compounds deposition and stimulates the growth of different bacteria, fungi, and nematodes. The mycorrhizosphere also regulates changes in soil pH, nutrient acquisition, suppression of pathogenic microorganisms, and soil structural changes. Mycorrhizal interactions occur both in the soil and within the roots. Mycorrhizal fungi can change root morphology by facilitating nutrient acquisition. The mycelial network of mycorrhizal fungi includes 20–30% of the total soil microbial biomass and plays a crucial role in nutrient cycling. The mycelial network in 1 g of soil is up to 100 m in AM fungi and 800 m in ectomycorrhizal fungi. The soil region under the influence of the extraradical mycelium of mycorrhizal fungi is known as the hyphosphere. It is critical for P acquisition by mycorrhizal fungi. Both the mycorrhizosphere and hyphosphere play an important role in plant co-existence. The microbial changes occurring in the mycorrhizosphere represent a distinct equilibrium compared with the rhizosphere of nonmycorrhizal plants. These microorganisms engage in direct interactions with each other and the extramatrical hyphae of mycorrhizal fungi. Moreover, the presence of one organism in the mycorrhizosphere can induce indirect responses in other organisms, affecting the host plant. This dynamic interplay may lead to selective population shifts, potentially influencing plant growth and development in collaboration with mycorrhizal fungi [37, 49]

Mycorrhizal fungi protect plants from both biotic and abiotic stresses. It increases the soil volume accessible for nutrient acquisition and plays a crucial role in nutrient cycling and maintenance of soil structure and biodiversity. Allelopathic effects on seedlings can be reduced in the presence of AM fungi. Common mycorrhizal networks can develop when both tree and crop component roots are colonized by the same fungi. Plant-to-plant nutrient transfer through common mycorrhizal networks has been postulated. Both arbuscular and ectomycorrhizal fungi are involved in nutrient transfer between the two plants [50, 51]. However, competition for P acquisition from common mycorrhizal networks between the component species was observed when mycorrhizal networks were attached to interspecific and size-asymmetric plants. The tree component can provide more carbon to AM and render most of the P service from this common network [50].

It is difficult to study and quantify nutrient transfer through the mycorrhizal network [39]. Apart from nutrient sharing, the mycorrhizal network also facilitates the sharing of water between the associated plants, which is crucial for mutual survival in water-limited environments [10].

The abundance and diversity of fungi are higher in a multicomponent ecosystem than in a monoculture because of wider host availability. AM fungi diversity was significantly higher in an AFS in which the spore density followed the decreasing order of tree rhizosphere, crop rhizosphere, and bulk soil [51]. The tree component helps maintain the active inoculum of AM, particularly under stress conditions, as the tree rhizosphere can act as a reservoir of mycorrhizal inoculum for the surrounding crops. The tree can increase the extent of mycorrhization and extra-radical hypha production in the crop component [50].

9.5.2.4 ROOT PLASTICITY

The basic idea of agroforestry regarding roots is that the different components will occupy different soil strata and obtain nutrients from different depths so that competition is avoided. But in a practical sense, it is hard to achieve. The spatial distribution of roots is altered when trees and crops are grown together, which plays a critical role in interspecific interactions in AFSs. In the competitive contact zone of the two root systems, plants attempt to avoid competition by avoiding further lateral root growth by favoring deeper vertical root growth. A stratified root system helps to avoid competition among the roots. For a stratified root system to develop, the root system of one-component species should be competitive enough to displace the root of the other component. The displaced root system should exhibit high plasticity, and the soil should support the growth of roots in deeper soil layers [38]. Root plasticity denotes the ability of the root system to adjust to local conditions and grow accordingly to avoid competition. Root plasticity enables better use of resources. The plasticity of tree roots in agroforestry depends on tree density, where closure spacing results in a greater plastic response of roots. Root plasticity is fundamentally related to plant interactions. Root system architecture depends on the inbuilt genotype, soil heterogeneity, and nutrient status of the plant. The plasticity of the roots of the tree component, as affected by the root density of the crop component, enables the tree root to grow in less competitive zones.

Trees such as walnut, oak, and apple have deeper roots when grown with annuals such as wheat and grasses. However, the root plasticity of crop components is limited [50].

9.5.2.5 SIGNIFICANCE OF ROOT–ROOT INTERACTION IN THE AFS

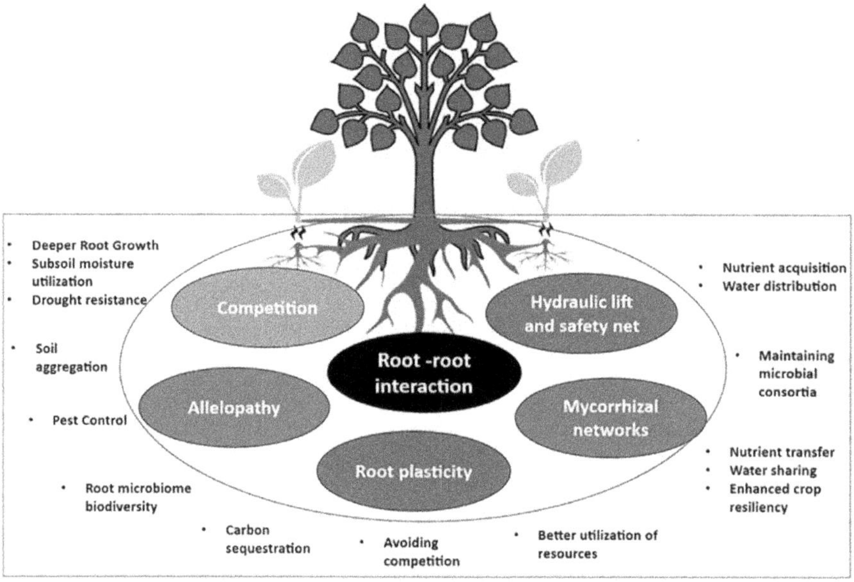

FIGURE 9.2 Significance of root–root interaction in the AFS.

Both positive and negative interactions between roots have their own significance in the AFS (Figure 9.2). Root competition in the top soil increases deeper root growth and better use of subsoil moisture and nutrients, thereby increasing drought resistance. In nitrogen-fixing trees, timely uptake of N by the crop component increases and maintains N fixation by the tree, as N will not accumulate in the soil system. Spatial and temporal overlap of roots may not always result in plant competition; however, it may help in nutrient availability of weaker components. Generally, the crop component derives benefit from the tree component because the extent of root exudation is higher under the tree component [28]. In AFSs such as *Acacia senegal*-wheat, better P availability was observed because of P mobilization by tree roots in the rhizosphere. In AFSs involving leguminous N-fixing trees, N supplementation of crops is expected. The pathway includes N release from litter decomposition, root exudates, root-to-root transfer, and common mycorrhizal networks. Leguminous trees, such as *Gliricidia sepium* and *Inga edulis*, have the ability to transfer or contribute to the N nutrition of crop components [39].

Allelopathy can be harnessed in AFSs for pest control. It requires careful selection of the component plant species. Allelochemicals frequently confer resistance to pathogens, insects, and nematodes in plants and contribute to weed reduction. Trees can regulate the germination and growth of weeds by releasing allelochemicals. For instance, scopolin and scopoletin, isolated from *Celtis laevigata*, inhibit the growth of *Amaranthus palmeri* [52]. *Eucalyptus citriodora* has inhibitory effects on the growth of several weeds, including *Bidens pilosa, Digitaria pertenuis, Eragrostis cilianensis, Setaria geniculate, and Parthenium hysterophorus* [53, 54].

The AFS helps to avoid microbial homogenization and loss of biodiversity. The varied substrate availability of the different components increases the diversity. Therefore, strong spatial heterogeneity in soil functioning can be expected under the AFS. The tree roots maintain the inoculum of AM in the soil, particularly under harsh environmental conditions. The presence of dense perennial roots in the tree component enables the rapid reestablishment of crops in case of external disturbances [50]. Rhizosphere interaction helps in the homogeneous exploration of soil in terms of resources. It will improve the resource use efficiency of the entire system. Tree roots can preferentially explore the soil volume not assessable to the crop roots. The safety net and hydraulic lift phenomenon will ensure better use of nutrients and water. Rhizosphere interaction plays a crucial role in water- and nutrient-limited environments [28].

9.6 HOW THE RHIZOSPHERE SHAPES THE DYNAMICS OF AFSS

Agroforestry has gained significant attention for its potential to address multiple challenges in modern agriculture, including soil degradation, biodiversity loss, and climate change mitigation. The rhizosphere emerges as a crucial and unique component within the intricate web of interactions that characterize AFSs. Rhizosphere plays a pivotal role in shaping the dynamics of agroforestry ecosystems. Its roles are briefed below.

Nutrient cycling: Rhizosphere chemistry improves nutrient availability to plants. The safety net, root plasticity, and nutrient pump hypotheses ensure better utilization of nutrients. The safety net hypothesis prevents nutrient leakage below the rooting depth, whereas the nutrient pump hypothesis ensures nutrient availability to shallow-rooted crops. Trees and crops may engage in mutualistic relationships with mycorrhizal fungi. The fungus assists in nutrient uptake, particularly of phosphorus from the soil, benefiting both trees and crops. The inclusion of leguminous trees in the AFS improves the nitrogen budget of the entire system.

Microbial diversity: One of the distinctive features of the rhizosphere in AFSs is its remarkable microbial diversity. Different plant species release unique compounds that attract specific microbial communities. The intricate interactions between plant roots and soil microorganisms create a dynamic environment in which beneficial symbiotic relationships thrive. Mycorrhizal fungi, for instance, form mutualistic associations with plant roots, thereby enhancing nutrient uptake and promoting plant growth. These symbiotic relationships contribute to increased nutrient cycling, fostering a more efficient use of resources in agroforestry landscapes.

Soil structure improvement: Tree roots can enhance soil structure by promoting the formation of soil aggregates. This can lead to improved water infiltration and retention, thereby reducing soil erosion. The root systems of trees and crops in AFSs contribute significantly to soil stability and erosion control. The rhizosphere acts as a binding agent, preventing soil erosion by anchoring the soil particles and reducing water runoff. Both tree and crop roots release organic compounds into the soil through rhizodeposition, which helps in soil aggregation.

Pest and disease suppression: Some trees release compounds that attract beneficial insects or microbes, which can act as natural enemies of pests or pathogens. This can contribute to pest and disease suppression in the AFS. Rhizospheres in AFSs serve as natural defense mechanisms against pathogens and pests. In addition, certain plant species release compounds through their roots that inhibit the growth of harmful soil-borne pathogens. This allelopathic effect, combined with the diverse microbial community in the rhizosphere, creates a robust defense system that reduces the reliance on synthetic pesticides. Therefore, agroforestry promotes a sustainable and ecologically friendly approach to pest and disease management.

Water-use efficiency: In the hydraulic lift phenomenon, water must pass through the rhizosphere, which improves water distribution in the soil profile. Trees and crops often have different root architectures and water-use patterns. This can lead to a more efficient use of water resources, with trees accessing deeper soil layers and crops using surface soil moisture.

Carbon sequestration: The rhizosphere plays a pivotal role in carbon sequestration, which is a critical component of climate change mitigation. Through the release of root exudates, plants stimulate microbial activity in the soil, leading to the formation of soil organic matter. This organic matter not only improves soil structure but also acts as a carbon sink, helping to mitigate the effects of greenhouse gas emissions. Agroforestry, with its diverse plant species, fosters a rich rhizosphere environment that enhances carbon sequestration and contributes to soil health. Continuous shedding of roots and organic matter from trees and crops contributes to carbon inputs in the soil. This can enhance carbon sequestration, potentially mitigating climate change.

Improved soil health: The diverse root exudates from trees and crops can stimulate microbial activity, leading to improved soil health and nutrient cycling. Additionally, the enhanced water absorption capacity of agroforestry plants helps in efficient water management, reducing the risk of drought and enhancing overall ecosystem resilience. This unique characteristic of the rhizosphere aligns with the principles of sustainable land management, making agroforestry a valuable strategy for maintaining soil integrity.

9.7 CONCLUSION

The AFS, which mainly focuses on the ecological and economic interactions among tree, crop, and livestock components, possesses a unique set of aboveground and belowground interactions. The belowground interactions of roots with other roots and soil microorganisms in the soil matrix play a crucial role in shaping the dynamics of the entire AFS. The rhizosphere zone acts as a biological hotspot in the soil and modifies the physical, chemical, and biological properties of the soil. The quantity and quality of organic matter input to the soil through litter input and rhizodeposition vary between the tree and crop components, which creates a spatiotemporal variation in the rhizosphere effect under the AFS. Tree components with better availability of resources can modify the linkage between crop and belowground communities by influencing the quality, quantity, and timing of release of plant-derived inputs. The interaction between tree roots and crop roots in soil involves both positive and negative interactions, each of which has ecological significance. The competition between roots takes place for nutrients and water. The root competition can alter the root morphology. Allelopathic interactions between the roots result in reduced germination, growth, and yield. The deep root system of tree crops recycles the leached nutrients below the crop rooting depth and facilitates water uptake and redistribution in the topsoil layer. The rhizosphere-mediated nutrient acquisition strategy of one crop component also influences the other component, particularly when both are closely associated, which is critical for immobile nutrients such as phosphorus. The common mycorrhizal networks between the tree and crop components facilitate nutrient and water sharing. Rhizosphere-mediated root interactions improve nutrient cycling, water-use efficiency, microbial diversity, plant resilience, and overall soil health.

KEYWORDS

- **rhizosphere**
- **agroforestry**
- **rhizodeposition**
- **plant growth promoting rhizobacteria**
- **mycorrhizal networks**
- **root plasticity**
- **root microbiome**
- **allelopathy**
- **hydraulic lift**
- **soil health**

REFERENCES

1. Betz, U. A. K., Betz, F., Kim, R., Monks, B., Phillips, F. (2019). Surveying the future of science, technology and business—A 35 year perspective. *Technological Forecasting and Social Change*, 144, 137–147. https://doi.org/10.1016/j.techfore.2019.04.005.
2. Boretti, A., Rosa, L. (2019). Reassessing the projections of the world water development report. *NPJ Clean Water*, 2(1), 15. https://doi.org/10.1038/s41545-019-0039-9.
3. Falcon, W. P., Naylor, R. L., Shankar, N. D. (2022). Rethinking global food demand for 2050. *Population and Development Review*, 48(4), 921–957. https://doi.org/10.1111/padr.12508.
4. Lal, R. (2019). Rights-of-soil. *Journal of Soil and Water Conservation*, 74(4), 81A–86A. https://doi.org/10.2489/jswc.74.4.81A.
5. Chander, K., Goyal, S., Nandal, D. P., Kapoor, K. K. (1998). Soil organic matter, microbial biomass and enzyme activities in a tropical agroforestry system. *Biology and Fertility of Soils*, 27(2), 168–172. https://doi.org/10.1007/s003740050416.
6. Dosskey, M. G. (2002). Setting priorities for research on pollution reduction functions of agricultural buffers. *Environmental Management*, 30(5), 0641–0650. https://doi.org/10.1007/s00267-002-2755-y.
7. Schroth, G., Fonseca, G. A. B., Harvey, C. A., Vasconcelos, H. L., Gascon, C., Izac, A. M. N. (2004). Introduction: The role of agroforestry in biodiversity conservation in tropical landscapes. In: *Agroforestry and Biodiversity Conservation in Tropical Landscapes,* Island Press, Washington.
8. Dixon, R. K. (1995). Agroforestry systems: Sources of sinks of greenhouse gases? *Agroforestry Systems*, 31(2), 99–116. https://doi.org/10.1007/BF00711719.
9. Leakey, R. (2006). Trees and markets for agroforestry tree products: Targeting poverty reduction and enhanced livelihoods. In: *World Agroforestry into the Future*.
10. Cardinael, R., Mao, Z., Chenu, C., Hinsinger, P. (2020). Belowground functioning of agroforestry systems: Recent advances and perspectives. *Plant and Soil*, 453, 1–13. https://doi.org/10.1007/s11104-020-04633-x.
11. Berendsen, R. L., Pieterse, C. M. J., Bakker, P. A. H. M. (2012). The Rhizosphere Microbiome and Plant Health. *Trends in Plant Science*, 17(8), 478–486. https://doi.org/10.1016/j.tplants.2012.04.001.
12. Mendes, R., Kruijt, M., De Bruijn, I., Dekkers, E., Van Der Voort, M., Schneider, J. H. M., Piceno, Y. M., DeSantis, T. Z., Andersen, G. L., Bakker, P. A. H. M. (2011). Deciphering the rhizosphere microbiome for disease-suppressive bacteria. *Science*, 332(6033), 1097–1100. https://doi.org/10.1126/science.1203980.
13. Marschner, H., Kirkby, E. A., Cakmak, I. (1996). Effect of mineral nutritional status on shoot-root partitioning of photoassimilates and cycling of mineral nutrients. *Journal of Experimental Botany*, 47(SPEC. ISS.), 1255–1263. https://doi.org/10.1093/jxb/47.special_issue.1255.
14. Van Der Heijden, M. G. A., Bardgett, R. D., Van Straalen, N. M. (2008). The unseen majority: Soil microbes as drivers of plant diversity and productivity in terrestrial ecosystems. *Ecology Letters*, 11(3), 296–310. https://doi.org/10.1111/j.1461-0248.2007.01139.x.

15. Sharma, P., Sharma, M. M. M., Malik, A., Vashisth, M., Singh, D., Kumar, R., Singh, B., Patra, A., Mehta, S., Pandey, V. (2021). Rhizosphere, rhizosphere biology, and rhizospheric engineering. In: *Plant Growth-Promoting Microbes for Sustainable Biotic and Abiotic Stress Management*. https://doi.org/10.1007/978-3-030-66587-6_21.
16. Bruand, A., Cousin, I., Nicoullaud, B., Duval, O., Bégon, J. C. (1996). Backscattered electron scanning images of soil porosity for analyzing soil compaction around roots. *Soil Science Society of America Journal*, 60(3), 895–901. https://doi.org/10.2136/sssaj1996.03615995006000030031x.
17. Czarnes, S., Hallett, P. D., Bengough, A. G., Young, I. M. (2000). Root- and microbial-derived mucilages affect soil structure and water transport. *European Journal of Soil Science*, 51(3), 435–443. https://doi.org/10.1046/j.1365-2389.2000.00327.x.
18. Grose, M. J., Gilligan, C. A., Spencer, D., Goddard, B. V. D. (1996). Spatial heterogeneity of soil water around single roots: Use of CT-scanning to predict fungal growth in the rhizosphere. *New Phytologist*, 133(2), 261–272. https://doi.org/10.1111/j.1469-8137.1996.tb01893.x.
19. Harris, K., Young, I. M., Gilligan, C. A., Otten, W., Ritz, K. (2003). Effect of bulk density on the spatial organisation of the fungus rhizoctonia solani in soil. *FEMS Microbiology Ecology*, 44(1), 45–56. https://doi.org/10.1016/S0168-6496(02)00459-2.
20. Riley, D., Barber, S. A. (1971). Effect of ammonium and nitrate fertilization on phosphorus uptake as related to root-induced Ph changes at the root-soil interface. *Soil Science Society of America Journal*, 35(2), 301–306. https://doi.org/10.2136/sssaj1971.03615995003500020035x.
21. Mudge, S. R., Smith, F. W., Richardson, A. E. (2003). Root-specific and phosphate-regulated expression of phytase under the control of a phosphate transporter promoter enables arabidopsis to grow on phytate as a sole P source. *Plant Science*, 165(4), 871–878. https://doi.org/10.1016/S0168-9452(03)00286-3.
22. Burns, R. G. (2010). Albert Rovira and a half-century of Rhizosphere research. In: Gupta, V. V. S. R., Ryder, M., Radcliffe, J., (Eds.), *The Rovira Rhizosphere Symposium* (pp. 1–10), Springer New York LLC: New York.
23. Lebeis, S. L. (2015). Greater than the sum of their parts: Characterizing plant microbiomes at the community-level. *Current Opinion in Plant Biology*, 24, 82–86. https://doi.org/10.1016/j.pbi.2015.02.004.
24. Gaiero, J. R., McCall, C. A., Thompson, K. A., Day, N. J., Best, A. S., Dunfield, K. E. (2013). Inside the root microbiome: Bacterial root endophytes and plant growth promotion. *American Journal of Botany*, 100(9), 1738–1750. https://doi.org/10.3732/ajb.1200572.
25. Richardson, A. E., Kawasaki, A., Condron, L. M., Ryan, P. R., Gupta, V. V. S. R. (2021). Root microbiome structure and microbial succession in the rhizosphere. In: *Rhizosphere Biology: Interactions Between Microbes and Plants* (pp. 109–128). https://doi.org/10.1007/978-981-15-6125-2_5.
26. Donn, S., Kirkegaard, J. A., Perera, G., Richardson, A. E., Watt, M. (2015). Evolution of bacterial communities in the wheat crop rhizosphere. *Environmental Microbiology*, 17(3), 610–621. https://doi.org/10.1111/1462-2920.12452.
27. Kawasaki, A., Donn, S., Ryan, P. R., Mathesius, U., Devilla, R., Jones, A., Watt, M. (2016). Microbiome and exudates of the root and rhizosphere of brachypodium

distachyon, a model for wheat. *PLoS One*, 11(10), e0164533. https://doi.org/10.1371/journal.pone.0164533.
28. Schroth, G., Schaller, M., Jiménez, F. (2007). Belowground interactions in tree-crop agroforestry: Need for a new approach. In: *Ecological Basis of Agroforestry,* CRC Press.
29. Awotoye, O. O., Atayese, M. O., Osonubi, O., Mulongoy, K., Okali, D. U. U. (1992). Response of Some Tropical Nitrogenfixing Woody Legumes to Drought and Inoculation with Mycorrhiza. *Biological Nitrogen Fixation And Sustainability of Tropical Agriculture*. Ibadan, Nigeria, 24–28 September 1990. Ibadan, Nigeria: IITA (pp. 67–77).
30. Huang, R. S., Smith, W. K., Yost, R. S. (1985). Influence of vesicular-arbuscular mycorrhiza on growth, water relations, and leaf orientation in leucaena leucocephala (LAM.) De wit. *New Phytologist*, 99(2), 229–243. https://doi.org/10.1111/j.1469-8137.1985.tb03652.x.
31. Mommer, L., Kirkegaard, J., van Ruijven, J. (2016). Root-root interactions: Towards a rhizosphere framework. *Trends in Plant Science*, 21(3), 209–217. https://doi.org/10.1016/j.tplants.2016.01.009.
32. Banerjee, S., Baah-Acheamfour, M., Carlyle, C. N., Bissett, A., Richardson, A. E., Siddique, T., Bork, E. W., Chang, S. X. (2016). Determinants of bacterial communities in canadian agroforestry systems. *Environmental Microbiology*, 18(6), 1805–1816. https://doi.org/10.1111/1462-2920.12986.
33. Radhakrishnan, S., Mohan, V. (2016). Status of microbial diversity in agroforestry systems in Tamil Nadu, India. *Journal of Basic Microbiology*, 56(6), 662–669. https://doi.org/10.1002/jobm.201500639.
34. Zhang, Q., Zhang, M., Zhou, P., Fang, Y., Ji, Y. (2018). Impact of tree species on barley rhizosphere-associated fungi in an agroforestry ecosystem as revealed by 18S RDNA PCR-DGGE. *Agroforestry Systems*, 92(2), 541–554. https://doi.org/10.1007/s10457-017-0086-5.
35. Vetterlein, D., Carminati, A., Kögel-Knabner, I., Bienert, G. P., Smalla, K., Oburger, E., Schnepf, A., Banitz, T., Tarkka, M. T., Schlüter, S. (2020). Rhizosphere spatiotemporal organization–A key to rhizosphere functions. *Frontiers in Agronomy*. https://doi.org/10.3389/fagro.2020.00008.
36. Ben zineb, A., Barkaoui, K., Karray, F., Mhiri, N., Sayadi, S., Mliki, A., Gargouri, M. (2022). Olive agroforestry shapes rhizosphere microbiome networks associated with annual crops and impacts the biomass production under low-rainfed conditions. *Front Microbiology*, 13. https://doi.org/10.3389/fmicb.2022.977797.
37. Sanon, A., Andrianjaka, Z. N., Prin, Y., Bally, R., Thioulouse, J., Comte, G., Duponnois, R. (2009). Rhizosphere microbiota interfers with plant-plant interactions. *Plant and Soil*, 321, 259–278. https://doi.org/10.1007/s11104-009-0010-5.
38. Schroth, G. (1998). A review of belowground interactions in agroforestry, focussing on mechanisms and management options. In: *Agroforestry Systems* (Vol. 43). https://doi.org/10.1023/a:1026443018920.
39. Isaac, M. E., Borden, K. A. (2019). Nutrient acquisition strategies in agroforestry systems. *Plant and Soil*, 444, 1–19. https://doi.org/10.1007/s11104-019-04232-5.
40. Zhang, W., Wang, B. J., Gan, Y. W., Duan, Z. P., Hao, X. D., Xu, W. L., Lv, X., Li, L. H. (2017). Competitive interaction in a jujube tree/wheat agroforestry system in

Northwest China's Xinjiang Province. *Agroforestry Systems*, 91(5), 881–893. https://doi.org/10.1007/s10457-016-9962-7.
41. Devi, M. (2017). Allelopathy in agroforestry: A review. *Journal of Pharmacognosy and Phytochemistry*, 6(3), 686–688.
42. Hossain, M. M., Miah, G., Ahamed, T., Sarmin, N. S. (2012). Allelopathic effect of moringa oleifera on the germination of vigna radiate. *International Journal of Agriculture and Crop Sciences*, 4(3), 114–121.
43. Thakur, M. K. (2014). Studies on allelopathic effects of some agroforestry tree species on soybean. *International Journal of Farm Sciences*, 4(2), 107–113.
44. Khan, M., Zakaria, M., Ali, F., Hussain, F., Imdadullah, S., Musharaf, S. (2016). Allelopathic effect of populus Nigra Bark on Zea Mays in agroforestry ecosystems. *Global Journal of Science Frontier Research*, 16(1), 21–28.
45. Walker, T. S., Bais, H. P., Grotewold, E., Vivanco, J. M. (2003). Root exudation and rhizosphere biology. *Plant Physiology*, 132(1), 44–51. https://doi.org/10.1104/pp.102.019661.
46. Bayala, J., Prieto, I. (2020). Water acquisition, sharing and redistribution by roots: Applications to agroforestry systems. *Plant Soil*, 453(1–2). https://doi.org/10.1007/s11104-019-04173-z.
47. Jobbágy, E. G., Jackson, R. B. (2004). The uplift of soil nutrients by plants: Biogeochemical consequences across scales. *Ecology*, 85, 2380–2389. https://doi.org/10.1890/03-0245.
48. Hinsinger, P., Betencourt, E., Bernard, L., Brauman, A., Plassard, C., Shen, J., Tang, X., Zhang, F. (2011). P for two, sharing a scarce resource: Soil phosphorus acquisition in the rhizosphere of intercropped species. *Plant Physiology*, 156(3), 1078–1086. https://doi.org/10.1104/pp.111.175331.
49. Priyadharsini, P., Rojamala, K., Koshila Ravi, R., Muthuraja, R., Nagaraj, K., Muthukumar, T. (2017). Mycorrhizosphere: The extended rhizosphere and its significance. In: *Plant-Microbe Interaction: An Approach to Sustainable Agriculture*. https://doi.org/10.1007/978-981-10-2854-0_5.
50. Battie-Laclau, P., Taschen, E., Plassard, C., Dezette, D., Abadie, J., Arnal, D., Benezech, P., Duthoit, M., Pablo, A. L., Jourdan, C. (2020). Role of trees and herbaceous vegetation beneath trees in maintaining arbuscular mycorrhizal communities in temperate alley cropping systems. *Plant Soil*, 453(1–2), 153–171. https://doi.org/10.1007/s11104-019-04181-z.
51. Ingleby, K., Wilson, J., Munro, R. C., Cavers, S. (2007). Mycorrhizas in agroforestry: Spread and sharing of arbuscular mycorrhizal fungi between trees and crops: Complementary use of molecular and microscopic approaches. *Plant Soil*, 294(1–2), 125–136. https://doi.org/10.1007/s11104-007-9239-z.
52. Lodhi, M. A. K., Rice, E. L. (1971). Allelopathic effects of celtis laevigata. *Bulletin of the Torrey Botanical Club*, 98, 83–89.
53. Luo, S. (2012). Allelopathy in South China agroecosystems. In: *Allelopathy: Current Trends and Future Applications* (pp. 145–157). Springer.
54. Singh, H. P., Batish, D. R., Setia, N., Kohli, R. K. (2005). Herbicidal activity of volatile oils from eucalyptus citriodora against parthenium hysterophorus. *Annals of Applied Biology*, 146(1), 89–94. https://doi.org/10.1111/j.1744-7348.2005.04018.x.

CHAPTER 10

Sustainable Agroforestry Practices: Mitigation and Adaptation to Climate Change

MAHADEV BERA,[1] SUMANTA DAS,[1,2] SUMAN DUTTA,[3] and KOUSHIK BANERJEE[4*]

[1]*School of Environment and Disaster Management, Ramakrishna Mission Vivekananda Educational and Research Institute, Narendrapur, Kolkata, India*

[2]*School of Agriculture and Food Sustainability, The University of Queensland, St Lucia, Queensland, Australia*

[3]*School of Genetics and Plant Breeding, Ramakrishna Mission Vivekananda Educational and Research Institute, Narendrapur, Kolkata, India*

[4]*ICAR-Mahatma Gandhi Integrated Farming Research Institute, Motihari, Bihar, India*

*Corresponding author

ABSTRACT

Agriculture is a major contributor to global greenhouse gas (GHG) emissions, both directly through farm activities and indirectly through land-use changes for agricultural expansion, with agriculture, forestry, and other land use responsible for nearly 22% of global anthropogenic GHG emissions. Agroforestry, which integrates trees or shrubs with crops and/or livestock at plot, farm, or landscape scales, has emerged as an effective strategy to cope with climate change. As an agroecological practice, it provides both

Agroforestry and Climate Change Mitigation: Harnessing Multitrophic Interactions for Sustainable Plant Protection and Ecosystem Resilience. Hanuman Singh Jatav, Vishnu D. Rajput, Ipsita Samal, Tanmaya Kumar Bhoi (Eds.)

DOI: 10.1201/9781779641397-10

socioeconomic and biophysical benefits while enhancing the resilience of farmers and agricultural systems. The Intergovernmental Panel on Climate Change has highlighted agroforestry's potential in climate change adaptation and mitigation, recognizing its capacity to reduce energy use, improve soil health, curb deforestation, enhance biodiversity, and strengthen food security. Agroforestry systems also improve water management, generate income opportunities, and provide ecosystem services, while simultaneously acting as a carbon sink through sequestration. By complementing other climate mitigation approaches, agroforestry contributes to reducing emissions while supporting sustainable livelihoods. Ultimately, sustainable agroforestry practices offer a multifaceted solution to climate challenges, combining resilience building, carbon storage, and flexible farming techniques, making them a crucial component in mitigating and adapting agricultural systems to the impacts of climate change.

10.1 INTRODUCTION

A protracted period during which there is a variation in the average climate or its variability that is statistically significant is referred to as "climate change" [1, 2]. Persistent modifications to the land use or atmosphere's composition, as well as natural processes, outside influences, or both, can all contribute to climate change [3]. The release of greenhouse gases (GHGs) is a key reason for concern due to the future estimates of global warming and its impact on biological life [4]. The atmospheric concentration of GHGs, including CO_2, has increased dramatically during the last century, and this trend is predicted to continue [5]. The increase in the mean temperature of the Earth's ocean and atmosphere, which is mostly caused by growing atmospheric concentrations of GHGs, is referred to as "global warming" [6]. One of the GHGs and the primary contributor to global warming is CO_2 [7]. It contributes between 9% and 26% of the greenhouse effect and makes up 72% of all anthropogenic GHGs [8]. Many countries are working to reduce GHG emissions at the source, but more efforts are needed to increase the sinks for these gases [9]. Reducing the need for energy, changing how it is used, and speeding up the process of atmospheric CO_2 removal through carbon sequestration are all ways to lower the concentration of CO_2 in the atmosphere [10]. Agroecosystems, which are vital to the global carbon cycle, contain about 12% of the terrestrial carbon in the world [11]. Given that trees can fix and store carbon from the atmosphere for a very long time, agroforestry is attractive when considering climate change and the global carbon cycle [12, 13]. Furthermore, all agroforestry systems were shown to be more capable of sequestering carbon than isolated agricultural land use systems [14].

One such method of adapting to climate change is agroforestry. It entails incorporating trees or bushes at various scales (plot, farm, or landscape) with crops and/or livestock [15]. Agroforestry has numerous advantages, including socioeconomic and biological ones, and strengthens farmers' resistance to climate-related dangers. Agroforestry is being highlighted as a potential agroecological solution by the Intergovernmental Panel on Climate Change in recent publications [15, 16]. In addition to helping adapt to climate change, agroforestry systems also preserve biodiversity, improve food security and income prospects, and create synergies between reducing climate change through carbon sequestration and providing ecological services [15, 17]. Agriculture is the main industry for rural development and livelihoods in low-income countries [14]. Food production in emerging nations is jeopardized by the unfavorable consequences of land pressure and climate change on agricultural systems [18, 19]. Despite being prevalent worldwide, intensive agriculture systems face challenges due to decreased production resulting from land degradation [20]. The production of food and agricultural intensification adversely affect soil and biodiversity, which has a negative knock-on effect on local food security, climate change, and farm revenue [21]. The "green revolution" paradigm, which involves subsidies and inputs such as fertilizers, has proven expensive and unsustainable since technology can never fully replace the functions that trees inherently perform [22]. The ongoing discussion about the sustainable intensification of agriculture emphasizes the value of diversity in enhancing crop and land management through the incorporation of trees into land use systems [23–25]. Combining higher yields with ecosystem services can lead to many sustainable agricultural goals, but few methods simultaneously improve farm productivity and agroecosystem diversity [26].

The Indian Council of Agricultural Research set a goal to increase India's forest cover from 23 to 33%, and agroforestry is helping the country reach that objective [26]. Agroforestry can address several issues and bring about a range of societal, environmental, and economic benefits, according to the third assessment report of the International Panel on

Climate Change [14, 27]. Systems of agroforestry can store carbon at rates between 0.7, 1.6, and 6.3 GT [28, 29]. Improved food availability, stable land tenure, higher agricultural revenue, preservation and restoration of carbon storage both above and below the surface, biodiversity preservation and recuperation, watershed hydrology upkeep, and soil conservation are examples of secondary environmental benefits [30, 31]. Sustainable adaptation to local and global change can be achieved by households and communities through the establishment of agroecosystems through agroforestry methods [17, 32]. This chapter focuses on mitigation and adaptation strategies for the negative impact of climate change through agroforestry practices.

10.2 AGROFORESTRY: A VIABLE PRACTICE TO MITIGATE THE IMPACTS OF CLIMATE CHANGE—FUNDAMENTALS AND MECHANISMS

Agroforestry is a land use management strategy that incorporates shrubs, crops, and cattle [33]. It has many positive effects on the environment, society, and economy. The overall resilience of the agroforestry system to changing climatic conditions can be improved by selecting tree species that are resilient to climate variations, including drought-resistant varieties [34]. Agroforestry naturally promotes biodiversity, and systems with greater diversity tend to be more resilient to the impacts of climate change [35]. Water management techniques that assist in adjusting to shifting precipitation patterns, such as contour planting to lessen soil erosion and water runoff, can be incorporated into agroforestry operations [36]. In agroforestry systems, trees store carbon in biomass and soil and absorb CO_2 from the atmosphere, thereby reducing the effects of climate change [37].

Agroforestry encourages sustainable land-use practices, which can help lower GHG emissions [14]. This approach can also reduce the use of artificial fertilizers, which are linked to GHG emissions. Agroforestry systems optimize the use of sun energy, nutrients, and water to increase the effectiveness of resources [38]. This can lessen the overall environmental impact of agricultural practices, which can help mitigate climate change. Silvopastoral systems, which combine trees with pastureland and cattle, improve carbon sequestration while also giving livestock shade, which enhances animal welfare and productivity [39]. Agroforestry is particularly advantageous in climate-vulnerable areas, such as those that are prone to soil erosion, flooding, or droughts [40]. The incorporation of trees can stabilize soils and create protective barriers. "Adaptation–mitigation" describes the combination of mitigating and adapting to climate change within the framework of agroforestry methods [41]. It is imperative to design and implement agroforestry systems that actively contribute to sequestering carbon to lessen the effects of climate change and sustainable land-use practices, while also being resilient to its effects [42]. The broader goals of sustainable agriculture and climate-smart practices align with this integrated strategy.

10.3 AGROFORESTRY: POTENTIAL FOR CARBON SEQUESTRATION

To reduce the amount of CO_2 in the atmosphere, we must move from lower biomass land uses, like grasslands, to tree-based ecosystems, including plantation forests, agroforestry, and forests [43]. Figure 10.1 illustrates the greater potential of agroforestry systems concerning annual systems concerning soil preservation, carbon sequestration, nutrient cycling, microclimate moderating, and overall production [38, 44]. The two primary agroforestry system components that are principally in charge of lowering CO_2 are trees and crops [38]. Capturing carbon in terrestrial reservoirs involves both above-ground biomass from plants, such as fuel wood and timber, and below-ground biomass from plants, such as the deep root zone, roots, soil microbes, and all forms of organic and inorganic carbon [45]. The crop species selected, the cropping strategy employed, the time of the following, and the kind and amount of residue that was put back into the soil all have a substantial influence on the stocks of organic carbon in the soil [46]. Considering the rational distribution of crops, 30 primary cropping systems have been identified in each district, even though over 250 double cropping systems are currently in use in India [47, 48]. The capacity for carbon sequestration of different crops was tested at the ICAR-Indian Institute of Soil Science, Bhopal. The results indicated that maize, sorghum, and pearl millet had better potential than rice, finger millet, and soybean [49].

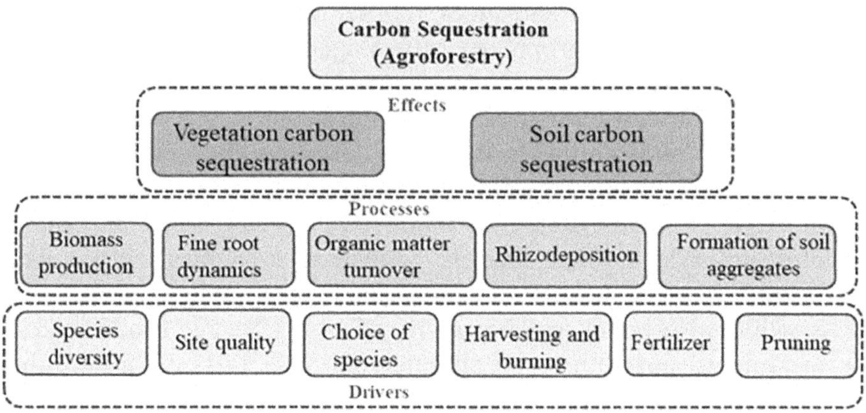

FIGURE 10.1 Effects and process of agroforestry carbon sequestration.

Furthermore, data on the potential for sequestering carbon in various Indian cropping systems was compiled utilizing long-term experiments conducted in various agroclimatic zones of the country. The study discovered that over 28 years in Jabalpur, Madhya Pradesh, the rate of carbon sequestration ranged from ~0.02 Mg C/ha/yr in cropping systems treated with NPK based on soybeans and wheat to 1.2 Mg C/ha/yr in NPK + FYM treatment for systems based on cassava over 13 years in Thiruvananthapuram, Kerala [50]. Most published research concurs that using some organic matter, such as stover, straw, green manure, FYM, or other tree material used to grow crops, is the only way to sequester carbon in pure cropping systems [51–53]. Integrated nutrient management has consistently outperformed NPK when it comes to cropping system CSP [54]. No-tillage techniques have been shown in numerous studies to increase the soil carbon pool in rice-wheat cropping systems [55, 56]. According to research conducted over eight years, the rice–wheat cropping system has an 8.9 Mg C/ha/yr potential for sequestering carbon when no-tillage is used [57, 58]. Table 10.1 shows the potential for sequestering carbon in different agroforestry systems.

According to Moreno and Rolo [66], agroforestry is a method of managing land in which crops and cattle are cultivated on the same plot of land alongside trees or shrubs [66]. Numerous potential advantages of agroforestry systems exist, including carbon sequestration and other economic and environmental advantages. Through photosynthesis, trees

TABLE 10.1 Potential for Sequestering Carbon in Different Agricultural Systems

Agroforestry System	Regions	Name of Species	CSP (Ng C ha1 yr1)	References
Agrihorticulture	Himachal	Fruit	12.15	[59]
Silviculture	Dehradun, Uttarakhand	*E. tereticornis*	4.40	[60]
Agrisilviculture	Chandigarh	*L. leucocephala*	10.48	[61]
Agrisilviculture	Jhansi, Uttar Pradesh	*A. procera*	3.70	[62]
Silviculture	Jhansi, Uttar Pradesh	*A. procera*	1.79	[63]
Agrisilviculture	Hyderabad, Andhra Pradesh	*L. leucocephala*	2.77	[64]
Agrisilviculture	Coimbatore, Tamil Nadu	*C. equisetifolia*	1.57	[65]
Silviculture	Hyderabad, Andhra Pradesh	*A. auriculiformis*	8.64	[64]
Silviculture	Hyderabad, Andhra Pradesh	*A. lebbeck*	0.62	[64]
Silviculture	Hyderabad, Andhra Pradesh	*A. albida*	0.82	[64]

in agroforestry systems store carbon in their biomass [67, 68]. The carbon stored in the tree's leaves, branches, trunk, and other above-ground parts is included in this. By adding organic carbon to the soil through root growth and death, trees also sequester carbon through their roots. Additionally, as leaves and other organic materials decompose, they release carbon into the soil [69]. In agroforestry systems, the inclusion of trees can boost soil microbial activity, which in turn increases the storage of organic carbon [70, 71]. Agroforestry techniques like contour planting and alley cropping can also help prevent soil erosion, which can cause the loss of topsoil that is rich in carbon and discourage soil carbon sequestration [72]. The roots of the trees in agroforestry systems improve the soil's structure, making it better able to hold onto organic matter, including carbon. Long-term carbon storage is possible for trees in agroforestry systems, which is critical for mitigating the effects of climate change [73]. The introduction of a diversity of types of plants found in agroforestry systems creates different carbon pools, which may improve the overall stability and resilience of the carbon sequestration process [68, 74]. Furthermore, agroforestry techniques may aid farmers in adjusting to shifting circumstances and promote carbon sequestration, making agriculture more resilient to climate change [75, 76]. In addition, the various studies on agroforestry carbon sequestration of different regions are given in Table 10.2.

TABLE 10.2 Various Studies on Agroforestry Carbon Sequestration of Different Regions

Potential of Agroforestry	Regions	Study Remarks	References
Biomass and carbon storage	North Western Himalayas, India	This research offers a useful technique for choosing alternative land use regimes and vegetation types to economically optimize carbon stocks, which can help mitigate the effects of global warming in other susceptible Himalayan catchments.	[77]
Fixation of atmospheric CO_2	Eastern Plateau Region of India	Compared to the open agriculture-based land use system, the agrihorticulture land use system had a carbon density that was six to seven times greater. There was a net increase in profit of 299.5 carbon credits, or 449,250 Indian rupees, from the agrihorticulture land use system.	[26]
Carbon sequestration	Indo-Gangetic Plains of India	The country's Indo-Gangetic regions saw great success with the commercial agroforestry system; however, more expansion is required with crops that are suitable and compatible with other regions.	[43]
Mitigation of Climate change through carbon sequestration	Pakistan (Semiarid Region)	A study reports that, concerning carbon sequestration, agroforestry systems are most promising. We propose that agroforestry system research and investment can help Pakistan successfully meet some of its targets for mitigating climate change.	[78]
Carbon sequestration and production of biomass	Northwestern Himalaya	According to the study, only soil and total carbon density affected biomass production and carbon storage capacity, even though the various agroforestry land use systems had a significant impact. Therefore, agroforestry might be a possible natural solution to climate change in temperate regions.	[79]
Sequestration of carbon and mitigation of climate change	Ethiopia	A recent study suggests that agroforestry techniques can effectively create carbon sinks in soils and biomass, thus providing a natural solution to mitigate the impacts of a changing climate. By reducing greenhouse gas emissions, these techniques also enhance the ability of current land uses to lessen and adjust to the consequences of climate change.	[80]

10.4 ADAPTATION OF AGROFORESTRY IN THE FACE OF CHANGING CLIMATE: RESEARCH GAP AND OPPORTUNITIES

Agroforestry holds immense promise for both mitigating and adapting to climate change [81]. Through photosynthesis, trees in agroforestry systems absorb carbon, lowering the atmospheric concentration of CO_2 [67, 82]. As a result, GHG concentrations are reduced worldwide. In place of conventional land-use methods, agroforestry offers sustainable options that possess the capacity to lower GHG emissions [83, 84]. It can take the place of slash-and-burn farming, which emits a lot of CO_2 into the atmosphere. According to Udawatta et al. [85], agroforestry promotes biodiversity by giving different plant and animal species somewhere to live. According to Gupta et al. [86], this improves ecosystem resilience and overall climate stability. Agroforestry techniques also improve the structure and health of the soil, reducing the likelihood of soil degradation [38]. This results in more fertile soils that can sustain plant growth and sequester more carbon [87]. Several crop and tree species are integrated with agroforestry systems [88]. Diversifying their sources of income helps farmers be more resilient to market swings and hazards associated with climate change.

Agroforestry also contributes to better control of water by lowering soil erosion, increasing water infiltration, and controlling water flow [89, 90]. This is crucial in reducing the impacts of floods and droughts and adapting to changing precipitation patterns [91]. Trees in agroforestry systems can also help control regional microclimates by reducing wind speed, providing shade, and regulating temperature [85]. Agroforestry techniques like windbreaks and shelterbelts protect against extreme weather events, prevent soil erosion, and shield crops from strong winds [92, 93]. The diverse nature of agroforestry systems makes agricultural landscapes more resilient to climate change [94]. Trees, crops, and livestock act as a buffer against the uncertainties brought on by climate change [24]. Water use efficiency can be improved through agroforestry systems like silvopasture and alley cropping, which maximize the use of water resources for crops and trees [95]. Trees that provide shade can protect crops from heat stress during hotter seasons. Overall, agroforestry is a flexible and sustainable approach to land use that addresses both the causes and effects of climate change [16]. Its potential lies in its ability to increase biodiversity, sequester carbon, and create resilient and adaptable agricultural systems [76].

Encouraging and adopting agroforestry techniques can make a significant contribution to global efforts to combat climate change [96]. In addition, risks related to climate change and the related services that agroforestry practices offer for adaptation (Figure 10.2).

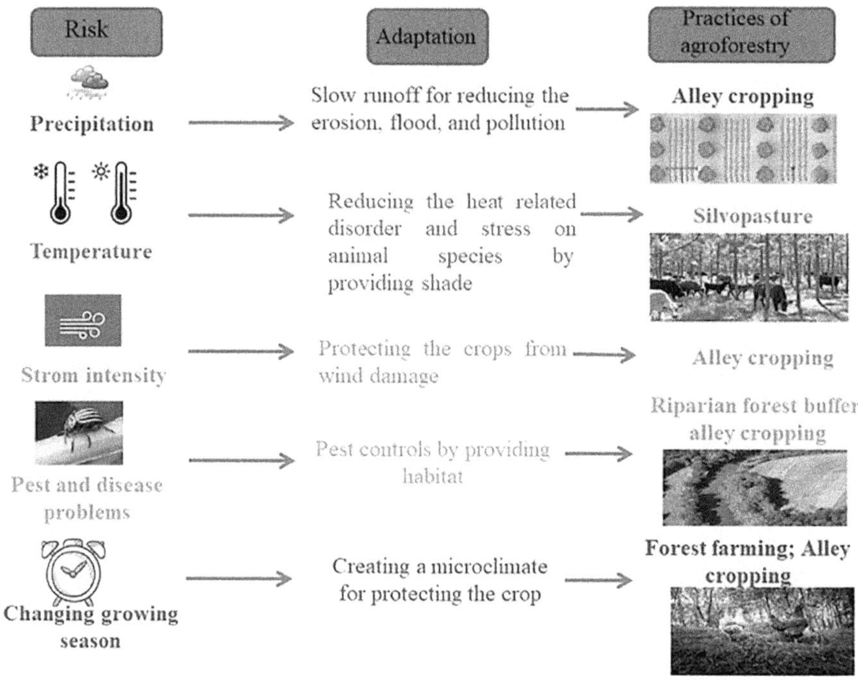

FIGURE 10.2 Components of agroforestry for adaptation to climate change.

Additionally, by raising livestock and/or tree crops alongside food crops on an identical plot of land, agroforestry is a sustainable land-use technique that boosts crop productivity [97]. This system can be put into place on wastelands, boundaries, and bunds that are available. Farmers preserve trees such as *Gmelina arborea*, *Mangifera indica*, *Dalbergia sissoo*, *Zizyphus mauritiana*, and *Acacia nilotica* on their agricultural land [43]. Agroforestry plays a vital role in mitigating climate change as trees sequester the most common GHG, CO_2, in their biomass [67, 98]. It contributes to both climate adaptation and mitigation, making it an environmentally and economically sound practice that improves overall farm productivity, enriches the soil, and preserves environmental services like phytoremediation, biodiversity conservation, watershed protection, and carbon sequestration [99, 100]. The ability of agroforestry to trap carbon has been acknowledged, but the lack of reliable information about the carbon reserves of AF interventions concerning forestry and agriculture has limited our understanding of carbon sinks in different AFS in the region [14, 101]. Even though agriculture and forestry together contribute to over 21% of emissions, neither national carbon accounts nor international carbon financing plans have included an objective assessment of AFS [102]. Carbon stocks in AFS have been documented by studies at the world, national, and zonal levels; nevertheless, their coverage is primarily limited to South Asia [37, 86].

Comprehensive data on trends in soil organic carbon and tree carbon stocks are missing from the majority of studies. It has proven to be very challenging to get insight into how different agroforestry systems might work as potential carbon sinks. The irregular trees found in AFS serve as wildlife corridors and let animals migrate across the landscape more quickly. This is essential for helping wildlife adapt to the changing weather patterns brought on by climate change [103]. Given that most of the countries in South Asia are agricultural, the region has a lot of potential to promote agroforestry as a technique for mitigating and adapting to climate change [104]. Agroforestry is still viable over 69% of South Asia's total land area, with 55% of that being even more so, according to a recent study [105, 106]. To maximize AFS's utility for mitigation and adaptation to the climate while limiting adverse consequences on the climate, strategically coordinated activities are needed [107]. Agroforestry techniques that assist in mitigating and adapting to climate change were also emphasized by the various research (Table 10.3).

10.5 AGROFORESTRY FOR SUSTAINABLE DEVELOPMENT

Agroforestry is a sustainable and multifunctional practice that can provide additional income sources for farmers through the cultivation of tree products [112, 113]. These products include timber, fruits, nuts, and medicinal plants, thus contributing to Goal 1 (no poverty). Agroforestry also enhances food security by diversifying agricultural production, improving soil fertility, and providing additional food sources through tree products, which aligns with Goal 2 (zero hunger). Furthermore, agroforestry

TABLE 10.3 Adaptation of Different Agroforestry Practices for Mitigating Climate Change

Agroforestry Strategies	Description	References
Intercropping 	The practice of planting trees alongside crops, or tree intercropping, raises the soil's carbon content and boosts agricultural output.	[16]
Silvopastoral 	To provide complementary uses and animal nourishment, agroforestry arrangements known as silvopastoral systems purposefully combine shrubs and trees with fodder plants, such as grasses and leguminous herbs. This kind of agroforestry aids in sequestering carbon and mitigating climate change.	[108]
Agrisilvicultural 	Agrisilvicultural systems, like home gardens and alley cropping, combine trees and crops. In silvopastoral systems, domesticated animals are grazed on pastures, rangelands, or on farms in conjunction with forestry practices.	[109, 110]
Random planting 	Different types of trees are planted randomly in the same field.	[78, 96]
Boundary planting 	Different types of trees are distributed around the edge of the field from the boundaries.	[111]

promotes healthy living by providing nutritious foods, medicinal plants, and a healthier environment, in line with Goal 3 (good health and well-being). Agroforestry can also help protect water resources by minimizing soil erosion, improving water retention, and reducing runoff, which contributes to better water quality, which is in line with Goal 6 (clean water and sanitation for all). Additionally, agroforestry can contribute to renewable energy sources, such as bioenergy, by utilizing tree biomass for fuel and energy production, which aligns with Goal 7 (affordable and clean energy). Agroforestry creates employment opportunities by involving farmers in the management and maintenance of tree-based systems, and it supports local economies, in line with Goal 8 (decent work and economic growth). Agroforestry supports sustainable land use and resource management, contributing to responsible consumption and production practices, which align with Goal 12 (responsible consumption and production). Furthermore, by improving resilience to climate impacts, supporting sustainable land management following Goal 13 (climate action), and sequestering carbon in trees and soil, agroforestry plays a critical role in mitigating the effects of climate change. In keeping with Goal 14 (life below water), it also supports ecological diversity in agricultural landscapes by serving as a habitat for a variety of plant and animal species. In general, agroforestry is a sustainable and all-encompassing strategy that incorporates social, economic, and environmental aspects, enhancing the general well-being of societies and the environment [114].

10.6 AGROFORESTRY: FUTURE PERSPECTIVES

Prospects for agroforestry practices are bright since they can address several global concerns and emerging trends [115]. For example, agroforestry can significantly aid in mitigating the effects of climate change by storing carbon in soil and trees [81]. Additionally, it fosters resilience through a variety of integrated farming systems, aiding agriculture's adaptation to changing climatic conditions [116]. With the growing threats to ecosystems, agroforestry provides a means of fostering biodiversity in agricultural landscapes by combining different tree species to offer habitat for a range of plant and animal species [117].

Agroforestry systems' ability to diversify agricultural output can help increase food security as diets with more nutritional value are those that include trees that yield fruits, nuts, and other edible products [118]. Sustainable agriculture is supported by the coexistence of trees in agroforestry systems with crops and livestock. This method strengthens overall farm resilience, lowers the need for synthetic inputs, and improves soil health [119]. By growing tree products like timber, fruits, nuts, and nontimber forest products, agroforestry can give farmers access to additional revenue streams, supporting rural livelihoods and economic expansion.

By decreasing soil erosion, increasing water retention, and decreasing runoff, agroforestry also aids in the management of water resources [120]. This enhances the water's quality and encourages agriculture to use water sustainably. Agroforestry systems can aid in the production of renewable energy in response to the growing demand for additional clean, sustainable energy sources by utilizing the biomass from trees to create bioenergy [121].

Technological developments, such as precision agriculture, remote sensing, and agroecological modeling, can improve agroforestry system planning and management. Integrating technology can raise productivity and optimize resource usage. As the advantages of agroforestry become more widely known, national and international policymakers are beginning to recognize and support this practice [122].

The potential of agroforestry to help organizations and governments achieve their sustainability goals is now being widely recognized [123]. There is a growing global cooperation and exchange of information regarding the optimal agroforestry techniques, which can help in advancing global sustainability initiatives and hasten the adoption of agroforestry worldwide [42]. The increasing awareness of consumers about eco-friendly and sustainable practices is driving the demand for agroforestry-produced goods, which may encourage farmers to use agroforestry techniques. The future of agriculture and sustainable development is expected to be significantly shaped by agroforestry, as the awareness of the practice increases and policies to support its development [124]. Agroforestry's potential for the future lies in its ability to simultaneously tackle several issues, providing a comprehensive and sustainable method of managing resources and land use [123, 125].

10.7 CONCLUSION

Sustainable agroforestry practices are essential for mitigating the consequences of and adjusting to climate change, as well as for addressing issues related to agriculture and the environment. Because agroforestry systems are capable of storing carbon, they function as carbon sinks and help to lower GHG emissions. Additionally, agroforestry contributes to a decrease in the demand for fuelwood and lumber from natural forests, hence lowering the rate of deforestation and the resulting carbon emissions.

Agroforestry encourages biodiversity by incorporating trees into agricultural landscapes, which offer diverse habitats for a range of plant and animal species. This, in turn, strengthens ecosystems and helps them resist changes caused by climate change. Agroforestry systems combine trees, crops, and livestock to enhance the cycling of nutrients, resulting in more fertile soil that requires fewer outside resources, such as synthetic fertilizers. By diversifying the agricultural landscape, agroforestry systems reduce their susceptibility to risks associated with climate change while providing multiple revenue streams that help farmers become less dependent on a single crop. The dynamic nature of agroforestry allows for adaptive management strategies that enable farmers to modify their practices in response to changing climatic conditions, ensuring long-term sustainability. Governments and institutions can promote sustainable agroforestry by creating policy frameworks, financial incentives, and capacity building. Incorporating sustainable agroforestry techniques into agricultural landscapes improves ecosystems' and communities' ability to adapt and endure while also lessening the effects of climate change. However, successful implementation requires coordinated efforts from policymakers, researchers, and local communities to achieve a harmonious balance between agricultural productivity and environmental conservation.

KEYWORDS

- **climate change**
- **agroforestry practices**
- **agricultural systems**
- **climate change mitigation**
- **sustainable development**

REFERENCES

1. Forzieri, G., Dakos, V., McDowell, N. G., Ramdane, A., Cescatti, A. (2022). Emerging signals of declining forest resilience under climate change. *Nature*, 608(7923), 534–539.
2. Qiao, L., Wang, X., Smith, P., Fan, J., Lu, Y., Emmett, B., Li, R., Dorling, S., Chen, H., Liu, S., Benton, T. G., Wang, Y., Ma, Y., Jiang, R., Zhang, F., Piao, S., Müller, C.,

Yang, H., Hao, Y., Li, W., Fan, M. (2022). Soil quality both increases crop production and improves resilience to climate change. *Nature Climate Change*, 12(6), 574–580.

3. Roy, P. S., Ramachandran, R. M., Paul, O., Thakur, P. K., Ravan, S., Behera, M. D., Sarangi, C., Kanawade, V. P. (2022). Anthropogenic land use and land cover changes—A review on its environmental consequences and climate change. *Journal of the Indian Society of Remote Sensing*, 50(8), 1615–1640.
4. Aakko-Saksa, P. T., Lehtoranta, K., Kuittinen, N., Järvinen, A., Jalkanen, J. P., Johnson, K., Jung, H., Ntziachristos, L., Gagne, S., Takahashi, C., Karjalainen, P., Rönkkö, T., Timonen, H. (2023). Reduction in greenhouse gas and other emissions from ship engines: Current trends and future options. *Progress in Energy and Combustion Science*, 94, 101055.
5. Cnossen, I. (2022). A realistic projection of climate change in the upper atmosphere into the 21st century. *Geophysical Research Letters*, 49(19), e2022GL100693.
6. Li, L., Zhang, Y., Zhou, T., Wang, K., Wang, C., Wang, T., Yuan, L., An, K., Zhou, C., Lü, G., Lü, G. (2022). Mitigation of China's carbon neutrality to global warming. *Nature Communications*, 13(1), 5315.
7. Letcher, T. M. (2022). Global warming, greenhouse gases, renewable energy, and storing energy. In: *Storing Energy* (pp. 3–12). Elsevier.
8. Gast, L., Cabrera Serrenho, A., Allwood, J. M. (2022). What contribution could industrial symbiosis make to mitigating industrial greenhouse gas (GHG) emissions in bulk material production?. *Environmental Science & Technology*, 56(14), 10269–10278.
9. Sonwani, S., Saxena, P. (Eds.). (2022). *Greenhouse Gases: Sources, Sinks and Mitigation*. Springer: Singapore.
10. Zhu, M., Liu, Y., Wu, X., Shen, J. (2023). Dynamic modeling and comprehensive analysis of direct air-cooling coal-fired power plant integrated with carbon capture for reliable, economic and flexible operation. *Energy*, 263, 125490.
11. Liu, X., Wang, S., Zhuang, Q., Jin, X., Bian, Z., Zhou, M., Meng, Z., Han, C., Guo, X., Jin, W., Zhang, Y. (2022). A review on carbon source and sink in arable land ecosystems. *Land*, 11(4), 580.
12. Sivaranjani, S., Panwar, V. P. (2023). Carbon sequestration in agroforestry systems. In *Agricultural Soil Sustainability and Carbon Management* (pp. 207–227). Academic Press.
13. Nandakishor, T. M., Gopi, G., Champatan, V., Sukesh, A., Aravind, P. V. (2022). Agroforestry in shade coffee plantations as an emission reduction strategy for tropical regions: Public acceptance and the role of tree banking. *Frontiers in Energy Research*, 10, 758372.
14. Sahoo, G., Swamy, S. L., Wani, A. M., Mishra, A. (2022). Agroforestry systems for carbon sequestration and food security: Implications for climate change mitigation. In: *Soil Health and Environmental Sustainability: Application of Geospatial Technology* (pp. 503–528). Springer International Publishing: Cham.
15. Quandt, A., Neufeldt, H., Gorman, K. (2023). Climate change adaptation through agroforestry: Opportunities and gaps. *Current Opinion in Environmental Sustainability*, 60, 101244.
16. Burgess, A. J., Cano, M. E. C., Parkes, B. (2022). The deployment of intercropping and agroforestry as adaptation to climate change. *Crop and Environment*, 1(2), 145–160.

17. Ntawuruhunga, D., Ngowi, E. E., Mangi, H. O., Salanga, R. J., Shikuku, K. M. (2023). Climate-smart agroforestry systems and practices: A systematic review of what works, what doesn't work, and why. *Forest Policy and Economics*, 150, 102937.
18. Kumar, L., Chhogyel, N., Gopalakrishnan, T., Hasan, M. K., Jayasinghe, S. L., Kariyawasam, C. S., Kogo, B. K., Ratnayake, S. (2022). Climate change and future of agri-food production. In *Future Foods* (pp. 49–79). Academic Press.
19. Farooq, M. S., Uzair, M., Raza, A., Habib, M., Xu, Y., Yousuf, M., Yang, S. H., Ramzan Khan, M. (2022). Uncovering the research gaps to alleviate the negative impacts of climate change on food security: A review. *Frontiers in Plant Science*, 13, 927535.
20. Chaudhuri, S., Roy, M., McDonald, L. M., Emendack, Y. (2023). Land degradation–desertification in relation to farming practices in India: An overview of current practices and agro-policy perspectives. *Sustainability*, 15(8), 6383.
21. Neuenschwander, P., Borgemeister, C., De Groote, H., Sæthre, M. G., Tamò, M. (2023). Food security in tropical Africa through climate-smart plant health management. *Heliyon*, 9(4), e15116.
22. Vishnoi, S., Goel, R. K. (2024). Climate smart agriculture for sustainable productivity and healthy landscapes. *Environmental Science & Policy*, 151, 103600.
23. Nguyen, T. T., Grote, U., Neubacher, F., Do, M. H., Paudel, G. P. (2023). Security risks from climate change and environmental degradation: Implications for sustainable land use transformation in the Global South. *Current Opinion in Environmental Sustainability*, 63, 101322.
24. van Noordwijk, M., Coe, R., Sinclair, F. L., Luedeling, E., Bayala, J., Muthuri, C. W., Cooper, P., Kindt, R., Duguma, L., Lamanna, C., Minang, P. A. (2021). Climate change adaptation in and through agroforestry: Four decades of research initiated by Peter Huxley. *Mitigation and Adaptation Strategies for Global Change*, 26(5), 18.
25. Silveira, J. G. D., Oliveira Neto, S. N. D., Canto, A. C. B. D., Leite, F. F. G. D., Cordeiro, F. R., Assad, L. T., Silva, G. C. C., Marques, R. d. O., Dalarme, M. S. L., Ferreira, I. G. M., Conceição, M. C. G. d., Rodrigues, R. D. A. R. (2022). Land use, land cover change and sustainable intensification of agriculture and livestock in the Amazon and the atlantic forest in Brazil. *Sustainability*, 14(5), 2563.
26. Kumar, R., Kumar, R., Karmakar, S., Kumar, A., Singh, A. K., Kumar, A., Singh, J. (2023). Impact of amide fertilizer on carbon sequestration under the agroforestry system in the eastern plateau region of India. *Sustainability*, 15(12), 9775.
27. Lee, H., Calvin, K., Dasgupta, D., Krinner, G., Mukherji, A., Thorne, P., ... & Park, Y. (2023). IPCC, 2023: Climate change 2023: Synthesis report, summary for policymakers. In: Core Writing Team, Lee, H., Romero, J., (Eds.), *Contribution of Working Groups I, II and III to the Sixth Assessment Report of the Intergovernmental Panel on Climate Change*. IPCC: Geneva, Switzerland.
28. Kumar, S., Dev, I., Kumar, N. (2023). Forestry and agroforestry. In: *Trajectory of 75 Years of Indian Agriculture after Independence* (pp. 497–519). Springer Nature Singapore: Singapore.
29. Udawatta, R. P., Walter, D., Jose, S. (2022). Carbon sequestration by forests and agroforests: A reality check for the United States. *Carbon Footprints*, 1, 8.
30. Kumawat, A., Yadav, D., Srivastava, P., Babu, S., Kumar, D., Singh, D., Vishwakarma, D. K., Sharma, V. K., Madhu, M. (2023). Restoration of agroecosystems with

conservation agriculture for food security to achieve sustainable development goals. *Land Degradation & Development*, 34, 3079–3097.
31. Mu, Y., Guo, Y., Li, X., Li, P., Bai, J., Linke, S., Cui, B. (2022). Cost-effective integrated conservation and restoration priorities by trading off multiple ecosystem services. *Journal of Environmental Management*, 320, 115915.
32. Amoak, D., Luginaah, I., McBean, G. (2022). Climate change, food security, and health: Harnessing agroecology to build climate-resilient communities. *Sustainability*, 14(21), 13954.
33. Elagib, N. A., Al-Saidi, M. (2020). Balancing the benefits from the water–energy–land–food nexus through agroforestry in the Sahel. *Science of The Total Environment*, 742, 140509.
34. Aurelle, D., Thomas, S., Albert, C., Bally, M., Bondeau, A., Boudouresque, C. F., Cahill, A. E., Carlotti, F., Chenuil, A., Cramer, W., Davi, H., De Jode, A., Ereskovsky, A., Farnet, A.-M., Fernandez, C., Gauquelin, T., Mirleau, P., Monnet, A.-C., Prévosto, B., Rossi, V., Sartoretto, S., Van Wambeke, F., Fady, B. (2022). Biodiversity, climate change, and adaptation in the Mediterranean. *Ecosphere*, 13(4), e3915.
35. Dinesha, S., Hosur, S. R., Toushif, P. K., Bodiga, D., Deepthi Dechamma, N. L., Ashwath, M. N., Pradhan, D. (2023). Sustaino-resilient agroforestry for climate resilience, food security and land degradation neutrality. In: *Land and Environmental Management through Forestry* (pp. 217–245).
36. Madhu, M. (2022). Soil and water conservation strategies for sustainable agriculture in changing climate scenario. *Community Based Climate Risk Management Through Watershed Development*, 1.
37. Ghale, B., Mitra, E., Sodhi, H. S., Verma, A. K., Kumar, S. (2022). Carbon sequestration potential of agroforestry systems and its potential in climate change mitigation. *Water, Air, & Soil Pollution*, 233(7), 228.
38. Fahad, S., Chavan, S. B., Chichaghare, A. R., Uthappa, A. R., Kumar, M., Kakade, V., Pradhan, A., Jinger, D., Rawale, G., Yadav, D. K., Kumar, V., Farooq, T. H., Ali, B., Sawant, A. V., Saud, S., Chen, S., Poczai, P. (2022). Agroforestry systems for soil health improvement and maintenance. *Sustainability*, 14(22), 14877.
39. Aryal, D. R., Morales-Ruiz, D. E., López-Cruz, S., Tondopó-Marroquín, C. N., Lara-Nucamendi, A., Jiménez-Trujillo, J. A., Pérez-Sánchez, E., Betanzos-Simon, J. E., Casasola-Coto, F., Martínez-Salinas, A., Sepúlveda-López, C. J., Ramírez-Díaz, R., La O Arias, M. A., Guevara-Hernández, F., Pinto-Ruiz, R., Ibrahim, M. (2022). Silvopastoral systems and remnant forests enhance carbon storage in livestock-dominated landscapes in Mexico. *Scientific Reports*, 12(1), 16769.
40. Datta, P., Behera, B. (2023). Climate change adaptation through agroforestry: empirical evidence from indian eastern himalayan foothills. In: *Strategizing Agricultural Management for Climate Change Mitigation and Adaptation* (pp. 167–181). Springer International Publishing: Cham.
41. Bogale, G. A., Bekele, S. E. (2023). Sustainability of agroforestry practices and their resilience to climate change adaptation and mitigation in Sub-Saharan Africa: A review. *Ekológia (Bratislava)*, 42(2), 179–192.
42. Octavia, D., Suharti, S., Murniati, Dharmawan, I. W. S., Nugroho, H. Y. S. H., Supriyanto, B., Rohadi, D., Njurumana, G. N., Yeny, I., Hani, A., Mindawati, N., Suratman, Adalina, Y., Prameswari, D., Hadi, E. E. W., Ekawati, S. (2022).

Mainstreaming smart agroforestry for social forestry implementation to support sustainable development goals in Indonesia: A review. *Sustainability*, 14(15), 9313.

43. Chavan, S. B., Dhillon, R. S., Sirohi, C., Uthappa, A. R., Jinger, D., Jatav, H. S., Chichaghare, A. R., Kakade, V., Paramesh, V., Kumari, S., Yadav, D. K., Minkina, T., Rajput, V. D. (2023). Carbon sequestration potential of commercial agroforestry systems in indo-gangetic plains of India: Poplar and eucalyptus-based agroforestry systems. *Forests*, 14(3), 559.
44. Hombegowda, H. C., Adhikary, P. P., Jakhar, P., Madhu, M. (2022). Alley cropping agroforestry system for improvement of soil health. In: *Soil Health and Environmental Sustainability: Application of Geospatial Technology* (pp. 529–549). Springer International Publishing: Cham.
45. Zanvo, S. M., Mensah, S., Salako, K. V., Kakaï, R. G. (2023). Tree height-diameter, aboveground and belowground biomass allometries for two West African mangrove species. *Biomass and Bioenergy*, 176, 106917.
46. Kan, Z. R., Chen, Z., Wei, Y. X., Virk, A. L., Bohoussou, Y. N. D., Lal, R., Zhao, X., Zhang, H. L. (2022). Contribution of wheat and maize to soil organic carbon in a wheat-maize cropping system: A field and laboratory study. *Journal of Applied Ecology*, 59(11), 2716–2729.
47. Brar, A. S., Kaur, K., Sindhu, V. K., Tsolakis, N., Srai, J. S. (2022). Sustainable water use through multiple cropping systems and precision irrigation. *Journal of Cleaner Production*, 333, 130117.
48. Shukla, S. K., Sharma, L., Jaiswal, V. P., Dwivedi, A. P., Yadav, S. K., Pathak, A. D. (2022). Diversification Options in Sugarcane-Based Cropping Systems for Doubling Farmers' Income in Subtropical India. *Sugar Tech*, 24(4), 1212–1229.
49. Kushwah, S. K., Dotaniya, M. L., Upadhyay, A. K., Rajendiran, S., Coumar, M. V., Kundu, S. and Subba Rao, A. (2014). Assessing carbon and nitrogen partition in kharif crops for their carbon sequestration potential. *National Academy Science Letter*, 37(3), 213–7.
50. Dhyan, S. K., Ram, A., Dev, I. (2016). Potential of agroforestry systems in carbon sequestration in India. *The Indian Journal of Agricultural Sciences*, *86*(9), 1103–1112.
51. Timsina, J. (2018). Can organic sources of nutrients increase crop yields to meet global food demand. *Agronomy*, 8(10), 214.
52. Powlson, D., Smith, P., Nobili, M. D. (2013). Soil organic matter. In: *Soil Conditions and Plant Growth* (pp. 86–131).
53. Mrunalini, K., Behera, B., Jayaraman, S., Abhilash, P. C., Dubey, P. K., Swamy, G. N., Prasad, J. V. N. S., Rao, K. V., Krishnan, P., Pratibha, G., Srinivasa Rao, C. (2022). Nature-based solutions in soil restoration for improving agricultural productivity. *Land Degradation & Development*, 33(8), 1269–1289.
54. Patil, R. R., Kumar, S. (2022). Rice transformer: A novel integrated management system for controlling rice diseases. *IEEE Access*, 10, 87698–87714.
55. Sharma, S., Singh, P. (2023). Tillage intensity and straw retention impacts on soil organic carbon, phosphorus and biological pools in soil aggregates under rice-wheat cropping system in Punjab, north-western India. *European Journal of Agronomy*, 149, 126913.

56. Sharma, S., Thind, H. S., Sidhu, H. S., Jat, M. L., Parihar, C. M. (2019). Effects of crop residue retention on soil carbon pools after 6 years of rice–wheat cropping system. *Environmental Earth Sciences*, 78, 1–14.
57. Rabach, V. O., Koske, J., Muna, M. M., Muriuki, J., Ngare, I. O. (2020). Carbon sequestration in agroforestry systems between conservation agriculture and conventional practice in the asal area of machakos county, Kenya. *Journal of Applied Agricultural Science and Technology*, 4(2), 118–133.
58. Kumari, E., Meena, R. S., Sen, A., Upadhyay, P. K., Meena, S. K., Nagargade, M., Tyagi, V. (2019). Biosequestration–A sustainable approach towards rejuvenation of soil organic carbon in India. *Sustainable Agriculture*, 190.
59. Goswami, S., Verma, K. S., Kaushalm, R. (2013). Biomass and carbon sequestration in different agroforestry systems of a Western Himalayan watershed. *Biological Agriculture & Horticulture.* https://doi.org/10.108 0/01448765.2013.855990
60. Dhyani, S. K., Puri, D. N., Narain, P. (1996). Biomass production and rooting behaviour of Eucalyptus tereticornis Sm. on deep soils and riverbed bouldery lands of Doon Valley, India. *Indian Forester*, 122(2), 128–36.
61. Mittal, S. P. & Singh, P. (1989). Intercropping field crops between rows of Leucaena leucocephala under rainfed conditions in northern India. *Agroforestry Systems*, 8(2), 165–172.
62. Newaj, R., Dhyani, S. K. (2008). Agroforestry for carbon sequestration: Scope and present status. *Indian Journal of Agroforestry*, 10, 1–9.
63. Rao, LGG., Joseph, B., Sreemannarayana, B. (2000). Growth and biomass production of some important multipurpose tree species on rainfed sandy loam soils. *Indian Forester*, 126(7), 772–781.
64. Rao, M. R., Ong, CK., Pathak, P., Sharma, MM. (1991). Productivity of annual cropping and agroforestry systems on a shallow Alfisol in semi-arid India. *Agroforestry Systems*, 15, 51–63.
65. Viswanath, S., Peddappaiah, RS., Subramoniam, V., Manivachakam, P., George, M. (2004). Management of Casuarina equisetifolia in wide-row intercropping systems for enhanced productivity. *Indian Journal of Agroforestry*, 6(2), 19–25.
66. Moreno, G., Rolo, V. (2019). Agroforestry practices: Silvopastoralism. In: *Agroforestry for Sustainable Agriculture* (pp. 119–164). Burleigh Dodds Science Publishing.
67. Tefera, Y., Hailu, Y., Siraj, Z. (2019). Potential of agroforestry for climate change mitigation through carbon sequestration. *Agricultural Research & Technology: Open Access Journal*, 22, 556196. Springer.
68. Dhyani, S. K., Ram, A., Newaj, R., Handa, A. K., Dev, I. (2020). Agroforestry for carbon sequestration in tropical India. In: *Carbon Management in Tropical and Sub-tropical Terrestrial Systems* (pp. 313–331).
69. Lu, C., Kotze, D. J., Setälä, H. M. (2021). Evergreen trees stimulate carbon accumulation in urban soils via high root production and slow litter decomposition. *Science of the Total Environment*, 774, 145129.
70. Rodríguez, L., Suárez, J. C., Rodriguez, W., Artunduaga, K. J., Lavelle, P. (2021). Agroforestry systems impact soil macroaggregation and enhance carbon storage in Colombian deforested Amazonia. *Geoderma*, 384, 114810.

71. Cardinael, R., Chevallier, T., Guenet, B., Girardin, C., Cozzi, T., Pouteau, V., Chenu, C. (2020). Organic carbon decomposition rates with depth and contribution of inorganic carbon to CO_2 emissions under a Mediterranean agroforestry system. *European Journal of Soil Science*, 71(5), 909–923.
72. Shinde, R., Sarkar, P. K., Thombare, N., Naik, S. K. (2019). Soil conservation: Today's need for sustainable development. *Agriculture & Food: e-Newsletter*, 1(5), 175–183.
73. Gupta, V. P. (2020). Role of agroforestry in soil conservation and soil health management: A review. *Journal of Pharmacognosy and Phytochemistry*, 9(4S), 555–558.
74. Khan, N., Jhariya, M. K., Raj, A., Banerjee, A., Meena, R. S. (2021). Soil carbon stock and sequestration: Implications for climate change adaptation and mitigation. In: *Ecological Intensification of Natural Resources for Sustainable Agriculture* (pp. 461–489).
75. Gifawesen, S. T., Tola, F. K., Duguma, M. S. (2020). Review on role of home garden agroforestry practices to improve livelihood of small scale farmers and climate change adaptation and mitigation. *Journal of Plant Sciences*, 8(5), 134–145.
76. Kay, S., Rega, C., Moreno, G., den Herder, M., Palma, J. H., Borek, R., Crous-Duran J., Freese D., Giannitsopoulos M., Graves A., Jäger M., Lamersdorf N., Memedemin D., Mosquera-Losada R., Pantera A., Paracchini M., Paris P., Roces-Díaz J., Rolo V., Rosati A., Sandor M., Smith J., Szerencsits E., Varga A., Viaud V., Wawer R., Burgess P., Herzog, F. (2019). Agroforestry creates carbon sinks whilst enhancing the environment in agricultural landscapes in Europe. *Land Use Policy*, 83, 581–593.
77. Sharma, H., Pant, K. S., Bishist, R., Gautam, K. L., Dogra, R., Kumar, M., Kumar, A. (2023). Estimation of biomass and carbon storage potential in agroforestry systems of north western Himalayas, India. *Catena*, 225, 107009.
78. Yasin, G., Nawaz, M. F., Zubair, M., Azhar, M. F., Mohsin Gilani, M., Ashraf, M. N., Qin, A., Ur Rahman, S. (2023). Role of traditional agroforestry systems in climate change mitigation through carbon sequestration: An investigation from the semi-arid region of Pakistan. *Land*, 12(2), 513.
79. Bhardwaj, D. R., Salve, A., Kumar, J., Kumar, A., Sharma, P., Kumar, D. (2023). Biomass production and carbon storage potential of agroforestry land use systems in high hills of north-western Himalaya: An approach towards natural based climatic solution. In: *Biomass Conversion and Biorefinery* (pp. 1–14).
80. Getnet, D., Mekonnen, Z., Anjulo, A. (2023). The potential of traditional agroforestry practices as nature-based carbon sinks in Ethiopia. *Nature-Based Solutions*, 4, 100079.
81. Matocha, J., Schroth, G., Hills, T., Hole, D. (2012). Integrating climate change adaptation and mitigation through agroforestry and ecosystem conservation. In: *Agroforestry-the Future of Global Land Use* (pp. 105–126).
82. Montagnini, F., Nair, P. R. (2004). Carbon sequestration: An underexploited environmental benefit of agroforestry systems. In: *New Vistas in Agroforestry: A Compendium for 1st World Congress of Agroforestry, 2004* (pp. 281–295). Springer: Netherlands.

83. Mutuo, P. K., Cadisch, G., Albrecht, A., Palm, C. A., Verchot, L. (2005). Potential of agroforestry for carbon sequestration and mitigation of greenhouse gas emissions from soils in the tropics. *Nutrient cycling in Agroecosystems*, 71, 43–54.
84. Sileshi, G. W., Gupta, S. R., Dagar, J. C. (2023). Synthesis: Agroforestry for sustainable intensification of agriculture in Asia and Africa. In: *Agroforestry for Sustainable Intensification of Agriculture in Asia and Africa* (pp. 781–790). Springer Nature Singapore: Singapore.
85. Udawatta, R. P., Rankoth, L. M., Jose, S. (2019). Agroforestry and biodiversity. *Sustainability*, 11(10), 2879.
86. Gupta, S. R., Dagar, J. C., Sileshi, G. W., Chaturvedi, R. K. (2023). Agroforestry for climate change resilience in degraded landscapes. In: *Agroforestry for Sustainable Intensification of Agriculture in Asia and Africa* (pp. 121–174).
87. van Zonneveld, M., Turmel, M. S., Hellin, J. (2020). Decision-making to diversify farm systems for climate change adaptation. *Frontiers in Sustainable Food Systems*, 4, 32.
88. Legesse, A., Negash, M. (2021). Species diversity, composition, structure and management in agroforestry systems: The case of Kachabira district, Southern Ethiopia. *Heliyon*, 7(3), e06477.
89. Kaushal, R., Mandal, D., Panwar, P., Kumar, P., Tomar, J. M. S., Mehta, H. (2021). Soil and water conservation benefits of agroforestry. In: *Forest Resources Resilience and Conflicts* (pp. 259–275). Elsevier.
90. Zhu, X., Chen, C., Wu, J., Yang, J., Zhang, W., Zou, X., Liu, W., Jiang, X. (2019). Can intercrops improve soil water infiltrability and preferential flow in rubber-based agroforestry system. *Soil and Tillage Research*, 191, 327–339.
91. Quandt, A. (2020). Contribution of agroforestry trees for climate change adaptation: Narratives from smallholder farmers in Isiolo, Kenya. *Agroforestry Systems*, 94(6), 2125–2136.
92. Mume, I. D., Workalemahu, S. (2021). Review on windbreaks agroforestry as a climate smart agriculture practices. *American Journal of Agriculture and Forestry*, 9(6), 342–347.
93. Ikromov, I. I., Kadamov, A. (2021). Agroforestry against wind erosion damage: A case study in Tajikistan. In: *Exploring and Optimizing Agricultural Landscapes* (pp. 693–705). Springer International Publishing: Cham.
94. Dagar, J. C., Gupta, S. R., Teketay, D. (Eds.). (2020). *Agroforestry for Degraded Landscapes*.
95. Jose, S., Udawatta, R. P. (2021). Agroforestry for ecosystem services: An introduction. In: *Agroforestry and Ecosystem Services* (pp. 1–17).
96. Oduniyi, O. S., Tekana, S. S. (2019). Adoption of agroforestry practices and climate change mitigation strategies in North West province of South Africa. *International Journal of Climate Change Strategies and Management*, 11(5), 716–729.
97. Pantera, A., Mosquera-Losada, M. R., Herzog, F., Den Herder, M. (2021). Agroforestry and the environment. *Agroforestry Systems*, 95(5), 767–774.
98. Cardinael, R., Cadisch, G., Gosme, M., Oelbermann, M., Van Noordwijk, M. (2021). Climate change mitigation and adaptation in agriculture: Why agroforestry should be part of the solution. *Agriculture, Ecosystems & Environment*, 319, 107555.

99. Bayu, T. (2020). Review on contribution of integrated soil fertility management for climate change mitigation and agricultural sustainability. *Cogent Environmental Science*, 6(1), 1823631.
100. Panchasara, H., Samrat, N. H., Islam, N. (2021). Greenhouse gas emissions trends and mitigation measures in Australian agriculture sector—A review. *Agriculture*, 11(2), 85.
101. Singh, N. R., Kumar, D., Rao, K. K., Bhatt, B. P. (2020). Agroforestry: Soil organic carbon and its carbon sequestration potential. *Climate Change and Agroforestry Systems: Adaptation and Mitigation Strategies*, 119.
102. Dhyani, S., Murthy, I. K., Kadaverugu, R., Dasgupta, R., Kumar, M., Adesh Gadpayle, K. (2021). Agroforestry to achieve global climate adaptation and mitigation targets: Are South Asian countries sufficiently prepared. *Forests*, 12(3), 303.
103. Akanwa, A. O., Mba, H. C., Ogbuene, E. B., Nwachukwu, M. U., Anukwonke, C. C. (2020). Potential of agroforestry and environmental greening for climate change minimization. In: *Climate Change and Agroforestry Systems* (pp. 47–86).
104. Aryal, J. P., Rahut, D. B., Sapkota, T. B., Khurana, R., Khatri-Chhetri, A. (2020). Climate change mitigation options among farmers in South Asia. *Environment, Development and Sustainability*, 22(4), 3267–3289.
105. Van Noordwijk, M., Ekadinata, A., Leimona, B., Catacutan, D., Martini, E., Tata, H. L., Öborn, I., Hairiah, K., Wangpakapattanawong, P., Mulia, R., Dewi, S., Rahayu, S., Zulkarnain, T. (2020). Agroforestry options for degraded landscapes in Southeast Asia. *Agroforestry for Degraded Landscapes: Recent Advances and Emerging Challenges*, 1, 307–347.
106. Huang, I. Y., James, K., Thamthanakoon, N., Pinitjitsamut, P., Rattanamanee, N., Pinitjitsamut, M., Yamklin, S., Lowenberg-DeBoer, J. (2023). Economic outcomes of rubber-based agroforestry systems: A systematic review and narrative synthesis. *Agroforestry Systems*, 97(3), 335–354.
107. Paukert, C., Olden, J. D., Lynch, A. J., Breshears, D. D., Christopher Chambers, R., Chu, C., Daly, M., Dibble, K. L., Falke, J., Issak, D., Jacobson, P., Jensen, O. P., Munroe, D. (2021). Climate change effects on North American fish and fisheries to inform adaptation strategies. *Fisheries*, 46(9), 449–464.
108. Bosi, C., Pezzopane, J. R. M., Sentelhas, P. C. (2020). Silvopastoral system with Eucalyptus as a strategy for mitigating the effects of climate change on Brazilian pasturelands. *Anais da Academia Brasileira de Ciências*, 92, e20180425.
109. Raj, A., Jhariya, M. K., Banerjee, A., Yadav, D. K. (2019). Agroforestry for climate mitigation and livelihood security in India. In: *Agroforestry and Climate Change* (pp. 189–207). Apple Academic Press.
110. Rijal, S. (2019). Agroforestry system: Approaches for climate change mitigation and adaptation. *Big Data In Agriculture (BDA)*, 1(2), 23–25.
111. Chavan, S. B., Dhillon, R. S., Ajit, Rizvi, R. H., Sirohi, C., Handa, A. K., Bharadwaj, K. K., Johar, V., Kumar, T., Singh, P., Daneva, V., Kumari, S. (2022). Estimating biomass production and carbon sequestration of poplar-based agroforestry systems in India. In: *Environment, Development and Sustainability* (pp. 1–29).
112. Raj, A., Jhariya, M. K., Banerjee, A., Meena, R. S., Nema, S., Khan, N., Yadav, S. K., Pradhan, G. (2022). Agroforestry a model for ecological sustainability. In: *Natural Resources Conservation and Advances for Sustainability* (pp. 289–307). Elsevier.

113. United Nations. (2017). The sustainable development goals report, 2017. https://unstats.un.org/sdgs/files/report/2017/thesustainabledevelopmentgoalsreport2017.pdf.
114. Mukhlis, I., Rizaludin, M. S., Hidayah, I. (2022). Understanding socio-economic and environmental impacts of agroforestry on rural communities. *Forests*, 13(4), 556.
115. Garrity, D. (2012). *Agroforestry and the Future of Global Land Use* (pp. 21–27). Springer: Netherlands.
116. Anantha, K. H., Garg, K. K., Barron, J., Dixit, S., Venkataradha, A., Singh, R., Whitbread, A. M. (2021). Impact of best management practices on sustainable crop production and climate resilience in smallholder farming systems of South Asia. *Agricultural Systems*, 194, 103276.
117. Nair, P. R., Kumar, B. M., Nair, V. D., Nair, P. R., Kumar, B. M., Nair, V. D. (2021). Agroforestry for biodiversity conservation. In: *An Introduction to Agroforestry: Four Decades of Scientific Developments* (pp. 539–562).
118. Kuyah, S., Sileshi, G. W., Luedeling, E., Akinnifesi, F. K., Whitney, C. W., Bayala, J., Kuntashula, E., Dimobe, K., Mafongoya, P. L. (2020). Potential of agroforestry to enhance livelihood security in Africa. *Agroforestry for Degraded Landscapes: Recent Advances and Emerging Challenges,* 1, 135–167.
119. Shrestha, J., Subedi, S., Timsina, K. P., Subedi, S., Pandey, M., Shrestha, A., Shrestha S., Hossain, M. A. (2021). Sustainable intensification in agriculture: An approach for making agriculture greener and productive. *Journal of Nepal Agricultural Research Council*, 7, 133–150.
120. Zhu, X., Liu, W., Chen, J., Bruijnzeel, L. A., Mao, Z., Yang, X., Cardinael, R., Meng, F.-R., Sidle, R. C., Seitz, S., Nair, V. D., Nanko, K., Zou, X., Chen, C., Jiang, X. J. (2020). Reductions in water, soil and nutrient losses and pesticide pollution in agroforestry practices: A review of evidence and processes. *Plant and Soil*, 453, 45–86.
121. Smith, L. G., Westaway, S., Mullender, S., Ghaley, B. B., Xu, Y., Lehmann, L. M., Pisanelli, A., Russo, G., Borek, R., Wawer, R., Borzęcka, M., Sandor, M., Gliga, A., Smith, J. (2022). Assessing the multidimensional elements of sustainability in European agroforestry systems. *Agricultural Systems*, 197, 103357.
122. Pandey, P. C., Pandey, M. (2023). Highlighting the role of agriculture and geospatial technology in food security and sustainable development goals. *Sustainable Development*, 31(5), 3175–3195.
123. Plieninger, T., Muñoz-Rojas, J., Buck, L. E., Scherr, S. J. (2020). Agroforestry for sustainable landscape management. *Sustainability Science*, 15, 1255–1266.
124. Lacerda, A. E. B., Hanisch, A. L., Nimmo, E. R. (2020). Leveraging traditional agroforestry practices to support sustainable and agrobiodiverse landscapes in Southern Brazil. *Land*, 9(6), 176.
125. van Noordwijk, M. (2021). Agroforestry-based ecosystem services: Reconciling values of humans and nature in sustainable development. *Land*, 10(7), 699.

CHAPTER 11

The Ecological Basis of Multitrophic Relationships

SATYABRATA SARANGI,* SUMAN SAMILITA DASH, and P BHAVANA

Department of Entomology, Odisha University of Agriculture and Technology, Bhubaneswar, Odisha, India

*Corresponding author

ABSTRACT

In agroforestry systems, the interaction of multiple trophic levels underpins ecological sustainability and productivity, making it essential to understand the ecological basis of multitrophic interactions. At the foundation of this web are primary producers—trees, shrubs, and crops—that establish energy and nutrient pathways. Phytophagous insects link autotrophic and heterotrophic levels by feeding on plants, while predatory insects regulate herbivores, providing natural biological control, a key principle of integrated pest management. Higher predators, such as birds and mammals, exert top-down control by feeding on herbivores and their predators, further balancing populations. This tri-trophic perspective is central to comprehending herbivore niches and adaptive evolution, as evidenced in systems like pea aphids, whose host specificity is influenced by predator avoidance, and the Blepharoneura fruit fly, which occupies highly specialized niches shaped by host plants and parasitoid interactions. Recognizing and managing these complex interactions reduces pesticide reliance, enhances pest control, improves soil health, and boosts productivity, while promoting biodiversity and ecosystem resilience. Thus, incorporating ecological and trophic dynamics into agroforestry design is vital for developing sustainable

Agroforestry and Climate Change Mitigation: Harnessing Multitrophic Interactions for Sustainable Plant Protection and Ecosystem Resilience. Hanuman Singh Jatav, Vishnu D. Rajput, Ipsita Samal, Tanmaya Kumar Bhoi (Eds.)

DOI: 10.1201/9781779641397-11

systems that mitigate risks, optimize natural processes, and foster multifunctional agricultural landscapes.

11.1 INTRODUCTION

For centuries, the natural world has captivated the imagination of naturalists, revealing an intricate tapestry of life woven within ecosystems. Moving beyond a narrow focus on individual species, the true essence lies in the dynamic interplay between trophic levels—a sophisticated ballet where herbivores waltz with plants, predators engage in pursuits of their prey, and decomposers set the bassline of decay. Embarking on an intellectual journey, this chapter endeavors to unravel the fascinating realm of multitrophic interactions, seeking to illuminate the ecological foundations orchestrating this grand symphony of life.

Traditional ecological investigations often adopted a myopic single-trophic level perspective, relegating herbivores to mere consumers, predators to top-down regulators, and decomposers to passive entities [62]. However, this reductionist lens fails to capture the intricate web of dependencies and dynamic interplay that defines life within an ecosystem [80]. Recent decades have witnessed a paradigm shift, with ecologists recognizing the critical role of multitrophic interactions in shaping communities, ecosystem functions, and the very fabric of biodiversity [131, 163].

At the heart of multitrophic interactions lies the dynamic interplay of top-down and bottom-up forces. Predators, perched at the apex of the food web, can exert cascading effects on lower trophic levels by suppressing herbivore populations, indirectly influencing plant growth and community composition [135]. Conversely, bottom-up forces, shaped by plant defense strategies, resource availability, and nutrient cycling, set the stage for herbivore abundance and predator performance [81]. This intricate dance between top-down and bottom-up forces creates a dynamic mosaic of community assemblages and ecosystem processes [133].

Moving beyond direct predator–prey and plant–herbivore interactions, the ecological dance floor pulsates with a myriad of indirect interactions. Plant traits, such as secondary metabolites, can attract natural enemies of herbivores, providing a subtle shield against herbivory through tritrophic interactions (TTIs) [174]. Moreover, the spatial and temporal heterogeneity of resources, driven by both abiotic and biotic factors, influences the connectivity and strength of multitrophic interactions, resulting in a patchwork of ecological outcomes across landscapes [30].

The unraveling of the ecological basis of multitrophic interactions is not merely an academic pursuit. It underpins the very foundation of sustainable management practices in agriculture, conservation, and pest control. By elucidating the intricate web of dependencies and indirect effects within food webs, we can design holistic management strategies that consider the cascading consequences of our actions across trophic levels.

This chapter aims to plunge into the multifaceted world of multitrophic interactions, unveiling the ecological principles that govern this complex choreography of life. Ultimately, this chapter seeks to illuminate the ecological symphony playing out within every ecosystem, urging us to listen closely and appreciate the profound interdependence that defines the tapestry of life on Earth.

11.1.1 WHAT IS THE MULTITROPHIC RELATIONSHIP?

Instead of linear chains, nature thrives on tangled webs. Multitrophic relationships weave the intricate among plants, herbivores, their predators, and beyond, across different "rungs" of the food chain. This intricate dance goes beyond simple predator–prey chases, revealing a symphony of connections and dependencies.

Plants, the primary producers, form the foundation, weaving energy into the fabric of life. Herbivores, like hungry threads, graze on this energy, while predators, the eagles of the ecosystem, keep the herbivores in check. But the tapestry is not a simple two-dimensional weave. It twists and turns, revealing intricate connections that extend beyond these basic interactions. Some scientists were among the first to unravel this intricate web, highlighting how the structure of an entire community, not just individual predator–prey pairs, influences population control [62]. Some further emphasized the importance of considering interactions across multiple trophic levels, recognizing the gradients of energy flow that shape exploitation within ecosystems [132].

The tapestry's strength lies in its dynamic interplay of top-down and bottom-up forces. Predators at the pinnacle, like fierce knots, can tighten their grip on herbivores below, influencing plant growth and shaping the entire community composition [135]. Conversely, the threads of plant defense, resource availability, and nutrient cycling, woven from the bottom-up, can influence herbivore abundance and even predator performance [81].

But the tapestry's beauty lies in its hidden threads, the indirect interactions. Plants, for example, can emit defensive cues that attract herbivores' natural enemies, creating a triangle of influence known as a TTI [174]. The ever-shifting patterns of resources, influenced by both the environment and the organisms themselves, further add to the complexity, creating a mosaic of ecological outcomes across landscapes [30]. Understanding multitrophic relationships is like appreciating the intricate beauty of this tapestry. It is about recognizing not just the obvious connections but also the subtle threads that weave life together, creating a vibrant and dynamic ecosystem.

11.1.1.1 PRIMARY TROPHIC LEVEL

1. **Plants:** Plants play a central role in multitrophic relationships by serving as the primary producers. They release volatile organic compounds (VOCs) and secondary metabolites in response to herbivore damage, influencing the behavior of herbivores and their natural enemies.

2. **Herbivorous insects**: Herbivorous insects feed on plants, triggering various plant defenses. The induced defenses, such as releasing plant volatiles, can attract natural enemies and indirectly affect herbivore populations.

11.1.1.2 SECONDARY TROPHIC LEVEL

1. **Predators:** Predators in agricultural ecosystems include insects like ladybugs, spiders, and predatory beetles. They play a vital role in managing herbivore populations. For instance, the predatory beetle *Hippodamia convergens* has been extensively studied for its predation on aphids [75].

2. **Parasitoids:** Parasitoids are insects that oviposit their eggs on or in the bodies of other insects, eventually killing their hosts. An example is the parasitoid wasp *Trichogramma* spp., which parasitizes the eggs of various agricultural pests [183].

11.1.1.3 CHEMICAL ECOLOGY AND COMMUNICATION

1. **Volatile organic compounds (VOCs):** Plants release VOCs in response to herbivore damage, acting as chemical signals that attract natural enemies. The phenomenon, known as tritrophic

communication, enhances the efficiency of natural enemies in locating their prey [44].

2. **Induced plant defenses:** Herbivore-induced plant defenses directly affect herbivores, coupled with indirectly influencing higher trophic levels. For example, the induction of extrafloral nectaries in plants can attract ants, providing indirect defense against herbivores [69].
3. **Implications for pest management:** Understanding multitrophic relationships is vital for developing integrated pest management (IPM) strategies. Conservation biological control, which involves enhancing the abundance and effectiveness of natural enemies, is an important aspect of sustainable agriculture (Table 11.1) [108].

TABLE 11.1 Some Key Interactions

Trophic Level	Organism	Chemical Signal	Effect on Next Trophic Level	References
Producer	Crop plant (e.g., maize)	A. Volatile terpenoids: Attract natural enemies of herbivores (e.g., ladybugs) B. Glucosinolates: Repel herbivores (e.g., cabbage moth)	Increased predation/ parasitism on herbivores, reduced herbivory	[43, 79]
Herbivore	Aphid	A. Honeydew: Attracts ants, which protect aphids from predators	Increased aphid survival and reproduction	[196]
	Colorado potato beetle	B. 2-tridecanone: Inhibits egg-laying by other potato beetles	Reduced population density of competing herbivores	[123]
Natural enemy	Ladybugs	A. Herbivore-induced plant volatiles: Attract ladybugs to herbivore-feeding sites	Increased predation on herbivores, reduced herbivory	[187]
	Trichogramma wasps	B. Egg kairomones: Attract wasps to lay eggs in herbivore eggs	Parasitism and reduction of herbivore populations	[94]

11.1.1.4 MULTITROPHIC INTERACTIONS IN AGRICULTURAL CROPS VIA CHEMICAL ECOLOGY

Chemical ecology plays a crucial role in mediating complex interactions between plants, herbivores, and natural enemies in agricultural ecosystems (Table 11.2). Plants release a diverse array of VOCs that serve as chemical cues, attracting beneficial insects like predators and parasitoids while deterring herbivores. This intricate interplay shapes pest control and crop health.

TABLE 11.2 Some Common Multitrophic Interactions *via* Chemical Ecology

Chemical Cues	Trophic Level 2: Herbivore	Trophic Level 3: Natural Enemy	References
Green leaf volatiles (GLVs)	Western corn rootworm (*Diabrotica virgiferavirgifera*)	Tachinid fly (*Lydella thompsoni*)	[142, 151, 202]
Linalool, limonene	Soybean aphid (*Aphis glycines*)	Minute pirate bug (*Orius insidiosus*)	[63]
Volatile esters, terpenes	Codling moth (*Cydia pomonella*)	Trichogramma wasps (*Trichogramma* spp.)	[109, 192]
Gossypol, terpenes	Bollworm (*Helicoverpa zea*)	Big-eyed bug (*Geocoris pallens*)	[20]

11.1.1.4 MECHANISMS OF INTERACTION

- **Herbivore attraction:** Green leaf volatiles emitted by damaged plants attract natural enemies like parasitoids and predators, leading to increased pest control.
- **Herbivore deterrence:** Certain VOCs, like linalool and limonene, repel herbivores, reducing their feeding and oviposition.
- **Olfactory cues:** Natural enemies use plant volatiles to locate herbivore infestations, improving their foraging efficiency.

11.1.1.4 BENEFITS OF UNDERSTANDING MULTITROPHIC INTERACTIONS

- **Developing sustainable pest management strategies**: By leveraging chemical ecology, we can promote natural enemy populations and reduce reliance on chemical insecticides.
- **Breeding pest-resistant crops:** Cultivating varieties that emit specific VOCs can attract beneficial insects and deter herbivores.
- **Optimizing crop management practices:** Understanding how cultural practices and environmental factors influence plant VOC

emissions can help us manipulate multitrophic interactions for better pest control (Figure 11.1).

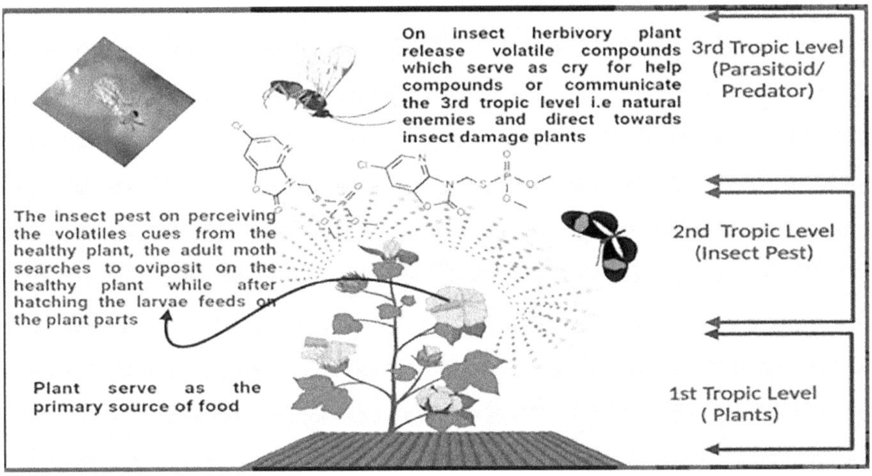

FIGURE 11.1 A schematic diagram depicting the multitrophic relationship and how chemical communication plays an intricate part in the ecosystem.

11.1.2 *ECOLOGY OF MULTITROPHIC INTERACTIONS*

The study of multitrophic interactions in agricultural entomology is a dynamic and interdisciplinary field that delves into the intricate relationships among plants, herbivores, predators, and parasitoids within agroecosystems. This review intends to provide a rundown overview of the ecology of multitrophic interactions in agricultural settings, shedding light on key concepts, influential studies, and their implications for sustainable pest management.

Trophic levels and interactions: Understanding multitrophic interactions involves examining the relationships between primary producers (plants), herbivores (insects), and their natural enemies, including predators and parasitoids. Pioneering work emphasized the significance of trophic interactions in ecological systems, laying the foundation for subsequent research in agricultural entomology [62].

Plant–insect interactions: Plants play a key role in multitrophic interactions, influencing both herbivores and their natural enemies. The phenomenon of induced plant defenses demonstrates how plants can respond to herbivore attacks by releasing VOCs, attracting predators and parasitoids to control herbivore populations [90].

Herbivore–natural enemy interactions: The dynamics between herbivores and their natural enemies are crucial for pest regulation. One classic study exemplifies the cascading effects of predatory behavior on herbivore populations and subsequently on plant fitness [144]. This work demonstrated that predators can exert top-down control, influencing the entire multitrophic system.

Conservation biological control: Conservation biological control centers on enhancing natural enemy populations to control pest species. Some researchers demonstrated the role of floral resources in promoting the survival and reproduction of natural enemies [190]. This approach emphasizes the importance of providing alternative food sources for predators and parasitoids, contributing to sustainable pest management strategies.

Application of molecular techniques: Advancements in molecular techniques have allowed researchers to explore multitrophic interactions at the genetic level. For instance, researchers investigated the signaling pathways involved in plant responses to herbivore attacks, providing insights into the molecular mechanisms underlying multitrophic interactions [96].

By understanding the intricate relationships between plants, herbivores, and their natural enemies, researchers can develop innovative strategies to enhance biological control and lessen the reliance on chemical pesticides.

11.1.3 THE CRUCIAL ROLE OF FOOD CHAINS AND FOOD WEBS IN MULTITROPHIC RELATIONSHIPS

Agricultural entomology explores the intricate interactions between insects, crops, and the environment. Among the key components shaping these relationships are food chains and food webs. This chapter delves into the pivotal role of food chains and food webs in multitrophic relationships within agricultural entomology, focusing on elucidating their significance in pest management and ecosystem stability (Figure 11.2).

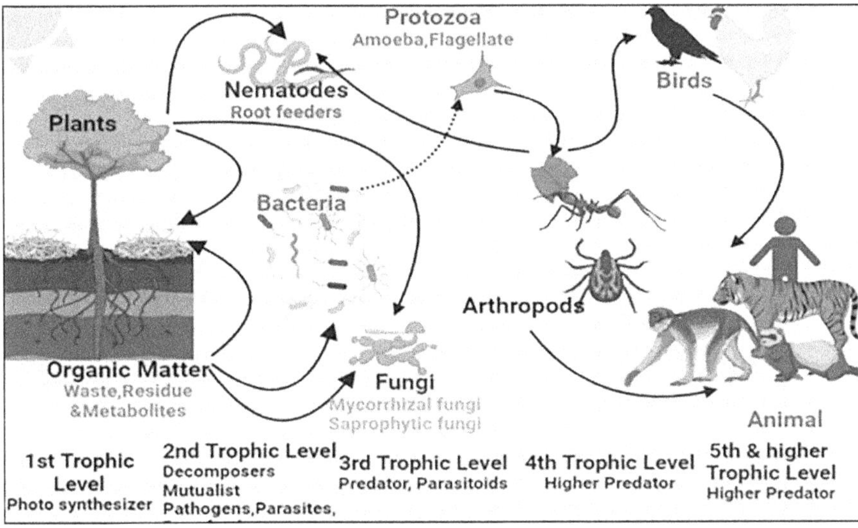

FIGURE 11.2 The intricate framework depicting the crucial role of food chains and food webs in multitrophic relationships.

Image courtesy: Soil and Water Conservation Society.

Conceptual framework of food chains and food webs: Grasping the structure and dynamics of food chains and food webs is fundamental to unraveling the multitrophic relationships in agricultural ecosystems. Within this framework, the interactions among herbivores, predators, and parasitoids, and their respective impacts on crop systems, are paramount.

Predator–prey dynamics and pest regulation: The presence of natural enemies, such as predators and parasitoids, plays a pivotal role in regulating pest populations. This is exemplified by the classic trophic cascade concept, where the impact of predators on herbivores indirectly affects plant productivity. Notable examples include the trophic cascades in aquatic ecosystems, demonstrating the far-reaching consequences of predator–prey interactions [130].

Indirect effects on crop health: Beyond direct predation, food chains and webs indirectly affect crop health. For instance, the presence of certain plant species can attract beneficial insects that function as natural enemies to pests, creating a diverse and resilient ecosystem. Some workers highlighted the significance of enhancing biodiversity to improve pest management in agroecosystems [6].

Case studies in agricultural entomology:

- **Ladybird beetles (Coccinellidae):** Coccinellids are well-known predators of aphids, as well as mites. The multitrophic interactions involving ladybird beetles, emphasizing their impact on pest suppression and crop protection, were also elucidated [74].
- **Parasitoid wasps (Hymenoptera):** Parasitoids, such as *Trichogramma* spp., have been extensively studied for their capability to control lepidopteran pests. Some researchers provide insights into the multitrophic relationships involving parasitoid wasps and their hosts, showcasing the potential for biological control in agricultural settings [67].

Implications for sustainable agriculture: Recognizing the significance of food chains and webs in agricultural entomology has profound implications for sustainable agriculture. IPM strategies that harness natural enemies and promote biodiversity contribute to reduced reliance on chemical pesticides, fostering ecologically sound farming practices.

By understanding and manipulating these ecological interactions, researchers and farmers can develop effective strategies for pest management and promote sustainable agriculture.

11.2 DIFFERENT TROPHIC-LEVEL INTERACTIONS

As we know, in an agroforestry (agriculture + forestry) ecosystem, several trophic levels operate in the environment, consisting of lower, middle, and higher trophic levels to maintain a sustainable equilibrium, to maintain constant productivity as well as plant protection, in the surrounding agroforestry ecosystem, and those elements are (Figure 11.3):

1. Primary trophic level/lower trophic level—Producers/plants
2. Secondary trophic level/middle trophic level—Herbivores/insect pests/nematodes/plant pathogens/pollinators
3. Tertiary trophic level/upper trophic level—Predators and parasitoids (natural enemies)
4. Quaternary trophic levels—Top predators/secondary predators

We have studied from much previous research that the agroforestry systems generally have a simpler structure than conventional forests and have a homogeneous species mixture. For these reasons, they need to

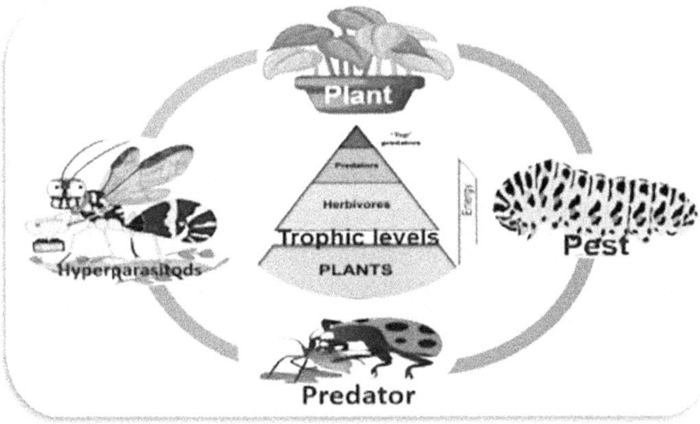

FIGURE 11.3 Pictorial illustration of multitrophic interactions in an agroforestry ecosystem.

support the impacts and processes of various impoverished types of species [5, 65] that provide ecosystem services, like species richness and species inter-relationships, in the agroforestry systems. Improving the maintenance of trophic interactions in plants by increasing the complexity and richness of the species they host [148], will benefit revenue production in plantation crops. Food webs in high-diversity environments connected by interconnected networks are expected to exhibit weaker trophic cascades compared to environments with low diversity in a diverse community and a greater number of species.

The producers (trees, shrubs, and herbs) are autotrophic as they produce the green matter, assisted by pollinators using pollination and photosynthesis as well. In contrast, insect pests are the detrimental elements of the crop ecosystem, devastatingly devouring the crop. On the other hand, the predators and parasitoids feed on the pest-saving plant products worth millions. The hyper-parasites parasitize the crop's intimate friend, that is, the natural enemies, which sometimes creates natural havoc in crop productivity, pest resurgence, as well as secondary pest outbreaks (the initial minor pest now becomes major, due to the death of natural enemies) (Figure 11.3).

11.2.1 TYPE OF MULTITROPHIC INTERACTIONS

The multitrophic interaction consists of two types [76, 171].

1. **Biotic interactions:** Interactions among living entities (plants–pests/pollinators–natural enemies, etc.)

2. **Abiotic interactions:** Interactions among nonliving entities (plants–climatic factors–soil parameters, etc.)

Of all the trophic-level interactions, the most widely studied is the TTI, which links trophic chains of primary producers, herbivores, and predators. The top-down sequence of predatory attacks on herbivores indirectly helps the plant, thereby creating biodiversity and community dynamics [161]. Through constitutive or herbivore-induced chemical signals, plants mediate TTIs, drawing predators to areas where herbivory has occurred [182]. They can also do this by physical traits, including plant structures, domatia, and/or extrafloral nectaries, which serve as habitats for predators [68].

11.2.2 PLANT–HERBIVORE–ANT INTERACTIONS

Interactions between plants, herbivores, and ants are one of the best-studied symbiotic multitrophic relationships in agroforestry ecosystems. Plant and ant symbioses are usually mediated by nectaries or domatia and are omnipresent in both, tropical [120] and temperate [34] habitats. Given the role of ants as ecosystem engineers linked to biogeochemical cycles and underlying processes, it is possible to link ant-driven trophic patterns top-down through other predators and plants, which treat TTI in different environments, having a significant effect. The Herbivores like plant aphids (sapsuckers) suck liquid food from the phloem and they secrete the honeydew as a nectar gift for their protector (the ant), and in return, the ant acts as their protector from different kinds of predators as well as parasitoids, which was well studied in pulse and mustard ecosystem, where aphid spp. *Aphis fabae* and *Lipaphis erysimi*, respectively, are protected from ants against various hymenopteran parasitoids and coleopteran predators [128]. In many agroforestry ecosystems, vertebrate predators like toads, frogs, lizards, birds, and bats play a main role in neutralizing many invertebrate pests' species, boosting natural crop growth and development [78, 92].

11.2.3 DIFFERENT HYPOTHESES IN MULTITROPHIC INTERACTIONS

The arthropod population in agroecosystems is known to be influenced by a variety of factors, including cropping diversity and agroforestry management, in addition to the biotic interactions that can be quantified by laboratory research [101]. Furthermore, field pests can be controlled by natural enemies [100]. To explain animal behavior in both basic and diverse settings, two theories were put forth [154].

1. **The Enemy hypothesis:** This predicts that the number of pests will decrease as there are more enemies in different cultures due to the accessibility of other foods and suitable microhabitats.
2. **The hypothesis of resource concentration:** It predicts that herbivores are more likely to reach large populations in uni/monocultures due to the exaggeration of the hosts.

Example–By citing a few examples, I want to justify the roles of multitrophic interactions:

1. The expanding coffee agroforestry [169], where upland coffee *Coffea arabica* L. (Rubiaceae) is a significant commercial cash crop, is a widespread agroforestry ecosystem in the Americas. Among the most common pests of coffee crops are the coffee spider mite, *Oligonychus ilicis* McGregor (Acari: Tetranychidae), and the coffee leaf miner, *Leucoptera coffeella* Guérin-Méneville (Lepidoptera: Lyonetiidae). The main adversaries of miners and spider mites are the predatory mite *Amblyseius herbicolus* Chant (Acari: Phytoseiidae) and an unidentified hymenopteran parasitoid of the family Eulophidae. Although these predatory mites consume red spider mites in all stages (adults, nymphs, larvae, and eggs) (Figure 11.4), the parasitoid targets the miner larvae [158].
2. A basic trophic chain model was used to examine how a food web of arthropods and vertebrate predators can be understood through experimental studies carried out in coffee agroecosystems in southern Mexico [138]. A small number of arthropod herbivores, like the coffee berry borer (CBB) *Hypothenemus hampei*, are consumed by both vertebrate predators (frogs, toads, lizards, bats, and birds) and arthropods (beetles, lacewings, wasps, robber flies, ants, and spiders) (Figure 11.4). Consequently, if the ecosystem's body structure is referred to as a trophic chain because arthropods are more frequently eaten by birds and bats than by herbivores, then efforts to eradicate these vertebrate carnivores should increase arthropod carnivores. However, because they feed on the berry borer's natural enemies, such as ants, eulophid wasps, and beetles, spiders in this situation function as hyper-parasites (secondary predators) [141].

Thus, the multitrophic interactions are necessary for the natural balance of the sustainable ecosystem and crop productivity, by bringing a barrier to the insect pests through natural control and artificial manipulation of the environment, which are discussed later.

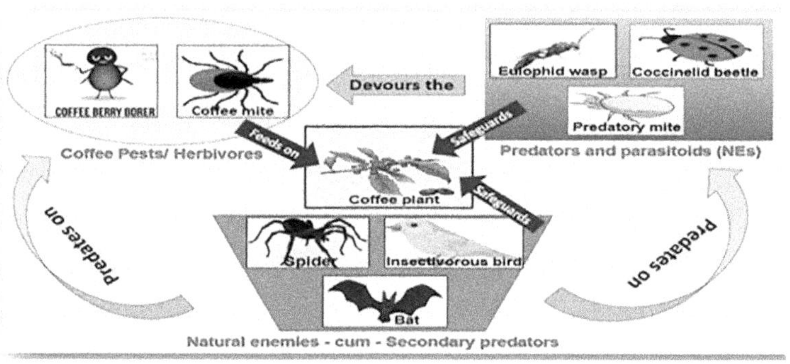

FIGURE 11.4 Multitrophic interactions in the coffee ecosystem among different trophic levels.

11.3 FACTORS AFFECTING MULTITROPHIC INTERACTIONS

The multitrophic interactions among different trophic levels are manipulated/affected by various factors like; (1) natural factors (birds, bats, ants, and pollinators, etc.), (2) climatic factors (temperature, rainfall, and humidity, etc.), and (3) anthropogenic factors (change in cropping pattern, change in agroforestry ecosystem, and other human interventions that are not naturally controlled).

11.3.1 NATURAL FACTORS AFFECTING MULTITROPHIC INTERACTIONS

11.3.1.1 BIRDS AND BATS

Bees, bumblebees, spiders, lizards, birds, and bats perform important ecosystem services such as dispersal of seeds, pollination, fruit set, pest management, etc., [106] (Figure 11.5). Especially, pest control with the help of bats, birds, and, in tropical agroforests, contributes to increased crop yields [93]. Ecologically sound management of tropical agricultural lands is a hopeful strategy for the conservation of NEs to improve human health by promoting food security and ecosystem resilience. A variety of arthropods are preyed upon by insectivorous birds and bats, which involves herbivorous pests [106, 198] as well as carnivorous arthropods such as ants and spiders [56] (Figure 11.5). Thus, while birds and bats often directly increase crop yields by killing insect herbivores, they can sometimes reduce the yield of the crops by acting as intra-guild predators by preying on both intermediate predators and herbivores [141].

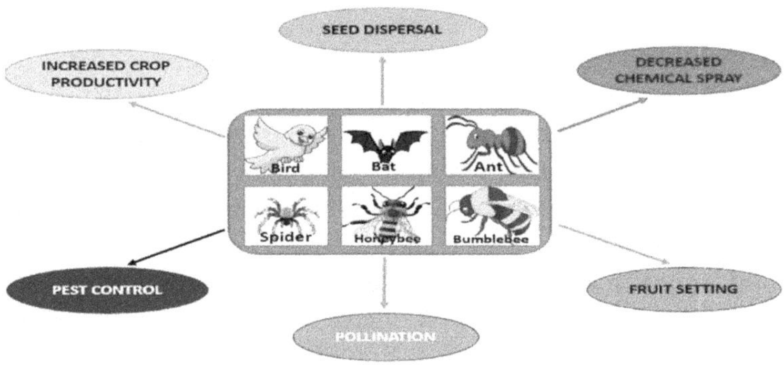

FIGURE 11.5 Schematic illustration of natural factors affecting various multitrophic interactions in agroforestry ecosystem.

The effects of predation by birds and bats on arthropod communities can be separated by placing fences during the day (only for birds), at night (only for bats and nighthawks), or during the day to assess the relationship between birds and bats (Figure 11.5). The richness and diversity of partly-natural habitats, especially trees, in agricultural areas, can make bird and bat populations rich and diverse [66], thus affecting crop pests for natural acclimatization (such as nests and breeding grounds, altered prey). However, birds and bats can negatively influence the control of pests by exploiting their natural enemies, namely intra-guild predation [181] (Figure 11.5).

Predators affect tree communities in different temperate climatic zones by preying on herbivores, indirectly affecting the composition of the plant community, age structure, diversification, crop productivity, and also nutrient cycling [114]. This trophic cascade reduces the negative impact on plants by decreasing the abundance of herbivores. In Southeast Asia, fruit and berry bats are considered pollinators of stink beans (*Parkia* spp.), durians (*Durio* spp.), and Indian trumpet

(*Oroxylum indicum*), a pivotal plant in the agroforestry industry. Bat pollination accounts for 80–100% of fruit set in crops [27]. They also contribute to seed germination [114].

11.3.1.1 ANTS

They are effective predators of various insect pests in various major plantation crop agroforestry ecosystems. They also act as symbiotic organisms with many agricultural pests (aphids, jassids, mealybugs, and other sucking pests, which provide sweet honeydew as a gift in return for protecting them from various natural enemies) in different agroecosystems.

Tropical ants are carnivores, detritivores, and generalist herbivores [77]. In many studies, in both natural and agricultural ecosystems, ants contribute by limiting the herbivore community and promoting plant growth and development, exhibiting top-down effects [162]. *Oecophylla longinoda* controlled the coffee antestia bugs on the West African side [111]. *Solenopsis geminata* Fabricius is an important bio-control agent of sun coffee plants in some areas. It ripens in Costa Rica but does not kill the CBB (*H. hampei* Ferrari) [186]. Numerous ant species manage fungal infections or pests related to cocoa. For instance, in Southeast Asia, the black cocoa ant (*Dolichoderus thoracicus* F. Smith) suppresses attacks on cocoa beans by black cocoa ants, cocoa bean borers, and mirid bugs. Mirid bugs and pod weevils are controlled by weaver ants (*Oecophylla smaragdina* Fabricius) [170].

11.3.2 CLIMATIC FACTORS

The impacts of climatic change on crop pests include the following:

1. Diversity and abundance
2. Geographical speciation
3. Increased in overwintering/aestivation
4. Rapid fecundity and volatility
5. Synchrony between herbivores and plants
6. Introduction of alternative autotrophs
7. Host plant resistance
8. Biotypes of herbivores
9. Tritrophic relationships
10. Changes in activity and relative abundance of natural enemies
11. Danger of invasion of new pests

11.3.2.1 SEASONAL VARIATION

Carnivores and pollinators can obtain a variety of resources from the diversity of agroforestry products, including food and shelter. The range and diversity of insects can be impacted by the agroforestry system's interactions with microclimatic elements such as temperature, relative humidity, light intensity, precipitation, wind, carbon dioxide, and water vapor [47]. Seasonal changes in weather conditions also play an important role in the reproduction, growth, development, and distribution of insects and affect their population parameters [41]. In some agricultural areas of Mizoram, the abundance of pests was greater and lesser in summer and winter, respectively [137]. Plants in agroforestry systems serve as safe havens for predators and pollinators. Combining trees with crops in agroforestry systems provides an environment for the natural growth and development of pests, animals, and pollinating species.

11.3.2.2 AGROFORESTRY IMPACT

The prediction of climate change's effects on forestry and agriculture will aid in determining how environmentally friendly the current systems are. In perennial environments, trees' shade effect can help keep pests at bay. It was found that both the number of herbivores and the proportion of herbivores in cocoa pods decreased when the tree canopy increased, which was found in a study [21]. Similarly, the CBB, *H. hampei* (Ferrari), was more abundant in coffee fruits in low shade (high temperature) compared to high shade (low temperature) systems [87]. Natural enemies of CBB, such as birds and insects, can survive better in the shade [93].

11.3.2.3 ABIOTIC STRESS PARAMETERS

Almost all insects are sensitive to some changes in temperature, which can have various effects on an insect's life history [10]. Species whose niche is determined by weather conditions may react more sensitively to climate change in temperate regions, where most insects grow during the warm part of the year. On the other hand, species whose niche is determined by climate conditions will respond more predictably to climate change in areas where abiotic or biotic limiting factors may be unpredictable [10].

Laboratory and experimental modeling supports the idea that pest biology may respond to higher temperatures [64].

Diapause is a period in which growth activity is stopped or delayed. It is a period, and its symptoms are controlled by environmental factors such as temperature, humidity, and photoperiod. In global warming, an increase in temperature between 1°C and 5°C will ensure that insects survive more in the winter with fewer fatalities, the formation of large populations, early disease occurrence, and crop damage from pests, influenced by global warming [64, 166].

11.3.3 ANTHROPOGENIC FACTORS

11.3.3.1 AGROFORESTRY INTENSIFICATION

Intensification of the agroforestry ecosystem [51] is essentially nothing but the "integration of trees into agriculture practices" [53], which is meant to improve the results of flower pollination and reduce honey bee loss. Agroforestry, which is a "win-win" in terms of production and environmental protection considered a "win" opportunity [176]. These are occasionally referred to as the Northern area. These include several agrosilviculture systems in the Himalayas for natural insecticides and pollination, as well as alley-cropping and agroforestry intercropping systems in North America (typically employing hardwoods for fruit and tree production) [129]. It is often necessary to utilize different machinery and different trees or crops in temperate climates, and it is unclear how silvo-forest systems affect pest management and pollination programs in these areas [146].

Predator luring can be increased through agroforestry by increasing microhabitat diversity and changing the microclimate. The uniqueness of agroforestry systems makes them a subject of particular interest for conservation biological control (CBC; 146). However, some studies on climatic agroforestry apple orchards have revealed that gradually enhancing plant diversity has marginal effects on arthropods and CBC [108], probably due to the inadequate tritrophic relationships.

Canopy management in agroforestry systems is a regulatory method to control microclimate for enhancing crop growth [86] and can influence pests and prey arthropods [50]. The diurnal and seasonal patterns of ground predators of the family Carabidae (e.g., tiger beetle) differ between open and closed habitats (e.g., forest and grassland) [105]. Since trees grown in agroforestry systems can adapt to microclimates and are not protected from extreme weather conditions, it is thought that agriculture will affect their daily and seasonal activities.

11.3.3.2 PLANTATION PATTERN AND CHANGE IN AGROFORESTRY PLANT DIVERSITY

1. **Sustainable tree incorporation:** Enhancing plant diversity and geographical complexity might hasten the efficacy of pests' natural enemies, such as predators and parasites, hence mitigating crop loss [14]. This conservative biological control works under this principle. Similar arguments have been made in favor of wild pollinators [201], which can increase the efficiency of crop pollination and reduce bee loss [51]. One solution to greater energy needs is agroforestry, which is essentially the "integration of trees into agriculture services" [53].
2. **Silvoagriculture:** Particularly interesting is silvoarable agroforestry, which intercrops arable crops with trees or shrubs for the benefits of pollination and pest control. The resources that the agroforestry agricultural system provides can be used to illustrate the advantages of natural enemies. For instance, intact trees can offer refugia, or places to hide over the winter, which is crucial for the survival of populations of terrestrial predators in other habitats [108]. Silvoarable farming systems can also develop nutrient-rich systems with proven benefits for parasitoids [29]. Additionally, it was shown that the number of bee nests can be increased by using silviculture systems in mountains in warm climates [54].
3. **Planting density:** The entomopathogenic fungus *Beauveria bassiana* (Bals-Criv.) Vuill., and the parasitic hymenopteran wasp *Cephalonomia stephanoideris* Betrem parasite managing the CBB *H. hampei* [13] has been seen in shady areas of coffee plantations. Many studies have shown that improvements in annual crops lead to increased microfauna abundance. More underground natural enemies (especially ants, carabid beetle larvae, and centipedes) have been recorded in maize-based agroforestry than in similar systems [167, 168].

These are the factors that indirectly as well as directly affect the multitrophic interactions in agroforestry ecosystems for sustainable plant protection at different trophic levels, which includes the plant–herbivores/pollinators-natural enemies.

11.4 INNOVATIVE TECHNIQUES USED IN AGROECOSYSTEMS USING MULTITROPHIC INTERACTION

11.4.1 APPLICATION OF MOLECULAR GUT CONTENT ANALYSIS (MGCA) IN AGRICULTURE

The predominant focus of recent MGCA studies has been on agronomic crops, with approximately half concentrating on cereal systems. A significant portion of these studies aimed to address a specific pest through PCR, while the rest utilized a combination of multiplex and sequencing methods. The more extensive investigations incorporated multiplex PCR alongside modeling techniques, enabling the anticipation of trophic structure and network parameters [22, 178], as well as the determination of pest host-plant diet breadth. Some studies also delved into the examination of how cropping systems influence predator–prey interactions [73, 88]. Within this section, we present a summary of the outcomes obtained through the application of MGCA to explore interactions between predators and pests within agronomic crop systems.

MGCA has been used in several recent research on agronomic crops to understand the involvement of generalist arthropod predators, which are prevalent in agroecosystems. As an illustration, *Orius insidiosus* (Hemiptera: Anthocoridae; Say), a generalist predator found in maize, proved effective in controlling *Helicoverpa zea* (Lepidoptera: Noctuidae; Boddie) as a biological control agent [139]. It is noteworthy that conservation efforts have been demonstrated to increase *H. zea* predation levels. Researchers discovered that foliar predators preyed more heavily on cereal leaf beetles than species that live on the ground. This suggests that conservation measures should be made to increase the number of foliar predators [98]. Several arthropod predators from nearby cotton-soybean fields tested positive for stink bug DNA, according to PCR-based gut content analysis used to investigate stink bug predation [178]. Although it is probably a predator of many species in soybeans, the common predator *Oxyopes salticus* (Araneae: Oxyopidae; Hentz) showed only a modest degree of predation on brown marmorated stink bugs, *Hyalomorpha halys*

(Hemiptera: Pentatomidae; Stål) [7]. On the other hand, in certain systems, such as cereal aphids in wheat, diversified populations led to higher levels of predation on a pest, and generalist predators tested positive for pests at high rates [33]. The idea that habitat shape in agricultural settings might change generalist predators' diets was validated more recently by diagnostic multiplex PCR research on food web structure [155, 156, 172].

It is essential to provide biodiversity for ecosystem services, and MGCA has helped to understand prey selection and whether generalist predators provide redundant or complementary predation services [155]. Numerous pest species exhibit a high degree of polyphagy, engaging in feeding activities across a diverse range of host plants. An examination of the digestive contents of the migratory *Apolygus lucorum* (Hemiptera: Miridae; Meyer-Dür) gathered from crop-free islands indicated that this herbivore consumes plant material from 17 different families, with a notable emphasis on cotton [195]. Nevertheless, there is evidence that *A. lucorum* exhibits a preference for mung beans over cotton [194], hinting at the potential of mung beans as a suitable trap crop within cotton systems. The promotion of crop diversity has been identified as a factor that augments arthropod populations and contributes to pest control [88]. A recent study utilizing MGCA revealed reduced feeding on crops in intercropping scenarios compared to monoculture systems [88].

Enzyme-linked immunosorbent assays (ELISA) represent a powerful tool in the study of trophic resource movement. In this method, prey is labeled with specific proteins, such as vertebrate egg protein, by [60], and the focal arthropod's gut is subsequently examined for the presence of these protein-marked food items using a standardized protein-specific ELISA. For instance, evidence of the linkage between buckwheat trophic interactions and grape production was revealed through the dispersal of natural enemies from buckwheat to grapes [82]. It is employed to assess the impact of trap crops and identify predation at specific pest life stages [60]. The utility of ELISA extends to tracking both predators and prey, demonstrating its cost-effectiveness and value as a tool in this context [59].

11.4.2 APPLICATION OF MOLECULAR GUT CONTENT ANALYSIS (MGCA) IN HORTICULTURE

Pests and the illnesses they spread can provide challenges for vegetable systems, but a new study indicates that trophic relationships may be better understood and that sustainability may be improved. Both *Frankliniella occidentalis* (Thysanoptera: Thripidae; Pergande) and *Nasonovia ribisnigri* (Hemiptera: Aphididae; Mosley) are common pests in Mediterranean lettuce and provide food for several local *Orius spp.* [52]. Examining these pests by diagnostic PCR together with a common substitute prey, Collembola, showed that predators were heavily preying on *N. ribisnigri* and *F. occidentalis*. Predators must have other prey for biological control to be effective, yet there are times when more resources can cause predators to stray from their intended prey. Multiple generalized canonical analysis, or MGCA, revealed that two predators in *Brassica oleracea* farms travel between non-crop vegetation and cropping fields: *Hippodamia variegate* (Coleoptera: Coccinellidae; Goeze) and *Micromus tasmaniae* (Neuroptera: Hemerobiidae; Walker). This implies that different resources might help with some pests' biological management [70].

11.4.3 INTRAGUILD PREDATION (IGP)

The evaluation of intricate IGP occurrences is essential to the design of biocontrol programs. For example, MGCA showed that in comparison to lacewing species (Neuroptera: Chrysopidae), there was a greater chance than previously thought of the invasive harlequin ladybeetle, *Harmonia axyridis*, preying on native ladybeetles [25]. Interestingly, no *Chrysoperla* individuals were found in the gut contents of *H. axyrids* that were collected in the wild. Therefore, it makes sense to conclude that *Chrysoperla* species may be less detrimental to native predators than the invasive *H. axyridis* in terms of biological control. On the other hand, in some systems, there was very little IGP between coccinellid species that were comparable [153]. In pear orchards, winter-active spider species participate in IGP events, although the frequency of these occurrences was found to be modest in comparison to the pests (pear psyllids) they eat.

Thus, in this instance, IGP's beneficial effects were outweighed by its ability to effectively suppress a specific pest through biological means [140]. The natural enemy hypothesis—which holds that a wider variety of natural enemies promotes more efficient biological control is supported by the findings of recent landscape research, which indicated that IGP was low and had no influence on aphid predation [134]. By analyzing how habitat management affects predation on other beneficial insects, MGCA can also determine the balance between benefits and drawbacks. For example, ambush predators associated with flowers consume a wide range of insects, such as bees and other naturally occurring adversaries [119]. Thus, depending on the agroecosystem and local community needs, the use of flowering habitats, typically associated with providing resources to beneficial arthropods, may be unfavorable (Figure 11.6).

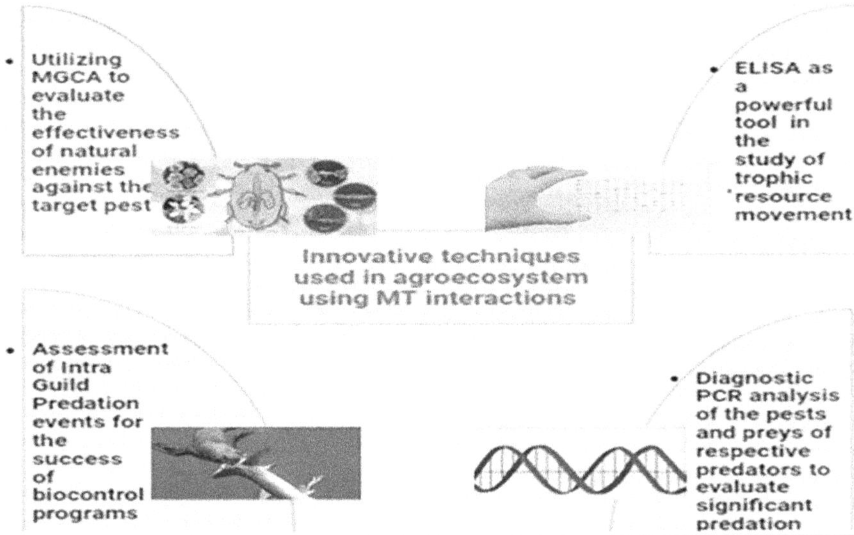

FIGURE 11.6 Innovative techniques used in agroecosystems using multitrophic interaction.

11.4.4 ROLE OF VERTEBRATES IN MULTITROPHIC INTERACTION

The analysis of gut content at the molecular level in vertebrates, especially insect-eating birds and bats, presents encouraging findings indicating that these groups could offer significant ecosystem services in various agroecosystems, as outlined in the review by [158]. Using the MGCA method, eight research on bats and four studies on birds were conducted in agricultural settings between 2015 and 2020. DNA analysis is typically performed on fecal samples or pellets obtained from either passively collecting fecal pellets from bat roosting places or catching individual birds or bats within the target crop system in vertebrate systems. study of vertebrate pest consumption ranges from targeted sampling of a single predator [4, 11] or numerous predator species or populations, to specialized research on specific prey species [118, 179], DNA metabarcoding study capturing the breadth of diet [38, 50, 95], and more.

Although further research is required, some general patterns of vertebrate pest ingestion have emerged, indicating both beneficial and negative effects as well as density-dependent foraging by birds and bats. Certain bat species exhibit seasonal or prey-abundance-dependent dietary adjustments, indicating their capacity to dynamically modify their foraging strategies and diets to take advantage of pest outbreaks [179]. For example, the chance of finding pink bollworm in the diet of Kuhl's pipistrelle (*Pipistrellus kuhlii*) increased as the pink bollworm (*Pectinophora gossypiella*; Lepidoptera: Gelechiidae; Saunders) grew in cotton [37]. In a similar vein, peak outbreaks in orchards correlated with bat activity and consumption of the pecan nut casebearer, *Acrobasis nuxvorella* (Lepidoptera: Pyralidae; Neunzig) [179]. In macadamia, insectivorous bats' ingestion of the four primary pest species was associated with pest outbreaks [197]. Lesser horseshoe bats in vineyards consume a variety of pest species and display seasonal diet variations [11]. Furthermore, it has been noted that bats eat pests that transmit disease in addition to herbivorous pests. Bats in Madagascar's rice fields were found to feed on economically significant pests and arthropods that carry rice-related diseases, according to fecal samples collected from six common bat species [95].

The common bent-wing bat's food habits are impacted by agricultural intensification in the landscape, according to a continent-wide landscape analysis conducted throughout southern Europe using DNA metabarcoding [4]. The kind of crop may determine how ecosystem services and disservices are distributed among birds. Exclusion tests and MGCA revealed that whereas prairie birds cause problems in soy crops, they contribute to net services in corn fields [50]. Although the diets of adult and nestling western bluebirds (Sialia mexicana) are different, both species offer benefits to vineyards in California [78]. In macadamia orchards, an analysis of 11 bird species revealed overlapping diet diversity, including the consumption of five major insect pest species [38]. A DNA-based approach detected minimal foraging on codling moths in organic apple orchards, with only one species, the brown-headed cowbird, found to consume codling moths [118].

11.5 ROLE OF POLLINATION IN MULTITROPHIC INTERACTION

11.5.1 FOOD SOURCES USED BY PLANT HERBIVORES

Besides plant tissue, herbivorous arthropods can also feed on a wide range of other plant-derived materials, including pollen, (extra) floral nectar, and (rotting) fruits; they can also feed on honeydew, dung, and bird droppings [84, 152]. Since plants directly influence plant–herbivore interactions and are important for a variety of adult herbivores, we have focused our research on plants as food sources. A sugar-rich plant secretion, nectar can also include proteins, lipids, amino acids, vitamins, secondary plant chemicals, and a variety of organic and mineral substances [9]. Pollen, in contrast, serves as a predominantly nitrogenous compound source, with protein levels ranging from 2.5% to 60% [157]. Additionally, pollen typically contains sterols, lipids, and carbohydrates (mainly starch), and its detailed chemical composition, nutritional value, and suitability as insect food have been extensively discussed elsewhere [157, 190].

Herbivores can be categorized into three groups based on the stages during which they feed on plant tissue and/or other food items:

- Species that, as larvae, limit their food to plant tissue; as adults, they switch to a diet of nectar or pollen. Several Hymenoptera (sawflies, *Tenthredinidae*) [15, 152] and Diptera (*Anthomyiidae*) are examples, along with the majority of Lepidoptera.
- Species that are primarily herbivorous as larvae but combine tissue eating with nectar/pollen feeding as adults. Coleoptera (e.g., *Diabrotica* spp.) [115] and Diptera (e.g., *Liriomyza* spp.) are two examples.
- Species combining tissue feeding with nectar/pollen feeding during both larval and adult stages. Examples include *Thysanoptera* [124], *Coleoptera* (e.g., pollen beetles of the genus *Meligethes*) [35], *Heteroptera* (e.g., *Lygus* spp.) [26, 39], and certain *Hymenoptera* (Tenthredinidae) [15].

Only a small number of species of adult *Lepidoptera* have been found to feed on pollen; most feed on sugar sources [152]. Many cyclorrhaphous flies belonging to the Diptera order feed on exposed fluids, including floral and extrafloral nectar, and can also eat pollen. Examples of these flies are *Tephritidae* (fruit flies), *Anthomyiidae* (root maggot flies), and Psilidae (rust flies). *Chrysomelidae* and *Bruchidae* are among the Coleoptera families that feed primarily on flowers [97, 159]. Direct nectar feeding is rare in flower-visiting beetles [103, 159].

Symphyta adult sawflies consume a variety of sugar-containing foods, such as nectar from flowers and other plants. According to reports, only a small number of species (such as those from the families Xyelidae,

Argidae, and Tenthredinidae) consume pollen [15, 84]. Heteroptera that feed on plants, including the Thyreocoridae, Lygaeidae, Rhopalidae, and Miridae families, visit flowers and appear to consume nectar [71, 73]. The tarnished plant insect, Lygus lineolaris, was discovered in much higher numbers in nectar-producing cotton than in nectariless cotton [164, 165], which provides indirect evidence that mirids use extrafloral nectar. Mirids have also been documented to feed on pollen [39]. Numerous herbivorous species of thrips (Thysanoptera) belonging to the families Thripidae and Phlaeothripidae consume nectar and pollen [99], including flower thrips of the genera Thrips, *Haplothrips*, and *Frankliniella*, as well as foliar-feeding thrips (e.g., *Thrips tabaci*) [124].

11.5.2 IMPLICATION OF POPULATION DYNAMICS OF ADULT HERBIVORES

Nectar and pollen ingestion affect the temporal dynamics of herbivore populations and their spatial distribution. At the population level, the impacts of adult feeding on traits specific to individual herbivores, such as fertility and lifespan, can be amplified over generations [191]. For example, the addition of pollen raises the rate of reproduction (rm) of *F. occidentalis* on cotton from 0.16 to 0.22 per day, leading to a significant increase in population. Similar to adding glucose to water, sugar feeding can similarly affect rm, as demonstrated by *Pieris brassicae* [191]. The potential impact of adult feeding on herbivore population growth highlights the significant influence of fluctuations in nectar or pollen availability on herbivore population dynamics.

Research comparing the herbivore pressure on cotton varieties that are nectariless and that produce extrafloral nectar shows that the latter have greater herbivore populations [1, 116, 165]. The dynamics of herbivore populations can also be influenced by pollen availability, as evidenced by the fluctuations in pear thrip populations and the harm done to sugar maple stands due to inconsistent spring flowering [110]. Many scientists show how important adult food sources are in determining the dynamics and composition of herbivore populations. While these interactions add complexity to multitrophic systems, it is essential to initially focus on the impact of nectar and pollen feeding in a bi-trophic (plant–herbivore) context before delving into more intricate multitrophic interactions involving pollen and nectar [191].

11.5.3 EFFECT OF ADULT FOOD ON OVIPOSITION PATTERNS

The spatial distribution of female herbivores' egg placement can be impacted by the proximity of adult food sources, just as the distribution of feeding sites affects the patterns of distribution of adult herbivores. When female herbivores graze near oviposition patches or choose hosts near adult food plants, they often save time and energy, lower the risk of predation, and streamline their search process [119]. Herbivore decision-making accuracy and speed may be limited by neural constraints, which could account for oviposition on less-than-ideal nectar plants [17, 83]. The choice of oviposition sites in various insect species, including *Lepidoptera, Diptera, Coleoptera*, and *Thysanoptera*, is influenced by the availability of adult food sources [2, 8, 24, 35, 55, 83, 90, 125, 126, 145]. The impact of floral food on oviposition decisions may depend on the spatial scale of interactions [83].

The choice of flowering plants may not just be driven by adult food needs but may also involve the greater visibility of flowering plants [48, 200]. Research indicates that adjusting the amount of nectar available, as opposed to eliminating flowers, can affect the likelihood of obtaining eggs, demonstrating how food resources impact oviposition habits [2, 136]. When female feeding is incorporated into the oviposition sequence, the effect of adult feeding on oviposition is especially noticeable [23]. Although herbivores are traditionally thought to choose oviposition sites based on larval development suitability, studies suggest that females may prioritize their survival, considering the proximity and quality of adult food sources [121]. In some cases, herbivores may choose flowering plants even if they are relatively unsuitable as larval hosts, indicating potential parent–offspring conflicts in host plant choice [32, 37, 49]. Dioecious plants with sexual dimorphism offer a unique system to study the effect of floral rewards on plant–herbivore interactions. Herbivory in dioecious plants is skewed toward plants that bear pollen, and many studies have failed to consider the potential that herbivores focus their efforts on staminate plants to maximize pollen exploitation [36].

In summary, the distribution of feeding sites and adult food sources significantly influences the spatial patterns of oviposition in herbivores. Oviposition decisions may involve a balance between optimizing offspring survival and maximizing female survival and fecundity, and this balance can be influenced by the availability and quality of adult food sources. Additionally, dioecious plants with sexual dimorphism provide a unique context for studying the impact of floral rewards on plant–herbivore interactions.

11.6 IMPLICATION IN BIOCONTROL

The text discusses the impact of floral resources on ecosystems involving herbivores, plants, and their natural enemies. Initially, the focus is on herbivores and plants relying on nectar or pollen, assuming the absence of a third trophic level. The presence of common enemies among herbivores may lead to indirect interactions, either negative or positive [28, 185]. Furthermore, herbivores' top-down control and the direct effects of floral resources can be altered by predators and parasitoids of herbivores [189, 190]. The relevance of flowering noncrop elements is highlighted by the fact that the survival and reproduction of natural enemies can be impacted by the lack of flowering plants in simpler agricultural systems [57]. Floral resources, while enhancing biological pest control, may also benefit herbivorous pests, as demonstrated in studies on nectar-feeding herbivores.

The text emphasizes that when examining the effects of biodiversity on herbivore levels, nectar and pollen feeding must be taken into account [2, 116, 164]. The intricate relationships found in agricultural systems are also covered, such as how flowering vegetation affects insect concentrations in cabbage crops [145, 175]. This chapter emphasizes how difficult it is to separate the role that nectar plays in these kinds of field investigations. There is an introduction of a study providing a theoretical framework to analyze how floral resources affect systems where both herbivores and their natural enemies feed on flowers. It emphasizes the Adult Food Fecundity Index as a crucial aspect affecting the balance between herbivores and natural enemies [199]. The relative impact of floral resources on their life histories varies, with natural enemies generally obtaining larger fecundity benefits. Finally, it acknowledges that not all flowers are equally suitable as food sources, discussing factors such as availability, apparency, accessibility, and nutritional suitability. Understanding these factors can help select food sources that benefit natural enemies without providing suitable food for herbivores.

11.7 PRODUCTIVITY INCREASES DUE TO MULTITROPHIC INTERACTION

Crop productivity is a critical factor in ensuring global food security, and sustainable agricultural practices are essential for achieving this goal. One promising avenue for enhancing crop productivity involves the manipulation and optimization of multitrophic relationships within agricultural ecosystems. This chapter explores the intricate interactions between plants, insect pests, and their natural enemies, shedding light on the potential benefits of increasing crop yields.

11.7.1 THE TRIANGULAR INTERACTIONS

1. **Plant–insect interactions:** Plants and insect pests engage in complex relationships, with the latter often posing threats to crop health. These interactions have been extensively studied by researchers to provide insights into the mechanisms of plant defense and pest adaptation [3, 91].
2. **Insect–pest–natural enemy interactions**: Natural enemies, including predators and parasitoids, play a crucial role in regulating insect pest populations. Some classic case studies have elucidated the dynamics of these interactions, emphasizing the potential for biological control in agriculture [58, 143].
3. **Plant–natural enemy interactions**: The relationship between plants and their natural enemies has also been investigated, revealing that plants can release VOCs to attract natural enemies [46]. This indirect defense mechanism enhances the effectiveness of biological control strategies.

11.7.2 MULTITROPHIC RELATIONSHIPS IN AGRICULTURE

Conservation biological control: The concept of conservation biological control involves fostering habitats that support natural enemies. Perfect examples of this approach are provided by some studies demonstrating the positive impact of diversified landscapes on pest suppression [108, 112].

Augmentative biological control: Augmentative biological control involves releasing natural enemies to enhance pest suppression. Some notable research showcases successful applications of this approach in various agricultural settings [127, 184].

11.7.3 ADVANTAGES OF MULTITROPHIC RELATIONSHIPS

Enhanced pest control: By leveraging the interactions between plants, insect pests, and their natural enemies, agricultural systems can benefit from a more sustainable and environmentally friendly approach to pest management.

Increased crop yields: The effective regulation of insect pest populations through natural enemies results in reduced crop damage, ultimately leading to increased yields. This has been demonstrated in studies [113, 180].

11.8 CONCLUSION

Studying multitrophic interactions among plants, insect pests, and their natural enemies in agroecosystems is crucial for understanding and managing ecological dynamics in agriculture. This intricate web of relationships has far-reaching implications for crop health, productivity, and sustainability. In conclusion, several key points emerge from the literature. First, the dynamic nature of multitrophic interactions underscores the complexity of agroecosystems. The interplay between plants, insect pests, and natural enemies involves a myriad of chemical, physical, and ecological factors. As highlighted by some researchers, the multifaceted nature of these interactions requires a holistic approach to pest management that considers not only individual species but also the broader ecosystem [85]. Second, the role of plant defenses in shaping multitrophic interactions cannot be overstated. Plants employ a variety of strategies, including the production of secondary metabolites and the induction of defensive proteins, to deter herbivores and attract natural enemies [45].

Some work emphasizes the importance of understanding plant-mediated interactions as a means to develop sustainable pest control strategies that harness the natural defenses of crops [90]. Furthermore, the impact of anthropogenic factors, such as pesticide use and habitat modification, cannot be ignored. In their study, they demonstrated the influence of landscape structure on the abundance and effectiveness of natural enemies [177].

Human-induced changes in the environment can disrupt these delicate interactions, potentially leading to increased pest outbreaks and reduced biodiversity. The concept of IPM involves the coordinated use of biological, cultural, and chemical control methods to minimize the impact of pests while promoting ecosystem health [102]. By incorporating natural enemies and enhancing plant defenses, IPM strategies aim to create a balanced and resilient agroecosystem [203–220]. In conclusion, the study of multitrophic interactions in agroecosystems is a dynamic and evolving field that requires interdisciplinary collaboration. As our understanding of these interactions deepens, so too will our ability to cultivate resilient and productive agricultural systems.

11.8.1 FUTURE ASPECT

Multitrophic interactions in agroecosystems involve the intricate relationships between plants, insect pests, and their natural enemies. Understanding these interactions is crucial for sustainable agriculture as they directly influence crop health, productivity, and overall ecosystem stability. This part explores the future aspects of multitrophic interactions, focusing on the dynamics between plants, insect pests, and their natural enemies.

1. **Enhanced understanding of chemical ecology**: Future research will likely delve deeper into the chemical ecology of multitrophic interactions. Studying the plant's chemical signaling to attract beneficial insects or repel pests can provide insights into developing eco-friendly pest management strategies [42].
2. **Microbial mediation of interactions**: Advances in microbial ecology will play a pivotal role in understanding how microorganisms influence multitrophic interactions. Beneficial microbes associated with plants can act as indirect defenders against pests by enhancing plant defenses or attracting natural enemies [16].
3. **Integration of molecular techniques**: Molecular techniques, such as genomics and transcriptomics, will contribute to a more comprehensive understanding of multitrophic interactions. Identifying specific plant genes involved in defense responses and the molecular mechanisms underlying pest attraction to certain plants will inform targeted pest control strategies [72].

4. **Climate change impacts**: Future studies should address the influence of climate change on multitrophic interactions. Altered temperature and precipitation patterns may affect the phenology and behavior of insects, plants, and natural enemies, ultimately shaping the dynamics of pest populations and their control [10].

5. **Integrated pest management (IPM) strategies**: Future research will likely focus on refining and expanding IPM strategies that leverage multitrophic interactions. Incorporating natural enemies into crop management practices can contribute to sustainable pest control while minimizing reliance on chemical pesticides [12].

The future of research on multitrophic interactions in agroecosystems holds great promise for advancing sustainable agriculture. Continued collaboration between ecologists, entomologists, and plant scientists will be essential in unlocking the full potential of multitrophic interactions for the benefit of agricultural ecosystems.

KEYWORDS

- **agroforestry**
- **herbivores**
- **predator-prey relationships**
- **tritrophic interactions**
- **trophic levels**

REFERENCES

1. Adjei-Maafol, K., Wilson, L. T. (1983). Factors affecting the relative abundance of arthropods on nectaried and nectariless cotton. *Environmental Entomology*, 12, 349–52
2. Adler, L. S., Bronstein, J. L. (2004). Attracting antagonists: Does floral nectar increase leaf herbivory? *Ecology*, 85, 1519–26
3. Agrawal, A. A. (2000). Overcompensation of plants in response to herbivory and the optimal defense hypothesis. *Ecology*, 81(7), 2059–2070.

4. Aizpurua, O., Budinski, I., Georgiakakis, P., Gopalakrishnan, S., Ibanez, C., Mata, V., Rebelo, H., Russo, D., Szodoray-Paradi, F., Zhelyazkova, V., et al. (2018). Agriculture shapes the trophic niche of a bat preying on multiple pest arthropods across Europe: Evidence from DNA metabarcoding. *Molecular Ecology*, 27, 815–825.
5. Allen, R., Platt, K., Wiser S (1995). Biodiversity in New Zealand plantations. *New Zealand Journal of Forestry*, 39(4), 26–29
6. Altieri, M. A. (1999). The ecological role of biodiversity in agroecosystems. *Agriculture, Ecosystems & Environment*, 74(1–3), 19–31.
7. Athey, K. J., Sitvarin, M. I., Harwood, J. D. (2017). Laboratory and field investigation of biological control for brown marmorated stink bug (*Halyomorphahalys* (Stal) (Hemiptera: Pentatomidae)). *Journal of the Kansas Entomological Society*, 90, 341–352.
8. Averill, A. L., Prokopy, R. J. (1993). Foraging of *Rhagoletis pomonella* flies about interactive food and fruit resources. *Entomologia Experimentalis et Applicata*, 66, 179–85
9. Baker, H. G., Baker, I. (1983). A brief historical review of the chemistry of floral nectar. In: Bentley, B., Elias, T. (Eds.), *The Biology of Nectaries* (pp. 126–52), Columbia University Press: New York.
10. Bale, J. S., Masters, G. J., Hodkinson, I. D., Awmack, C., Bezemer, T. M., Brown, V. K., Butterfield, J., Buse, A., Coulson, J. C., Farrar, J., Good, J. E. G., Harrington, R., Hartley, S., Hefin Jones, T., Lindroth, R. L., Press, M. C., Symrnioudis, I., Watt, A. D., Whittaker, J. B. (2002). Herbivory in global climate change research: Direct effects of rising temperature on insect herbivores. *Global Change Biology*, 8(1), 1–16.
11. Baroja, U., Garin, I., Aihartza, J., Arrizabalaga-Escudero, A., Vallejo, N., Aldasoro, M., Goiti, U. (2019). Pest consumption in a vineyard system by the lesser horseshoe bat (*Rhinolophus hipposideros*). *PLoS One*, 14, e0219265.
12. Barzman, M., Bàrberi, P., Birch, A. N. E., Boonekamp, P., Dachbrodt-Saaydeh, S., Graf, B., Hommel, B., Erik Jensen, J., Kiss, J., Kudsk, P., Lamichhane, J. R., Messéan, A., Moonen, A. C., Ratnadass, A., Ricci, P., Sarah, J.-L., Messean, A. (2015). Eight principles of integrated pest management. *Agricultural Systems*, 182, 1–10.
13. Beer, J., Muschler, R., Kass, D., Somarriba, E. (1998). Shade management in coffee and cacao plantations. *Agroforestry Systems,38*, 139–164
14. Begg, G. S., Cook, S. M., Dye, R., Ferrante, M., Franck, P., Lavigne, C., Lövei, G. L., Mansion-Vaquie, A., Pell, J. K., Petit, S. (2017). A functional overview of conservation biological control. *Crop Protection*, 97, 145–158.
15. Benson, R. B. (1950). An introduction to the natural history of British sawflies. *Transactions of the Society of British Entomologists*, 10, 45–142
16. Berendsen, R. L., Pieterse, C. M. J., Bakker, P. A. H. M. (2018). The rhizosphere microbiome and plant health. *Trends in Plant Science*, 23(3), 245–258.
17. Bernays, E. A. (2001). Neural limitations in phytophagous insects: Implications for diet breadth and evolution of host affiliation. *Annual Review of Entomology,* 46, 703–27
18. Bezemer, T. M., Wagenaar, R., van Dam, N. M., W¨ackers, F. L. (2003). Interactions between above- and belowground insect herbivores as mediated by the plant defense system. *Oikos*, 101, 555–62

19. Bigger, D. S., Chaney, W. E. (1998). Effects of *Iberis umbellate* (Brassicaceae) on insect pests of cabbage and potential biological control agents. *Environmental Entomology*, 27, 161–67
20. Birch, M. C., Young, M. C. (1998). Gossyplure and related terpenes: Behavioral and chemosensory
21. Bisseleua, H. B. D., Fotio, D., Yede Missoup, A. D., Vidal, S. (2013). Shade tree diversity, cocoa pest damage, yield compensating inputs, and farmers' net returns in West Africa. *PLoS One*, *8*, e56115.
22. Bowers, C., Toews, M., Liu, Y., Schmidt, J. M. (2020). Cover crops improve early season natural enemy recruitment and pest management in cotton production. *Biological Control*, 141, 104149.
23. Brantjes, N. B. M. (1976). Riddles around the pollination of *Melandrium album* (Mill.) Garcke (Caryophyllaceae) during the oviposition by *Hadenabicruris Hufn.* (Noctuidae, Lepidoptera). II. *Proceedings of the Royal Netherlands Academic Science Series, C*, 1, 127–41
24. Brommer, J. E., Fred, M. S. (1999). Movement of the Apollo butterfly (*Parnassius apollo*) related to host plant and nectar plant patches. *Ecological Entomology*, 24, 125–31
25. Brown, P. M. J., Ingels, B., Wheatley, A., Rhule, E. L., de Clercq, P., van Leeuwen, T., Thomas, A. (2015). Intraguild predation by Harmonia axyridis (Coleoptera: Coccinellidae) on native insects in Europe: Molecular detection from field samples. *Entomological Science*, 18, 130–133.
26. Bugg, R. L., W¨ackers, F. L., Brunson, K. E., Phatak, S. C., Dutcher, J. D. (1990). Tarnished plant bug (Hemiptera: Miridae) on selected cool-season leguminous cover crops. *Journal of Entomological Science*, 25, 463–74
27. Bumrungsri, S., Harbit, A., Benzie, C., Carmouche, K., Sridith, K., Racey, P. (2008). The pollination ecology of two species of *Parkia* (Mimosaceae) in southern Thailand. *Journal of Tropical Ecology*, 24, 467.
28. Chaneton, E. J., Bonsall, M. B. (2000). Enemy-mediated apparent competition: Empirical patterns and the evidence. *Oikos*, 88, 380–94
29. Chaplin-Kramer, R., O'Rourke, M. E., Blitzer, E. J., Kremen, C. (2011). A meta-analysis of crop pest and natural enemy response to landscape complexity. *Ecology Letters*, 14, 922–932.
30. Chase, J. M., Leibold, M. A. (2003). *Ecological Niches: Linking Classical and Contemporary Approaches*. University of Chicago Press.
31. Cheng, H. H. (1972). Oviposition and longevity of dark-sided cutworm, *Euxoa messoria* (Lepidoptera: Noctuidae), in the laboratory. *Canadian Entomologist*, 104, 919–25
32. Chew, F. S., Robbins, R. K. (1984). Egg-laying in butterflies. In: Vane-Wright, R., Ackery, P. R. (Eds.), *The Biology of Butterflies* (pp. 65–79). Academic: London.
33. Choate, B. A., Lundgren, J. G. (2015). Invertebrate communities in spring wheat and the identification of cereal aphid predators through molecular gut content analysis. *Crop Protection*, 77, 110–118.
34. Clark, R. E., Farkas, T. E., Lichter-Marck, I., Johnson, E. R., Singer, M. S. (2016). Multiple interaction types determine the impact of ant predation of caterpillars in a forest community. *Ecology*, 97, 3379–3388.

35. Cook, S. M., Murray, D. A., Williams, I. H. (2004). Do pollen beetles need pollen? The effect of pollen on oviposition, survival, and development of a flower-feeding herbivore. *Ecological Entomology*, 29, 164–73
36. Cornelissen, T., Stiling, P. (2005). Sex-biased herbivory: A meta-analysis of the effects of gender on plant-herbivore interactions. *Oikos*, 111, 488–500
37. Courtney, S. P. (1981). Coevolution of pierid butterflies and their cruciferous food plants. *Oecologia*, 51, 91–96
38. Crisol-Martinez, E., Moreno-Moyano, L. T., Wormington, K. R., Brown, P. H., Stanley, D. (2016). Using next-generation sequencing to contrast the diet and explore pest-reduction services of sympatric bird species in macadamia orchards in Australia. *PLoS One*, 11, e0150159.
39. de Groot, A. T., van der Wal, E., Schuurman, A., Visser, J. H., Blommers, L. H. M., van Beek, T. A. (1998). Copulation behavior of *Lygocoris pabulinus under* laboratory conditions. *Entomologia Experimentalis et Applicata*, 88, 219–28
40. De Ruiter, P. C. (2002). Stability in real food webs: Weak links in long loops. *Science*, 296, 1120–23
41. Dhaliwal, G. S., Arora, R. (2001). *Integrated Pest Management Concepts and Approaches* (pp. 27–60). Kalyani Publishers: New Delhi, India.
42. Dicke, M. (2009). Ecological and evolutionary approaches to managing crop insect pests. *Annual Review of Entomology*, 54, 75–95.
43. Dicke, M., Baldwin, I. T. (1999). The evolutionary context for communication between plants and their arthropod enemies. *Annual Review of Ecology and Systematics*, 30(1), 111–131.
44. Dicke, M., Baldwin, I. T. (2010). The evolutionary context for herbivore-induced plant volatiles: Beyond the 'cry for help'. *Trends in Plant Science*, 15(3), 167–175.
45. Dicke, M., Baldwin, I. T., Preston, C. A. (2009). *The Language of Plants: Science, Philosophy, Literature*. University of Minnesota Press.
46. Dicke, M., et al. (2003). Volatile induced plant defense. *Trends in Plant Science*, 7(4), 201–204.
47. Dwivedi, S. C., Singh, S. M. K., Katiyar, R. R. (2003). Seasonal incidence of insect pests associated with mango crop. *Annals of Plant Protection Science*, 16, 159–162.
48. Feeny, P. (1976). Plant apparency and chemical defense. *Recent Advances in Phytochemistry*, 10, 1–40
49. Firempong, S., Zalucki, M. P. (1990). Host plant selection by *Helicoverpa armigera* (H¨ ubner) (Lepidoptera: Noctuidae); the role of certain plant attributes. *Australian Journal of Zoology*, 37, 675–83
50. Garfinkel, M. B., Minor, E. S., Whelan, C. J. (2020). Birds suppress pests in corn but release them in soybean crops within a mixed prairie/agriculture system. *Condor*, 122, duaa009.
51. Garibaldi, L. A., Steffan-Dewenter, I., Winfree, R., Aizen, M. A., Bommarco, R., Cunningham, S. A., Kremen, C., Carvalheiro, L. G., Harder, L. D., Afik, O. (2013). Wild pollinators enhance fruit sets of crops regardless of honey bee abundance. *Science*, 339, 1608–1611.
52. Gomez-Polo, P., Alomar, O., Castane, C., Aznar-Fernandez, T., Lundgren, J. G., Pinol, J., Agusti, N. (2016). Understanding trophic interactions of Orius spp. (Hemiptera:

Anthocoridae) in lettuce crops by molecular methods. *Pest Management Science*, 72, 272–279.

53. Gordon, A., Newman, S., Coleman, B., Thevathasan, N. (2018). Temperate agroforestry: An overview. In: Gordon, A., Newman, S., Coleman, B. (Eds.), *Temperate Agroforestry Systems* (pp. 1–6), 2nd edn. CABI: Wallingford, UK.
54. Graham, J. B., Nassauer, J. I. (2019). Wild bee abundance in temperate agroforestry landscapes: Assessing effects of alley crop composition, landscape configuration, and agroforestry area. *Agroforestry Systems*, 93(3), 837–850.
55. Grossmueller, D. W., Lederhouse, R. C. (1987). The role of nectar source distribution in habitat use and oviposition by the tiger swallowtail butterfly. *Journal of the Lepidopterists' Society*, 41, 159–65
56. Gunnarsson, B. (2007). Predation on spiders: Ecological mechanisms and evolutionary consequences. *Journal of Arachnology*, 35, 509–529.
57. Gurr, G. M., Wratten, S. D., Tylianakis, J., Kean, J., Keller, M. (2005). Providing plant foods for natural enemies in farming systems: Balancing practicalities and theory. In: Wäckers, F.L., van Rijn, P.C.J., Bruin, J. *Plant-Provided Food for Carnivorous Insects: A Protective Mutualism and Its Applications* (pp. 326–47), Cambridge University Press.
58. Hagen, K. S., Van den Bosch, R. (1968). Augmentative biological control of insects in California. *Proceedings of the Tall Timbers Conference on Ecological Animal Control by Habitat Management*, 8, 129–147.
59. Hagler, J. R. (2019). It's Gut Check Time! A Universal Food Immunomarking Technique for Studying Arthropod Feeding Activities. *Annals of the Entomological Society of America*, 112, 211–219.
60. Hagler, J. R., Mostafa, A. M. (2019). A gut analysis technique for pinpointing egg-specific predation events. *Journal of Insect Science*, 19(3), 15.
61. Hails, R. S., Crawley, M. J. (1991). The population dynamics of an alien insect: *Andricus quercuscalicis* (Hymenoptera, Cynipidae). *Journal of Animal Ecology*, 60, 545–62
62. Hairston, N. G., Smith, F. E., Slobodkin, L. B. (1960). Community structure, population control, and competition. *The American Naturalist*, 94(879), 421–425.
63. Hare, J. D., Obrycki, J. J. (2004) Linalool induces ovicidal activity in *Orius insidiosus* (Say) (Coleoptera: Anthocoridae), a predator of the soybean aphid, *Aphis glycines* matsumura (Hemiptera: Aphididae). *Biological Control*, 30(1), 185–191.
64. Harrington, R., Fleming, R. and Woiwod, I. (2001). Climate change impacts on insect management and conservation in temperate regions: Can they be predicted? *Agricultural and Forest Entomology*, 3, 233.240.
65. Hartley M (2002) Rationale and methods for conserving biodiversity in plantation forests. *Forest Ecology and Management*, 155, 81–95
66. Harvey, C. A., Medina, A., Sanchez, D. M., Vílchez, S., Hernandez, B., Saenz, J. C., Maes, J. M., Casanoves, F., Sinclair, F. L. (2006). Patterns of animal diversity in different forms of tree cover in agricultural landscapes. *Ecological Applications*, 16, 1986–1999
67. Hasan, F., Ansari, M. S., Ahmad, N., Ahmad, A. (2017). Biopesticides: An eco-friendly approach for pest control. *Journal of Biopesticides*, 10(2), 125–127.

68. Heil, M. (2015). Extrafloral nectar at the plant-insect interface: a spotlight on chemical ecology, phenotypic plasticity, and food webs. *Annual Review of Entomology*, 60, 213–232.
69. Heil, M. (2008). Indirect defense via tri-trophic interactions. *New Phytologist*, 178(1), 41–61.
70. Heimoana, V., Pilkington, L. J., Raman, A., Mitchell, A., Nicol, H. I., Johnson, A. C., Gurr, G. M. (2017). Integrating spatially explicit molecular and ecological methods to explore the significance of non-crop vegetation to predators of brassica pests. *Agriculture, Ecosystems & Environment*, 239, 12–19.
71. Hendrichs, J., Katsoyannos, B. I., Papaj, D. R., Prokopy, R. J. (1991). Sex differences in movement between natural feeding and mating sites and tradeoffs between food consumption, mating success, and predator evasion in Mediterranean fruit flies (Diptera, Tephritidae). *Oecologia,* 86, 223–31
72. Hilker, M., Fatouros, N. E. (2015). Plant responses to insect egg deposition. *Annual Review of Entomology*, 60, 493–515.
73. Hilty, J. (2006). *Description of Families of Flower-visiting Plant Bugs*. http://www.shout.net/~jhilty/files/plant bugs.htm
74. Hodek, I., Honek, A. (1996). *Ecology of Coccinellidae*. Springer Science & Business Media.
75. Hodek, I., van Emden, H. F., Honěk, A. (2012). *Ecology and Behavior of the Ladybird Beetles (Coccinellidae)*. John Wiley & Sons.
76. Hofsvang, T., Snoan, B., Andersen, A., Heggen, H., Anh, L. (2005) *Liriomyza sativa* (Diptera: Agromyzidae), an invasive species in South-East Asia: Studies on its biology in Northern Vietnam. *International Journal of Pest Management*, 51, 71–80.
77. Hölldobler, B., Wilson, E. O. (1990) *The Ants.* Harvard University Press: Cambridge, MA, USA.
78. Holt, R. D., Polis, G. A. (1997). A theoretical framework for intraguild predation. *American Naturalist*, 149(4), 745–64
79. Hopkins, R. J., Mithen, R. (2015). Glucosinolates: Bioactive metabolites in the Brassicaceae. In: *The Arabidopsis Book* (p. e0168). American Society of Plant Biologists.
80. Hunter, M. D., Price, P. W. (1992). Playing chutes and ladders: Heterogeneity and the relative roles of bottom-up and top-down forces in natural communities. *Ecology*, 73(3), 724–732.
81. Hunter, M. D., Schultz, J. C. (1993). Induced plant defenses breached? Phytochemical induction protects an herbivore from disease. *Oecologia*, 94(1), 195–203.
82. Irvin, N. A., Hagler, J. R., Hoddle, M. S. (2018). Measuring natural enemy dispersal from cover crops in a California vineyard. *Biological Control*, 126, 15–25.
83. Janz, N. (2005). The relationship between habitat selection and preference for adult and larval food resources in the polyphagous butterfly *Vanessa cardui* (Lepidoptera: Nymphalidae). *Journal of Insect Behavior*, 18, 767–79
84. Jervis, M. A., Vilhelmsen, L. B. (2000). Mouthpart evolution in adults of the basal 'symphytan' hymenopteran lineages. *Biological Journal of the Linnean Society*, 70, 121–46

85. Johnson, S. N., Erb, M., Hartley, S. E. (2018). Roots under attack: Contrasting plant responses to below- and aboveground insect herbivory. *The New Phytologist*, 219(2), 695–709.
86. Jones, M., Sinclair, F. L., Grime, V. L. (1998) Effect of tree species and crown pruning on root length and soil water content in semi-arid agroforestry. *Plant Soil*, 201, 197–207
87. Jonsson, M., Raphael, I. A., Ekbom, B., Kyamanywa, S., Karungi, J. (2015). Contrasting effects of shade level and altitude on two important coffee pests. *Journal of Pest Science*, *88*, 281–287.
88. Ju, Q., Ouyang, F., Gu, S. M., Qiao, F., Yang, Q. F., Qu, M. J., Ge, F. (2019). Strip intercropping peanut with maize for peanut aphid biological control and yield enhancement. *Agriculture, Ecosystems & Environment*, 286, 8.
89. Karban, R. (1997). Neighborhood affects a plant's risk of herbivory and subsequent success. *Ecological Entomology*, 22, 433–39
90. Karban, R., Baldwin, I. T. (1997). *Induced Responses to Herbivory*. University of Chicago Press.
91. Karban, R., Myers, J. H. (1989). Induced plant responses to herbivory. *Annual Review of Ecology and Systematics*, 20, 331–348.
92. Karp, D. S., Daily, G. C. (2014). Cascading effects of insectivorous birds and bats in tropical coffee plantations. *Ecology*, 95, 1065–74
93. Karp, D. S., Mendenhall, C. D., Sandí, R. F., Chaumont, N., Ehrlich, P. R., Hadly, E. A., Daily, G. C. (2013). Forest bolsters bird abundance, pest control, and coffee yield. *Ecology Letters*, 16, 1339–1347
94. Keller, M. A., Debach, P. (1969). Biological control of *Heliothis zea* (Lepidoptera: Noctuidae) in corn by *Trichogramma pretiosum* (Hymenoptera: Trichogrammatidae). *Journal of Economic Entomology*, 62(4), 885–890.
95. Kemp, J., Lopez-Baucells, A., Rocha, R., Wangensteen, O. S., Andriatafika, Z., Nair, A., Cabeza, M. (2019). Bats as potential suppressors of multiple agricultural pests: A case study from Madagascar. *Agriculture, Ecosystems & Environment*, 269, 88–96.
96. Kessler, A., Baldwin, I. T. (2002). Plant responses to insect herbivory: The emerging molecular analysis. *Annual Review of Plant Biology*, 53(1), 299–328.
97. Kevan, P. G., Baker, H. G. (1998). Insects on flowers. In: Huffaker, C. B., Gutierrez, A. P. (Eds.), *Ecological Entomology* (pp. 553–583). Wiley: New York.
98. Kheirodin, A., Sharanowski, B. J., Carcamo, H. A., Costamagna, A. C. (2020). Consumption of cereal leaf beetle, *Oulema melanopus*, by generalist predators in wheat fields detected by molecular analysis. *Entomologia Experimentalis et Applicata*, 168, 59–69.
99. Kirk, W. D. J. (1997). Feeding. In: Lewis, T. (Ed.), *Thrips as Crop Pests* (pp. 217–57). CABI: London.
100. Klein, A. M., Steffan-Dewenter, I., Tscharntke, T. (2002b). Predator-prey ratios on cocoa along a land-use gradient in Indonesia. *Biodiversity and Conservation*, 11, 683–693.
101. Klein, A. M., Steffan-Dewenter, I., Buchori, D., Tscharntke, T. (2002a). Effects of land-use intensity in tropical agroforestry systems on coffee flower-visiting and trap-nesting bees and wasps. *Conservation Biology*, 16, 1003–1014.

102. Kogan, M. (1998). Integrated pest management: Historical perspectives and contemporary developments. *Annual Review of Entomology*, 43, 243–270.
103. Koptur, S., Lawton, J. H. (1988). Interactions among vetches bearing extrafloral nectaries, their biotic protective agents, and herbivores. *Ecology*, 69, 278–83.
104. Kranz, A. J., Wolz, K. J., Miller, J. R. (2018) Effects of shrub crop interplanting on apple pest ecology in a temperate agroforestry system. *Agroforestry Systems*, 93, 1179–1189 https://doi.org/10.1007/ s10457-018-0224-8
105. Krumpalova, Z., Tuf, I. H. (2013). Circadian rhythms of ground living spiders: Mechanisms of coexistence strategy based on the body size. *Polish Journal of Ecology*, 61, 575–586
106. Kunz, T. H., Braun de Torrez, E., Bauer, D., Lobova, T., Fleming, T. H. (2011). Ecosystem services provided by bats. In: *Year in Ecology and Conservation Biology, Annals of the New York Academy of Sciences*, Ostfeld, R. S., Schlesinger, W. H. (Eds.), (Vol. 1223, pp. 1–38). New York Academy of Sciences: New York.
107. Landis, D. A., et al. (2000). Increased parasitism of early-season *Plutella xylostella* (Lepidoptera: Plutellidae) by *Trichogramma* spp. (Hymenoptera: Trichogrammatidae) associated with perimeter trap cropping. *Environmental Entomology*, 29(4), 876–883.
108. Landis, D. A., Wratten, S. D., Gurr, G. M. (2000). Habitat management to conserve natural enemies of arthropod pests in agriculture. *Annual Review of Entomology*, 45(1), 175–201.
109. Landolt, P. J., Guédot, M. (1993). Volatile semiochemicals influencing the oviposition behavior of the codling moth L. (Lep., Tortricidae). *Experientia*, 49(12), 1025–1030.
110. Leskey, T. C., Teulon, D. A. J., Cameron, E. A. (1997). Effects of temperature and sugar maple pollen on oviposition and longevity of pear thrips (Thysanoptera: Thripidae). *Environmental Entomology*, 26, 566–71
111. Leston, D. (1973) The ant mosaic-tropical tree crops and the limiting of pest and diseases. *Proceedings of the National Academy of Sciences USA,* 19, 311–341.
112. Letourneau, D. K., et al. (2011). Does plant diversity benefit agroecosystems? A synthetic review. *Ecological Applications*, 21(1), 9–21.
113. Letourneau, D. K., et al. (2015). Crop protection in tropical agroforestry systems: Contributions of agroecology to the paradigm of IPM. *Environmental Management*, 56(4), 933–949.
114. Letourneau, D. K., Jedlicka, J. A., Bothwell, S. G., Moreno, C. R. (2009). Effects of natural enemy biodiversity on the suppression of arthropod herbivores in terrestrial ecosystems. *Annual Review of Ecology, Evolution, and Systematics*, 40, 573–592.
115. Ludwig, K. A., Hill, R. E. (1975). Comparison of gut contents of adult western and northern corn rootworms in northeast Nebraska. *Environmental Entomology*, 4, 435–38
116. Lukefahr, M. J., Martin, D. F., Meyer, J. R. (1965). Plant resistance to five Lepidoptera attacking cotton. *Journal of Economic Entomology,* 58, 516–18
117. Lukefahr, M. J., Martin, D. F. (1964). Effects of various larval and adult diets on fecundity and longevity of bollworm, tobacco budworm, and cotton leafworm. *Journal of Economic Entomology,* 57, 233–35
118. Mangan, A. M., Piaggio, A. J., Hopken, M. W., Werner, S. J., Pejchar, L. (2018). A molecular analysis to assess codling moth *Cydia pomonella* L. (Lepidoptera: Tortricidae) predation by orchard birds. *Ecological Indicators*, 93, 1222–1225.

119. Masonick, P., Hernandez, M., Weirauch, C. (2019). No guts, no glory: Gut content metabarcoding unveils the diet of a flower-associated coastal sage scrub predator. *Ecosphere*, 10, e02712.
120. Mayer, V. E., Frederickson, M. E., McKey, D., Blatrix, R. Current issues in the evolutionary ecology of ant–plant symbioses. *New Phytologist*, 202(3), 749–764.
121. Mayhew, P. J. (1997). Adaptive pattern of host-plant selection by phytophagous insects. *Oikos*, 79, 417–28
122. Miller, W. E. (1996). Population behavior and adult feeding capability in Lepidoptera. *Environmental Entomology*, 25, 213–26
123. Mori, K., Pasteels, J. M. (1982). 2-Tridecanone, an egg-laying deterrent pheromone in the Colorado potato beetle, *Leptinotarsa decemlineata* (Say). *Journal of Chemical Ecology*, 8(10), 1349–1358.
124. Murai, T. (2000). Effect of temperature on development and reproduction of the onion thrips, *Thrips tabaci* Lindeman (Thysanoptera: Thripidae), on pollen and honey solution. *Applied Entomoloy and Zoology*, 35, 499–504
125. Murphy, D. D., Menniger, M. S., Ehrlich, P. R. (1984). Nectar source distribution as a determinant of oviposition host species in *Euphydrya schalcedona. Oecologia*, 62, 269–271.
126. Murphy, D. D. (1983). Nectar sources as constraints on the distribution of egg masses by the checkerspot butterfly, *Euphydrya schalcedona* (Lepidoptera: Nymphalidae). *Environmental Entomology*, 12, 463–66
127. Naranjo, S. E. (2001). Conservation and Evaluation of Natural Enemies in IPM Systems for *Bemisia tabaci. Crop Protection*, 20(9), 835–852.
128. Nelson, A. S., Pratt, R. T., Pratt, J. D., Smith, R. A., Symanski, C. T., Prenot, C., Mooney KA: Progressive sensitivity of trophic levels to warming underlies an elevational gradient in ant–aphid mutualism strength. *Oikos*, 2018, 0.
129. Newman, S. M., Gordon, A. (2018). Temperate agroforestry: Key elements, current limits and opportunities for the future. In: Gordon, A. M., Newman, S. M., Coleman, B. (Eds.), *Temperate Agroforestry Systems* (pp. 274–298), 2nd edn., CABI: Wallingford, UK.
130. Odum, E. P. (1959). *Fundamentals of Ecology*. Saunders.
131. Ohgushi, T. (2008). Herbivore-induced indirect interaction webs on terrestrial plants: The importance of non-trophic, indirect, and facilitative interactions. *Entomological Science*, 11(3), 249–263.
132. Oksanen, L., Fretwell, S. D., Arruda, J., Niemelä, P. (1981). Exploitation ecosystems in gradients of primary productivity. *The American Naturalist*, 118(2), 240–261.
133. Olff, H., Ritchie, M. E. (1998). Effects of herbivores on grassland plant diversity. *Trends in Ecology & Evolution*, 13(7), 261–265.
134. Ortiz-Martínez, S., Staudacher, K., Baumgartner, V., Traugott, M., Lavandero, B. (2020). Intraguild predation is independent of landscape context and does not affect the temporal dynamics of aphids in cereal fields. *Journal of Pest Science*, 93, 235–249.
135. Pace, M. L., Cole, J. J., Carpenter, S. R., Kitchell, J. F. (1999). Trophic cascades are revealed in diverse ecosystems. *Trends in Ecology & Evolution*, 14(12), 483–488.

136. Parsons, F. S. (1940). Investigation on the cotton bollworm, *Heliothis armigera* Hub. 3. The relationship between oviposition and the flowering curves of flowering plants. *Bulletin of Entomological Research*, 31, 147–77
137. Paul, D., Lalnunsangi. (2011). Insect natural enemy complex in some agroforestry systems of Mizoram, India. *Science Vision*, 11, 90–95.
138. Perfecto, I., Vandermeer, J. (2014). *Coffee Agroecology: A New Approach to Understanding Agricultural Biodiversity, Ecosystem Services and Sustainable Development* (p. 352). Earthscan: London.
139. Peterson, J. A., Burkness, E. C., Harwood, J. D., Hutchison, W. D. (2018). Molecular gut-content analysis reveals high frequency of *Helicoverpa zea* (Lepidoptera: Noctuidae) consumption by *Orius insidiosus* (Hemiptera: Anthocoridae) in sweet corn. *Biological Control*, 121, 1–7.
140. Petrakova, L., Michalko, R., Loverre, P., Sentenska, L., Korenko, S., Pekar, S. (2016). Intraguild predation among spiders and their effect on the pear psylla during winter. *Agriculture, Ecosystems & Environment*, 233, 67–74.
141. Philpott, S. M., Soong, O., Lowenstein, J. H., Pulido, A. L., Lopez, D. T., Flynn, D. F., DeClerck, F. (2009). Functional richness and ecosystem services: Bird predation on arthropods in tropical agroecosystems. *Ecological Applications,* 19, 1858–1867.
142. Poveda, V. A., Steffan, J. J. (2008). Green leaf volatiles attract and arrest the tachinid fly *Lydella thompsoni*, a parasitoid of the western corn rootworm. *Journal of Chemical Ecology*, 34(2), 203–212.
143. Price, P. W. (1970). Evolutionary biology of parasitoids and predators with implications for biological control. *Environmental Entomology*, 1(4), 289–293.
144. Price, P. W., Bouton, C. E., Gross, P. (1986). Interactions among three trophic levels: Influence of plants on interactions between insect herbivores and natural enemies. *Annual Review of Ecology and Systematics*, 17(1), 337–364.
145. Prokopy, R. J., Duan, J. J., Vargas, R. I. (1996). Potential for host range expansion in *Ceratitis capitata* flies' impact of the proximity of adult food to egg-laying sites. *Ecological Entomology*, 21, 295–99
146. Pumariño, L., Sileshi, G. W., Gripenberg, S., Kaartinen, R., Barrios, E., Muchane, M. N., Midega, C., Jonsson, M. (2015). Effects of agroforestry on pest, disease and weed control: A meta-analysis. *Basic and Applied Ecology*, 16, 573–582
147. Pyke, G. H. (1991). What does it cost a plant to produce floral nectar? *Nature*, 350, 58–59
148. Ramı´rez, P. A., Simonetti, J. A. (2011) Conservation opportunities in commercial plantations: The case of mammals. *Journal for Nature Conservation*, 19, 351–355
149. Rana, R. L., Charlet, L. D. (1997). Feeding behavior and egg maturation of the red and gray sunflower seed weevils (Coleoptera: Curculionidae) on cultivated sunflower. *Annals of the Entomological Society of America*, 90, 693–99
150. Rand, T. A., Louda, S. M. (2004). Exotic weed invasion increases the susceptibility of native plants to attack by a biocontrol herbivore. *Ecology*, 85, 1548–54
151. Robert, C. M. P., Raffa, K. F., Townsend, P. A. (2011). Diurnal and seasonal variation in green leaf volatile emissions from maize: Implications for host location by herbivores and their natural enemies. *Journal of Chemical Ecology*, 37(1), 31–43.
152. Romeis, J., St¨adlerE, W¨ackers, F. L. (2005). Nectar- and pollen-feeding by adult herbivorous insects. In: Wäckers, F.L., van Rijn, P.C.J., Bruin, J. *Plant-Provided*

Food for Carnivorous Insects: A Protective Mutualism and Its Applications (pp. 178–219), Cambridge.
153. Rondoni, G., Fenjan, S., Bertoldi, V., Ielo, F., Djelouah, K., Moretti, C., Buonaurio, R., Ricci, C., Conti, E. (2018). Molecular detection of field predation among larvae of two ladybird beetles are partially predicted from laboratory experiments. *Scientific Reports*, 8, 2594.
154. Root, R. B. (1973) Organization of a plant-arthropod association in simple and diverse habitats: The fauna of collards (*Brassica oleracea*). *Ecological Monographs*, 43, 95–124.
155. Roubinet, E., Jonsson, T., Malsher, G., Staudacher, K., Traugott, M., Ekbom, B., Jonsson, M. (2018). High Redundancy as well as Complementary Prey Choice Characterize Generalist Predator Food Webs in Agroecosystems. *Scientific Reports*, 8, 8054.
156. Roubinet, E., Straub, C., Jonsson, T., Staudacher, K., Traugott, M., Ekbom, B., Jonsson, M. (2015). Additive effects of predator diversity on pest control caused by few interactions among predator species. *Ecological Entomology*, 40, 362–371.
157. Roulston, T. H., Cane, J. H. (2000). Pollen nutritional content and digestibility for animals. In: Dafni, A., Hesse, M., Pacini, E. (Eds.), *Pollen and Pollination* (pp. 187–211). Springer: Vienna.
158. Russo, D., Bosso, L., Ancillotto, L. (2018). Novel perspectives on bat insectivory highlight the value of this ecosystem service in farmland: Research frontiers and management implications. *Agriculture, Ecosystems & Environment*, 266, 31–38.
159. Samuelson, G. A. (1994). Pollen consumption and digestion by leaf beetles. In: Jolivet, P. H., Cox, M. L., Petitpierre, E. (Eds.), *Novel Aspects of the Biology of Chrysomelidae* (pp. 179–83), Kluwer Academic Publisher: Dordrecht.
160. Schimdt, J. M., Doria, A. A., Blaauw, B., Kheirodin, A., Pandey, S., Lennon, K., Kaldor, A. D., Toledo P. F. S and Grabarczyk, E. E. (2021). Identifying molecular-based trophic interactions as a resource for advanced integrated pest management. *Insects*, 12, 358.
161. Schmitz, O. J. (2006). Predators have large effects on ecosystem properties by changing plant diversity, not plant biomass. *Ecology*, 87, 1432–1437.
162. Schmitz, O. J, Hamback, P. A., Beckerman, A. P. (2000) Trophic cascades in terrestrial systems: A review of the effects of carnivore removals on plants. *American Naturalist*, 155, 141–153.
163. Schmitz, O. J. (2008). Effects of predator hunting mode on grassland ecosystem function. *Science*, 319(5865), 952–954.
164. Schuster, M. F., Lukefahr, M. J., Maxwell, F. G. (1976). Impact of nectarless cotton on plant bugs and natural enemies. *Journal of Economic Entomology*, 69, 400–2
165. Scott, W. P., Snodgrass, G. L., Smith, J. W. (1988). Tarnished plant bugs (Hemiptera, Miridae) and predaceous arthropod populations in commercially produced selected nectaried and nectariless cultivars of cotton. *Journal of Entomological Science*, 23, 280–86
166. Sharma, H. C., Dhillon, M. K., Kibuka, J., Mukuru, S. Z. (2005). Plant defense responses to sorghum spotted stem borer, *Chilo partellus* under irrigated and drought conditions. *Institute Sorghum Millets News*, 46, 49.52.

167. Sileshi, G., Mafongoya, P. L. (2006a). Long-term effects of improved legume fallows on soil invertebrate macrofauna and maize yield in eastern Zambia. *Agriculture, Ecosystems and Environment, 115*, 69–78.
168. Sileshi, G., Mafongoya, P. L. (2006b). Variation in macrofaunal communities under contrasting land use systems in eastern Zambia. *Applied Soil Ecology, 33*, 49–60.
169. Somarriba, E., Harvey, C. A., Samper, M., Anthony, F., González, J., et al. (2004) Biodiversity conservation in neotropical coffee (*Coffea arabica*) plantations. In: Schroth, G., Fonseca, G. A. B., Harvey, C. A., Gascon, C., Vasconcelos, H. L., Izac, A.-M. N. (Eds.), *Agroforestry and Biodiversity Conservation in Tropical Landscapes* (pp. 198–226). Island Press: Washington, DC, USA.
170. Stapley, J. H. (1973) Insect pests of coconuts in the Pacific Region. *Outlook on Agriculture*, 7, 17–22.
171. Stastny, M., Battisti, A., Petrucco-Toffolo, E., Schlyter, F., Larsson, S. (2006) Host-plant use in the range expansion of the pine processionary moth, *Thaumetopoea pityocampa*. *Ecological Entomology*, 31, 481–490.
172. Staudacher, K., Rubbmark, O. R., Birkhofer, K., Malsher, G., Sint, D., Jonsson, M., Traugott, M. (2018). Habitat heterogeneity induces rapid changes in the feeding behavior of generalist arthropod predators. *Functional Ecology*, 32, 809–819.
173. Strauss, S. Y., Watson, W., Allen, M. T. (2003). Predictors of male and female tolerance to insect herbivory in *Raphanus raphanistrum*. *Ecology*, 84, 2074–82
174. Strauss, S. Y. (1991). Indirect effects in community ecology: Their definition, study, and importance. *Trends in Ecology & Evolution*, 6(7), 206–210.
175. Swirski, E., Izhar, Y., Wysoki, M., Gurevitz, E., Greenberg, S. (1980). Integrated control of the long-tailed mealybug, *Pseudococcus longispinus* (Hom., Pseudococcidae), in avocado plantations in Israel. *Entomophaga*, 25, 415–26
176. The Woodland Trust. (2018). Agroforestry in England: Benefits, barriers and opportunities. In: *The Woodland Trust Policy Paper*. The Woodland Trust: UK.
177. Thies, C., Tscharntke, T. (1999). Landscape structure and biological control in agroecosystems. *Science*, 285(5429), 893–895.
178. Tillman, P. G., Greenstone, M. H., Hu, J. S. (2015). Predation of stink bugs (Hemiptera: Pentatomidae) by a complex of predators in cotton and adjoining soybean habitats in Georgia, USA. *Florida Entomologist*, 98, 1114–1126.
179. Torrez de, E. C. B., Brown, V. A., McCracken, G. F., Kunz, T. H. (2019). Sympatric bat species prey opportunistically on a major moth pest of pecans. *Sustainability*, 11, 6365.
180. Tscharntke, T., et al. (2007). Landscape moderation of biodiversity patterns and processes - eight hypotheses. *Biological Reviews*, 82(3), 309–335.
181. Tschumi, M., Ekroos, J., Hjort, C., Smith, H. G., Birkhofer, K. (2018). Rodents, not birds, dominate predation-related ecosystem services and disservices in vertebrate communities of agricultural landscapes. *Oecologia*, 188, 863–873.
182. Turlings, T. C. J., Erb, M. (2018). Tritrophic interactions mediated by herbivore-induced plant volatiles: Mechanisms, ecological relevance, and application potential. *Annual Review of Entomology*, 63, 433–452.
183. Van Driesche, R. G., Hoddle, M. S., Center, T. D. (2008). *Control of Pests and Weeds by Natural Enemies: An Introduction to Biological Control*. John Wiley & Sons.

184. Van Lenteren, J. C. (2012). The state of commercial augmentative biological control: Plenty of natural enemies, but a frustrating lack of uptake. *BioControl*, 57(1), 1–20.
185. van Veen, F. J. F., Morris, R. J., Godfray, H. C. J. (2006). Apparent competition, quantitative food webs, and the structure of phytophagous insect communities. *Annual Review of Entomology*, 51, 187–208
186. Varon, E. H. (2002) *Distribuciónespacio-temporal de hormigas con potencialcomodepredadoras de Hypothenemus hampei e Hypsipyla grandella, ensistemasagroforestales de café, en Costa Rica. Ms. thesis*. CATIE: Turrialba, Costa Rica.
187. Vet, L. E. M., Dicke, M. (1992). Infochemical communication in tri-trophic interactions between plants, herbivores, and natural enemies: Relevance for pest management. *Entomologia Experimentalis et Applicata*, 64(1), 71–84.
188. W¨ackers, F. L., van Rijn, P. C. J., Bruin, J. (Eds.) (2005). *Plant-Provided Food for Carnivorous Insects: A Protective Mutualism and Its Applications*. Cambridge University Press: Cambridge, UK.
189. W¨ackers, F. L., van Rijn, P. C. J. (2005). Food for protection: An introduction In: Wäckers, F.L., van Rijn, P.C.J., Bruin, J. *Plant-Provided Food For Carnivorous Insects: A Protective Mutualism and Its Applications* (pp. 1–14), Cambridge: UK.
190. W¨ackers, F. L. (2005). *Suitability of (extra-) Floral Nectar, Pollen, and Honeydew as Insect Food Sources* (pp. 17–74)
191. Wackers, F. L., Romeis J and Rijn, P. V. (2007). Nectar and pollen feeding by insect herbivores and implications for multitrophic interactions. *Annual Review of Entomology*, 52, 301–23
192. Wäckers, F. L., Vet, L. E. M. (1990). How does the egg parasitoid *Trichogramma maidli* distinguish between host and non-host eggs? Role of herbivore-induced plant volatiles in discrimination. *Journal of Chemical Ecology*, 16(10), 2955–2971.
193. Wang, Q., Bao, W., Yang, F., Yang, Y., Lu, Y. (2018). A PCR-based analysis of plant DNA reveals the feeding preferences of *Apolygus lucorum* (Heteroptera: Miridae). *Arthropod-Plant Interact*, 12, 567–574.
194. Wang, Q., Bao, W., Zhang, Q., Fu, X., Yang, Y., Lu, Y. (2019). Host plant use of a polyphagous mirid, *Apolygus lucorum*: Molecular evidence from migratory individuals. *Ecology and Evolution*, 9, 11518–11528.
195. Wang, Q., Bao, W. F., Yang, F., Xu, B., Yang, Y. Z. (2017). The specific host plant DNA detection suggests a potential migration of *Apolygus lucorum* from cotton to mungbean fields. *PLoS One*, 12, e0177789.
196. Way, M. J. (1963). Mutualism in cotton aphid-ant relationships. *Biological Reviews*, 38(3), 344–371.
197. Weier, S. M., Moodley, Y., Fraser, M. F., Linden, V. M. G., Grass, I., Tscharntke, T., Taylor, P. J. (2019). Insect pest consumption by bats in macadamia orchards established by molecular diet analyses. *Global Ecology and Conservation*, 18, e00626.
198. Whelan, C. J., Wenny, D. G., Marquis, R. J. (2008). Ecosystem services provided by birds. In: Ostfeld, R. S., Schlesinger, W. H. (Eds.), *Year in Ecology and Conservation Biology, Annals of the New York Academy of Sciences*, (Volume 1134, pp. 25–60). Blackwell Publishing: Boston, MA.

199. Winkler, K., W¨ackers, F. L., Bukovinszkine-Kiss, G., van Lenteren, J. C. (2006). Nectar resources are vital for *Diadegma semiclausum* fecundity under field conditions. *Basic and Applied Ecology*, 7, 133–40
200. Wolfe, L. M. (1997). Differential flower herbivory and gall formation on males and females of *Neeap sychotrioides*, a dioecious tree. *Biotropica*, 29, 169–174
201. Woodcock, B., Bullock, J., McCracken, M., Chapman, R., Ball, S., Edwards, M., Nowakowski, M., Pywell, R. (2016). Spill-over of pest control and pollination services into arable crops. *Agriculture, Ecosystems & Environment*, 231, 15–23.
202. Zhu, J., Chen, H., Wäckers, F. L. (2012). Orientation of the minute pirate bug *Orius insidiosus* to soybean volatiles induced by the soybean aphid *Aphis glycines*. *Journal of Chemical Ecology*, 38(4), 434–444.
203. Bhoi, T. K., Samal, I., Majhi, P. K., Komal, J., Mahanta, D. K., Pradhan, A. K., Saini, V., Nikhil Raj, M., Ahmad, M. A., Behera, P. P., Ashwini, M. (2022). Insight into aphid mediated Potato Virus Y transmission: A molecular to bioinformatics prospective. *Frontiers in Microbiology*, *13*, 1001454.
204. Mahanta, D. K., Jangra, S., Ghosh, A., Sharma, P. K., Iquebal, M. A., Jaiswal, S., Baranwal, V. K., Kalia, V. K., Chander, S. (2022). Groundnut bud necrosis virus modulates the expression of innate immune, endocytosis, and cuticle development-associated genes to circulate and propagate in its vector, Thrips palmi. *Frontiers in Microbiology*, *13*, 773238.
205. Bhatnagar, S., Mahanta, D. K., Vyas, V., Samal, I., Komal, J., Bhoi, T. K. (2023). Storage pest management with nanopesticides incorporating silicon nanoparticles: A novel approach for sustainable crop preservation and food security. *Silicon*, 1–13.
206. Mahanta, D. K., Komal, J., Samal, I., Bhoi, T. K., Dubey, V. K., Pradhan, K., Nekkanti, A., Gouda, M. N. R., Saini, V., Negi, N., Bhateja, S., Jat, H. K., Jeengar, D. (2023a). Nutritional aspects and dietary benefits of "Silkworms": Current scenario and future outlook. *Frontiers in Nutrition*, *10*, 1121508.
207. Samal, I., Bhoi, T. K., Vyas, V., Majhi, P. K., Mahanta, D. K., Komal, J., Sangeeta, S., Kumar, P.V.K., Acharya, L. K. (2023). Resistance to fungicides in entomopathogenic fungi: Underlying mechanisms, consequences, and opportunities for progress. *Tropical Plant Pathology*, 49, 5–17.
208. Mahanta, D. K., Bhoi, T. K., Komal, J., Samal, I., Nikhil, R. M., Paschapur, A. U., Singh, G., Kumar, P. V. D., Desai, H. R., Ahmad, M. A., Singh, P. P., Majhi, P. K., Mukherjee, U., Singh, P., Saini, V., Shahanaz, Srinivasa, N., Yele, Y. (2023b). Insect-pathogen crosstalk and the cellular-molecular mechanisms of insect immunity: Uncovering the underlying signaling pathways and immune regulatory function of non-coding RNAs. *Frontiers in Immunology*, 14, 1169152.
209. Samal, I., Bhoi, T. K., Majhi, P. K., Murmu, S., Pradhan, A. K., Kumar, D., Saini, V., Paschapur, A. U., Raj, M. N., Ankur, Manik, S., Behera, P. P., Mahanta, D. K., Komal, J., Alam, P., Balawi, T. A. (2023b). Combatting insects mediated biotic stress through plant associated endophytic entomopathogenic fungi in horticultural crops. *Frontiers in Plant Science*, *13*, 1098673.
210. Komal, J., Desai, H. R., Samal, I., Mastinu, A., Patel, R. D., Kumar, P. D., Majhi, P. K., Mahanta, D. K., Bhoi, T. K. (2023). Unveiling the genetic symphony: Harnessing CRISPR-Cas genome editing for effective insect pest management. *Plants*, *12*(23), 3961.

211. Samal, I., Bhoi, T. K., Mahanta, D. K., Komal, J. (2023c). Establishing the role of silicon (Si) in plant resistance to insects: A bibliometric approach. *Silicon*, 16, 2119–2128.
212. Majhi, P. K., Samal, I., Bhoi, T. K., Pattnaik, P., Pradhan, C., Gupta, A., Mahanta, D. K., Senapati, S. K. (2023). Climate-smart agriculture: An integrated approach to address climate change and food security. In: *Climate Change and Insect Biodiversity* (pp. 221–245). CRC Press.
213. Mahanta, D. K., Samal, I., Komal, J., Bhoi, T. K., Majhi, P. K., Ahmad, M. A. (2023c). Understanding anthropogenic climate change, its consequences on insect pests, and techniques in forecasting and monitoring pest dynamics: A contemporary scenario. In: *Climate Change and Insect Biodiversity* (pp. 44–67). CRC Press.
214. Bhoi, T. K., Samal, I., Mahanta, D. K., Komal, J., Jinger, D., Sahoo, M. R., Achary, G. C., Nayak, P., Sunani, S. K., Saini, V., Raghuraman, M., Singh, S. (2023). Understanding how silicon fertilization impacts chemical ecology and multitrophic interactions among plants, insects and beneficial arthropods. *Silicon, 15*(6), 2529–2549.
215. Bhoi, T. K., Samal, I., Mahanta, D. K., Komal, J., Majhi, P. K., Ankur. (2024). Exploring Melatonin's potential as an alternative strategy for protecting plants from biotic stresses. In: *Melatonin in Plants: A Pleiotropic Molecule for Abiotic Stresses and Pathogen Infection* (pp. 223–242). Springer Nature Singapore: Singapore.
216. Samal, I., Bhoi, T. K., Pradhan, A. K., Mahanta, D. K. (2023d). Plantix app: A success story of artificial intelligence in plant protection. *Van Sangyan*, 24.
217. Mahanta, D. K., Bhoi, T. K., Komal, J., Samal, I., Mastinu, A. (2024). Spatial, spectral and temporal insights: Harnessing high-resolution satellite remote sensing and artificial intelligence for early monitoring of wood boring pests in forests. *Plant Stress*, 11, 100381.
218. Samal, I., Mahanta, D. K., Bhoi, T. K., Komal, J., Jatav, H. S., Jatav, S. S., Sathyanarayana, E. (2024). *Approaches of Biochar in Ecosystem Management: Current Scenario and Future Perspectives*. IntechOpen
219. Vyas, V., Bhoi, T. K., Samal, I., Singh, S., Mahanta, D. K. (2024). Revitalizing degraded soils with agroforestry interventions: Opportunities, challenges, and future direction. In: *Agroforestry to Combat Global Challenges: Current Prospects and Future Challenges* (pp. 529–549).
220. Samal, I., Bhoi, T. K., Mahanta, D. K., Komal, J., Singh, S. (2024). Biorational pest management: Potentials, unintended consequences, and future concerns. In: *Biorationals and Biopesticides: Pest Management* (p. 47).

CHAPTER 12

Plant–Microbes Synergy in Agroforestry: Microbial Partnerships

MUHAMMAD DANISH TOOR,[1,2] MOHSIN ALI,[3*] SYEDA DUAA ZAHRA,[3] and SALEEM SAJJAD[4]

[1]*Department of Soil Science and Plant Nutrition, Faculty of Agriculture, Ondokuz Mayis University, Samsun, Turkey*

[2]*Department of Agro-chemistry and Soil Science, Faculty of Agronomy, Agricultural University, Plovdiv, Bulgaria*

[3]*School of Chemistry and Chemical Engineering, Hubei University, Wuhan, PR China*

[4]*College of Agriculture, University of Sargodha, Sargodha, Punjab, Pakistan*

*Corresponding author

ABSTRACT

Agroforestry systems, which integrate trees with crops, are increasingly recognized as sustainable agricultural practices capable of addressing soil degradation, nutrient depletion, and climate change. Central to their success are the complex plant–microbe interactions that enhance ecosystem functioning and resilience. Microbes such as mycorrhizal fungi, rhizobia, and beneficial bacteria form symbiotic associations with plants, supporting nutrient acquisition, stress tolerance, and disease resistance. In agroforestry, these interactions are particularly significant due to the deep and extensive root systems of trees, which foster diverse and dynamic microbial communities. Mycorrhizal associations, for example, facilitate improved nutrient uptake and carbon sequestration, contributing to soil

Agroforestry and Climate Change Mitigation: Harnessing Multitrophic Interactions for Sustainable Plant Protection and Ecosystem Resilience. Hanuman Singh Jatav, Vishnu D. Rajput, Ipsita Samal, Tanmaya Kumar Bhoi (Eds.)

DOI: 10.1201/9781779641397-12

fertility and climate change mitigation. Similarly, nitrogen-fixing rhizobia enrich the soil with essential nitrogen, while beneficial bacteria help suppress pathogens and improve plant health. By enhancing plant growth, productivity, and resilience, these microbial partnerships play a crucial role in optimizing agroforestry systems. However, managing these interactions is vital to minimize challenges such as competition for resources. Harnessing plant–microbe synergies offers an opportunity to advance sustainable agriculture, improve ecosystem services, and strengthen the adaptability of agroforestry systems to environmental changes.

12.1 INTRODUCTION

Soil ecosystems are significantly impacted by the diverse range of land use activities that utilize soil resources to further human economic or social objectives [1]. The various land use types are the outcome of how natural and human factors interact, which can have an ecological impact on the soil [2]. Changes in land use have a significant impact on the physical and chemical characteristics of soil, as well as the biodiversity and ecological functions of soil [3]. While unreasonable land use management can result in a sharp decline in soil biodiversity and the degradation of ecological functions, which can impact the sustainable development of soil resources, reasonable land use management can improve soil quality and increase the resistance to external disturbance [4, 5]. In soil biochemistry, which is concerned with humus formation, nutrient cycling, and the breakdown of soil organic matter, soil microorganisms play a significant role. Simultaneously, soil microorganisms serve as the link between soil and plants, sustaining ecosystem function [6]. As a result, microbial diversity is frequently employed to assess land quality and soil fertility [7–10]. The ability of soil bacteria, one of the major classes of soil microorganisms, to modify community structure and diversity in response to changes in land use may be crucial to the ecosystem's ability to function [11]. Changes in land use may have an impact on the diversity of soil biomass by changing the soil's nutrient cycling [12, 13]. Comprehending the data regarding microbial structure and functional diversity is crucial in elucidating the function of microbial communities across various land use categories. This, in turn, can offer theoretical direction for assessing and refining soil quality and land use management.

The majority of microorganisms cannot be grown in a lab, making the traditional methods of bacterial culture and identification difficult to use. This makes it impossible to fully explore the diversity of bacterial communities throughout the entire micro ecosystem from a macroscopic perspective [14]. Conserved section-specific amplification of V3–V4 16S rRNA has been a widely used method of sequencing analysis in recent years due to the advancements in next-generation sequencing and molecular biology comprehensive analysis [15, 16].

12.1.1 DIVERSITY AND COMPOSITION OF SOIL MICROORGANISMS

Improper land management techniques and incorrect land use are the causes of soil deterioration. The quality of the soil has gradually declined as a result of repeated tillage, overgrazing, shifting agriculture, excessive cutting of vegetation, destruction of sensitive land, and uneven fertilizer application [17]. In order to minimize negative environmental effects and preserve soils, sustainable agriculture practices that preserve soil biological processes and rely less on mechanical cultivation and outside inputs are required [18, 19]. Considered an alternative approach, agroforestry systems (AFSs) [20] enable the continued sustainable use of land for the production of crops and livestock while also aiding in the prevention of land degradation [21–23]. AFS is a type of multicropping in which a crop is grown alongside at least one woody perennial species, creating interactions between the two that are both economically and ecologically significant [24]. Because of the constant deposition of plant biomass and turnover of leaf litter, these systems are good at preserving and improving the quality of the soil. In a large part of the forest, leaf litter components comprising organic matter are approached by deeply ingrained roots of plants. This becomes a source that continuously supplies the soil with organic matter, boosting soil organic matter stocks [25, 26], and the possibility for sequestering carbon [27–31]. Additionally, biological qualities in the soil under AFS can be enhanced [32]. According to several authors, the AFS has higher soil microbial biomass and microbial diversity because of the benefits of trees and organic matter inputs, as well as variations in the quantity and quality of litter and root exudates [33]. Any agroecosystem's productivity depends on the existence of a sizable and varied soil microbial community. Furthermore, according to Cardoso and Kuyper [34], plant species mixtures in AFS typically permit a greater variety and/or abundance of mycorrhizae fungi than monocultures and improve the efficiency of nitrogen biological fixation, particularly in tropical soils. While the literature lists a number of advantages of agroforestry practices for soil microorganisms, further study is required to close important knowledge gaps and gain a thorough understanding of buffer effects on overall environmental quality [35, 36].

12.1.2 AGROFORESTRY SOIL

Depending on the crop type, climate, and geography, agroforestry can have varying effects on soil quality through changes in ecosystem functions and services brought about by the direct and indirect effects of trees. By pumping back nutrients that have been leached through their deep roots, trees contribute significantly to the cycling of nutrients by acting as a "safety net" against nutrient losses from the cycle. In tree-based systems, trees aid in dry deposition and also absorb atmospheric nutrients. Agroforestry presents a promising way to replenish the carbon lost in the soil as a result of heavy tillage, fertilizer use, and agricultural intensification. Because the components obtained in a mixed cropping canopy are more diverse in structure and function than in mono-cropping, agroforestry encourages more efficient use of resources [37].

12.1.3 MONOCULTURE

In the scientific literature on agriculture, the term "monoculture" is frequently used. Even though the concept of monoculture is also used in the social sciences, "agricultural and biological sciences" (949 papers) and/or "environmental sciences" (412 papers) account for the majority of these papers. Even with its broad appeal, it is exceedingly challenging to come up with a consensus definition for this term. This is most likely due to the fact that the concept of monoculture seems to be fairly obvious; after all, it might seem pointless to define a term that everyone knows what it means. But if we look closely at this idea, we find that there are a few points that merit more research. Indeed, there are important scientific and communication-related ramifications to having a vague notion of what an effective monoculture is. This is especially true with regard to the unfavorable connotation attached to the term "monoculture" in connection with the potential negative effects of agriculture on the environment, primarily the danger of biodiversity loss. Conversely, as the monoculture suggests, productive specialization is linked to the financial benefits brought about by economies of agglomeration, supply concentrations, returns to scale, and the fortification of regional supply chains.

Regarding this matter, the various dictionaries' definitions of monoculture are excessively broad and unhelpful. Fortunately, there is a more helpful definition of crop monoculture given by, who defines it as "the practice of replanting the same crop species in the same field, with no break to a different crop." The authors go on, though, saying that "large areas planted to the same crop species is also known as crop monoculture."

"We mean crop monoculture in the temporal and not the spatial sense," they conclude. Thus, based on this definition, one could contend that a monoculture can exist in both space and time. Numerous other writers adopt a similar viewpoint, concentrating on fields or farms and, implicitly, restricting the monoculture condition to the temporal dimension. The statement "cultivation of a single crop over a large area for consecutive years is the agronomic definition of monoculture" does not align with this viewpoint. There are two major implications to the notion that a monoculture condition is defined by an inherent spatial connotation that extends beyond the single field/farm dimension. First, since all crops are by definition "monocultures in time," the idea can be applied to multiyear crops like tree crops. The second has to do with the ability to lessen a monoculture state by increasing diversity through the introduction of crop rotations, the introduction of alternative crops, or inter- or mixed-cropping [38].

12.2 ROLE OF MICROBES

12.2.1 ROLE OF MICROBES IN NUTRIENT CYCLING

Most of the biological activity in soil is found in the topsoil, which can range in depth from a few to about 120 cm. The biological elements (soil microorganisms) comprise less than 0.5% of the total soil volume and approximately 10% of soil organic matter in the topsoil [22]. They are crucial to the breakdown of organic residues and the cycling of nutrients. In this sense, the organic waste is transformed into biomass or mineralized to produce CO_2, H_2O, and nutrients [39, 40].

An aspect of soil sustainability that cannot be obtained through physical/chemical measurements and/or analyses of the diversity of higher organisms is provided by the roles of these effective rhizosphere microbes. Microbes adapt to their surroundings quickly because they react quickly to changes. Changes in microbial populations and activities may thus serve as a highly useful indicator of shifts in soil health, as this adaptation may enable microbial analyses to be discriminating in soil health assessment. A diverse range of life forms can be found in the dynamic habitat which is soil, which also gives plants a mechanical support system from which they can draw nutrients. The interactions between soil organisms are ongoing and can occasionally be symbiotic, competitive, or mutually beneficial. Because the "good" bacteria, fungi, nematodes, and insects can outcompete the "bad" ones and because they play a major role in breaking down organic materials to release nutrients for plants, it improves the health of the soil. Microorganisms play a crucial role in the recycling of nutrients. Thus, it is discovered that distinct biological communities are supported by the same fundamental soil structure in various locations [41].

12.2.2 ROLE OF MICROBES IN DISEASE SUPPRESSION

Numerous plant pathogens, including nematodes, bacteria, viruses, and fungi, are the cause of different plant diseases and are managed by biocontrol agents. In order to suppress the various plant diseases and their pathogens, bio-agents employ a variety of mechanisms of action. These mechanisms include the pathogens' direct parasitism, the pathogens' inhibition through the use of antibiotics and extracellular cell wall-degrading enzymes, the competition of the pathogens for different nutrients (such as iron, nitrogen, or carbon) in colonization sites, plant defense mechanisms development, and stimulation [42, 43]. Beneficial bacteria may use multiple methods, which may be triggered simultaneously, for the biocontrol of plant diseases. By overexpressing one or more of these features, genetically modified biocontrol agents can be utilized to improve biocontrol by enabling many strains with distinct antipathogen properties to cooperate. In addition to promoting plant growth, the combined application of various biocontrol agents can also inhibit plant fungal diseases [44, 45].

The *microflora* in the soil aid in the uptake of nutrients, improving yield and reducing or suppressing disease. Numerous types of soil *rhizobacteria* exist that aid in the suppression of disease. *Bacillus subtilis* produces over 20 different antibiotics and has the potential to reduce disease. *Bacillus* species have been shown to be effective in controlling a variety of pathogens, including *Colletotrichum acutatum*, *Colletotrichum capsici*, *Colletotrichum gloeosporioides*, *Pythium aphanidermatum*, and *Rhizoctonia solani*, in a variety of crop plants, including tomato, chilli, and brinjal. Under both *in vitro* and field conditions, *Pseudomonas* spp. show antifungal activity against *Pyricularia oryzae*, *R. solani*, *Xanthomonas oryzae* pv. *oryzae*, and *Fusarium oxysporum* f. sp. *udum* [46, 47]. It has been reported that a number of soil-borne antagonists, such as *Trichoderma* spp., can prevent tomato fungal wilt brought on by *F. oxysporum* f. sp. *lycopersici* [48, 49].

12.2.3 SOIL MICROBES IMPROVING SOIL HEALTH

An integrated assessment of soil health can be obtained from microbes, which is not possible with physicochemical estimation. Microbes adapt quickly to new environments and respond very quickly to changes in the surrounding environment. Microbes that are dominant and thrive in a novel environment will be the most suitable. Due to their adaptability, microbial analyses are sensitive when assessing the health of the soil; any changes in their behavior could be interpreted as a clear indication of a shift in the health of the soil [50]. Because of their large surface area relative to soil volume, they also respond to environmental stress more quickly than higher organisms. An indicator of improved soil or a sign of degraded soil can be found in the measurable changes in soil characteristics caused by changes in the microbial population or activities. Microbial biomass converts much more quickly than the total amount of organic matter in the soil. The majority of microbial indicators have a sensitive power relationship with different soil types [51, 52]. A study revealed a direct correlation between soil respiration and the amount of microbial biomass found in the soil [53, 54]. An important factor that is directly related to soil health is the biodegradation of heavy metals, toxic chemicals, and pesticides. The effects of these harmful substances on the health of the soil are significantly impacted by microbial activity. Over time, the biodegradation of heavy metals in the soil may occur even though their presence will not. Seasons may affect polyhydrocarbon biodegradation because certain seasons have higher microbial activity than others [20]. Therefore, it is unquestionably impossible to use the total amount of hazardous chemicals present in the soil as a reliable indicator of both soil health and biodegradation. Instead, biodegradation needs to be calculated in reference to specific microbial processes. Furthermore, microbes react in order to integrate the effects of various chemical component mixtures that are not possible to obtain through independent analysis of the chemical mixtures [55, 56]. A microbial indicator is a parameter that represents aspects of environmental influences or factors that may not be included in the data gathered by experiential parameters [57]. An assessment of environmental processes that collectively indicate whether the soil is functioning normally is what is meant to be indicated by soil health indicators. According to Moghimian et al. [58], microbial indicators can estimate a variety of soil microbial processes as well as other related parameters.

12.3 SYMBIOTIC PARTNERSHIPS

The synergy between plants and microbes in AFSs is the backbone of sustainable agriculture via promoting environmental health. In nature, symbiotic relationships are exemplified by the interactions between certain microbes and plants. This relationship gives both in-touch partners a chance to benefit from each other, that is, food, shelter, specific nutrients, etc.

12.3.1 MYCORRHIZAL FUNGI

Inadequate nutrient and water availability limit crop production. In this regard, Mycorrhizal fungi earn importance among the diversity of symbiotic fungi as they are the most prevalent symbiotic partners for quite a few species of trees and crops to enhance the availability of nutrients and water [59]. Moreover, mycorrhizal fungi are capable of growing under multiple stress factors, such as pH, moisture, salinity, heavy metals, nutrients, pathogens, allelopathic chemicals, and so on [60]. Mycorrhizal fungi are habitual to tap into the photosynthates present in the plants to get fed. Meanwhile, the fungi expand their mycelial network to a vast area around the plant where the roots of the plants are unable to reach. It boosts the uptake of nutrients and water that are later supplied to the plants per demand for better growth and needs. The courtesy of mycorrhizal fungi saves the plant from dehydration and deficiency of rare nutrients, that is, phosphorus (P) and nitrogen (N) [61].

Bowles et al. [62] observed symbiotic relationships of mycorrhizal fungi with tomato plants (*Solanum lycopersicum* L.). The shoots and roots of these modified tomato species showed better growth as they succeeded in comparatively high nutrient uptake of P, N, and NH_3^+ even under low availability of water. In addition, AMF reduced nutrient leaching NO_3^- up to 54% in their partner tomato plants, concluding that AMF improves nutrient and water uptake and reduces nutrient leaching too. He et al. [63] reported an important study of the acid rain effect on *Lespedeza formosa* shrub modified with an AMF symbiotic relationship. He observed this modified *L. formosa* under varying pH of acid rain (5.6, 4.0, and 2.5) and concluded that acid rain reduces water uptake in plants but mycorrhizal *L. formosa* was not only resistant to acid rain but showed useful attributes also with this AMF partnership like enhanced water uptake in dry conditions, photosynthetic ability, better nitrogen and phosphorus (also K^+, Mg^{2+}) uptake in shoot and root, protection against Al phytotoxicity, and importantly provided low N/P ratio that is a useful value to spot nutrient limitations in plant. The collective effect of all said unique properties of *L. formosa* and AMF partnership is considered the reason behind acid rain tolerance ability; however, there can be some other hypotheses [64, 65]. Plants grow better in partnership with mycorrhizal fungi, as it increases their environmental stress tolerance ability [66] in the following ways. (1) Mycorrhizal fungi become a barrier acting as the first line of defense for their partner plant against soil-borne pathogens [67]. (2) Mycorrhizal fungi generously facilitate their associated plant by improving soil structure, better water retention, and root aeration [68].

Birhane et al. [69] reported results of a greenhouse experiment on a symbiotic partnership of *Commiphora myrrha* (*C. myrrha*) seedlings and arbuscular mycorrhizal fungi (AMF), unveiling those seedling bearing said association grew better in topsoil with adaptive long roots compared to subsoil in self-generated drought conditions taking up high concentrations of nutrients (N, P, and K), making more root colonization, confirming that AMF assists and enhances the chances of survival for the partner plants during stressful drought situations.

The facts behind how mycorrhizal symbiotic association assist plants against environmental stress tolerance can be summarized as (see Table 12.1):

TABLE 12.1 The Impact of Mycorrhizal Fungi on Plant Against Environmental Stress Tolerance

(Stress) Factors/Nutrients		Plant Species		References
		Without MFP*	With MFP*	
Photosynthetic ability		Ordinary	Increase	[63, 76, 78, 79]
Acid rain[a] (pH 2.5–5.6)		High	Very Low	
Heavy metal toxicity effect (Al^{3+})		Very high (2 mg/g)	Low (1.4 mg/g)	
Root colonization		None (generally)	Many	
Nitrogen-to-phosphorus ratio (N/P)		Average (9.5)	Reduce (6.9)	
Shoot and root	Nitrogen uptake	Ordinary (15.8 mg/g)	Increase (16.7 mg/g)	[73, 74]
	Phosphorus uptake	Ordinary (1.6 mg/g)	Increase (2.3 mg/g)	

Notes: *MFP = mycorrhizal fungi partnership; [a]adverse effect.

Mycorrhizal fungi (MF) being habitual of expanding their hyphal network long increase the absorptive surface area of plant roots allowing plant to access water from deeper soil layers and farther distances during drought conditions [70].

It also strengthens plants against soil salinity tolerance by improving ion uptake regulation, for instance, excluding harmful ions like sodium and absorbing fruitful ones like potassium. It also improves osmotic balance in plant cells through compatible solutes, either synthesized or transported.

MF helps plants in dealing with heavy metal toxicity either by immobilizing or sequestering such metals in the soil, resulting a reduction in their uptake. MF can offer compartmentalization, another solution to accumulate such metals in actual or less toxic form within the plant tissue, preventing from damage.

MF offers pathogen defense to plants either producing antibiotics or offering induced systemic resistance (ISR) by playing a role outside and inside the plant, forming a mantle around roots and stimulating the innate immune system of plant, respectively [71].

The soil around plant roots becomes less compact, improving the soil structure with the courtesy of mycorrhizal hyphae because it anchors deep into the soil, gripping the soil particles to prevent soil erosion, ensuring good aeration, and healthy root growth [72].

MF usually never lets the partner plant suffer from nutrient stress even in nutrient-deficient soil, as it spreads its hyphal network wide into the soil to enhance the uptake of essential nutrients (UOEN), for example, N, P, Mg, K, Zn, Cu, and so on [73, 74].

MF somewhat helps plants to withstand against temperature stress as it regulates the ion balance and water supply to the plant via its hyphae [75].

AMF partnership with plants also enhances photosynthetic capability [76] even under stressful conditions like acid rain, which affect photosynthesis in plants [77].

In addition, MF helps to decrease heavy metal (i.e., Al^{3+}) toxicity effect [78, 79] and decreases acid rain stress [80].

Plants enjoy a number of lifesaving advantages in AMF symbiotic partnerships, so this association is quite fruitful in agroforestry in terms of obtaining better plant growth in the fields following the massive crop yield (see Figure 12.1).

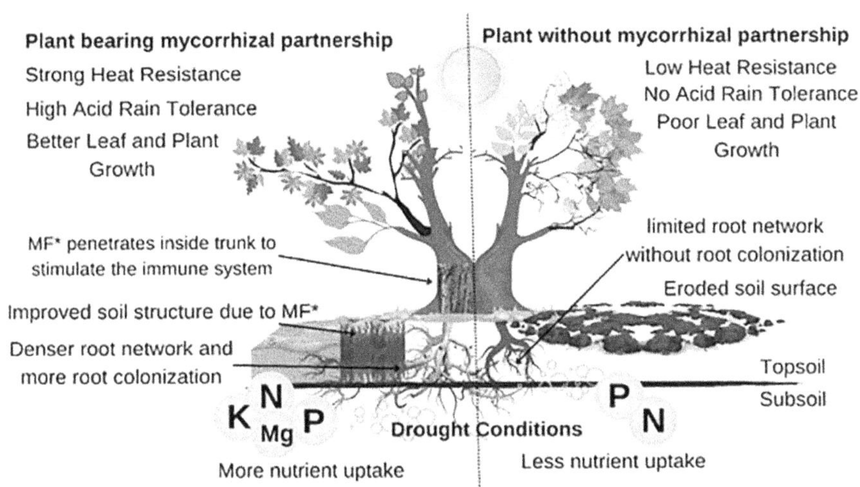

FIGURE 12.1 Comparison of mycorrhizal and ordinary plants (*MF = mycorrhizal fungi).

12.3.2 NITROGEN-FIXING BACTERIA

Nitrogen is of immense importance as it is the structural part of life-essential biomolecules, that is, proteins (amino acids), nucleic acids, chlorophyll, and so on, in plants but, despite the presence of 78% in the air; yet, nitrogen (N_2) is inaccessible to most plants. This is where plants seek help from nitrogen-fixing bacteria (NFB). So, importance of NFB lies in their function when they come into play with varying roles such as converting atmospheric nitrogen (N_2) to its useable form for plant [81], ammonia (NH_3) this is done by a number of bacteria (see Table 12.2) to increase soil fertility, crop productivity at agricultural level, reduce the need of synthetic fertilizers, and balance the ecosystem. Here it is to elaborate on why soil fertility and other factors are crucial.

TABLE 12.2 Summary of Nitrogen-Fixation Bacteria, Their Association, and Importance

Type of Bacteria	Associated Plants	Living Nature	Importance
Rhizobium	*Leguminous plants* (e.g., beans, peas)	Symbiotic	Enhances soil fertility and plant yield
Bradyrhizobium	Soybeans and some *legumes*	Symbiotic	Enhances soil fertility and plant yield
Frankia	*Nonleguminous* plants (e.g., alder)	Symbiotic	Enhances soil fertility in nonlegume hosts
Azospirillum	Grasses (e.g., wheat, rice)	Associative	Aids in plant growth without forming nodules
Azotobacter	None	Free-living	Enhances soil nitrogen content without plant association
Clostridium	None	Free-living	Enhances soil nitrogen in waterlogged conditions
Anabaena (cyanobacteria)	Freshwater systems and *Azolla fern*	Photosynthetic	Improves nitrogen balance in water bodies
Nostoc (cyanobacteria)	Freshwater systems	Photosynthetic	Contributes to nitrogen balance in water bodies

12.3.2.1 SOIL FERTILITY

NFB not only enriches the nitrogen content in the soil that is crucial nutrient needed to plants for growth but also replenish the soil fertility by releasing N_2 back into the soil if the host plant dies, reducing the need for synthetic nitrogen fertilizers, which have a serious harmful environmental impact upon intense application [82].

12.3.2.2 AGRICULTURAL PRODUCTIVITY

Plants that establish symbiotic association with NFB are highly benefited from them, resulting in high crop yield that is demanded. It happens because NFB provides associated plants with nutrient-rich soil to grow.

Adeleke and Babalola [83] documented a report on the effect of endospheric microbes in crop production, claiming the growth factors for plants important in crop yield are (induced) systemic resistance, growth hormone synthesis, antibiosis, siderophore production, and bioremediation. In addition, other researchers have reported microbes involved in the increment of a certain crop yield. *Azotobacter* present in the soil structure are useful for rice as they enhance crop yield by aiding in nitrogen fixation and phosphate solubilization [84]. *Bacillus circulans*, *Azotobacter chroococcum*, and *Cladosporium* are beneficial for wheat crop (*Triticum aestivum*) as they promote nitrogen fixation, enhance biocontrol agents, phosphate solubilization, *phytothormones*, and lytic enzymes [85]. *Pseudomonas*, *Bacillus*, *Azospirillum*, *Rhizobium*, and *Xanthomonas* are important for maize (*Zea mays*) because they provide protection against pathogens, enhance nutrient uptake, and nitrogen fixation [86]. *Rhizobium* and *Methylobacterium* are useful for soybean (*Glycine max*) as they stimulate the production of metabolites that favors plant growth followed by a better crop yield [87]. *Azospirillum* and *Pseudomonas* affect tomato (*S. lycopersicum*) production positively by improving phosphate and nitrogen uptake resulting in better plant growth [88]. *Pseudomonas* increases phosphate uptake in black pepper (*Piper nigrum*) giving good crop yield [89].

12.3.2.3 ECOSYSTEM BALANCE

Being crucial to nutrient recycling, NFB are key to balancing ecosystem. NFB supports food webs indirectly by ensuring the surplus presence of nitrogen in the soil texture, facilitating the growth of young seedlings followed by amplified productivity. Herbivores graze on nitrogen-rich plants and become available to carnivores. Eventually, when organisms die again bacteria come into play for decomposition, followed by nutrient recycling. In addition, nutrient-rich soil allows a variety of plants to grow, encouraging biodiversity of *flora* species that consumes more carbon dioxide from the atmosphere that help in mitigating the effects of climate change. Nitrogen-fixing plants in association with NFB assist afforestation and are impactful prior to succession as nitrogen depletes from the forest area after fire or clear-cutting [90] (see Figure 12.2).

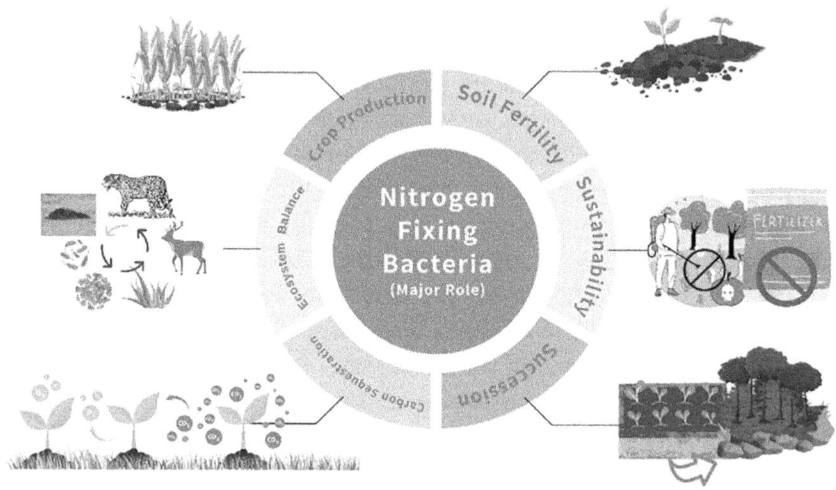

FIGURE 12.2 An overview of the importance of nitrogen-fixing bacteria.

12.3.2.4 AGRICULTURAL SUSTAINABILITY

The courtesy of these NFB paves the way for sustainable agriculture to some extent as they make the soil nutrient-rich reducing the need for synthetic fertilizers that are harmful on misuse and increase the toxicity of cultivable land. In this regard, NFB also limits *eutrophication.*

Let us have a bird's-eye view of the types of NFB that contribute toward the above-mentioned important aspects and how they manage to do that.

1. *Symbiotic NFB*: It shows a close association with plants living in their root nodules. *Rhizobium*, *Bradyrhizobium*, and *Frankia* fall in this category. *Rhizobium* [91] begins the formation of nodules at the root surface of *Leguminous flora* species like beans, peas, and clover, whereas *Bradyrhizobium*, being similar to *Rhizobium*, most often shows association with soybeans. In contrast to *Rhizobium* and *Bradyrhizobium*, *nonleguminous* plants permit their roots to structure nodules by the courtesy of *Frankia* [92] and the examples of such plants are alder, bayberry, and so on.
2. *Associative NFB:* It lives closely associated to the host plants without forming any special structures, that is, nodules on their roots. The most important bacteria in the slot of this category are *Azospirillum* [93] that is associated with grasses and important crops, such as wheat and rice.

3. *Free-living NFB*: It lives independently in the loam stuff without showing association with plants. The examples are *Azotobacter* [94], a habitant of neutral and alkaline soils, and *Clostridium*, an anaerobic bacteria habitant of waterlogged bodies.
4. *Cyanobacteria*, well-known photosynthetic blue-green algae (BGA), come with nitrogen-fixing ability as well. *Anabaena* and *Nostoc* are prime examples of BGA that can easily be found in freshwater bodies, and it is no surprise to find them making symbiotic relationships with aquatic plants like *Azolla*, a fern [95].

Conclusively, NFB play a vital role in maintaining life regarding the aforementioned aspects; however, some NFB have different significance, that is, NFB as NO_2 source or sink [96], NFB for quorum sensing [97], and so on.

12.4 ORGANIC MATTER DECOMPOSITION

Conversion of organic matter into its components through the decomposition process in dead species plays a fundamental role in nutrient recycling, which is an integral process to maintain soil health and ensure better plant growth in AFSs.

12.4.1 DECOMPOSERS

Talking about decomposers brings two microbes to mind fungi and bacteria that are the prime decomposers of organic matter breakdown, but some protozoans are decent contributors in the decomposition of dead organic matter. Their mode of decomposition may vary from each other, let us have a look.

12.4.1.1 BACTERIA

Bacillus and *Pseudomonas* species decompose organic material in soil and release nutrients in soil making them available to the plants. *Clostridium* species of bacteria decompose dead organisms (animals and plants) aerobically, releasing gases, that is, methane in the atmosphere, but *Clostridium* species HSAI-1 found in the subsurface of deep terrestrial environments are exceptionally decomposes humic substances anaerobically [98]. *Nitrobacter* and *Nitrosomonas* species are specialized for nitrification in which they convert ammonia (NH_3) to nitrites, followed by final conversion into nitrates, an absorbable form for plants.

Now, it is considered that exceptionally many bacteria, that is, *Acidobacteria, Streptomyces, Pedobacter, Luteibacter*, and so on, are actively involved in cellulose decomposition [99].

12.4.1.2 FUNGI

Aspergillus, *Penicillium*, and *Trichoderma* species decompose hard and complex materials in dead plants like lignin and cellulose. Another hypothesis states that saproxylic bacteria assist fungi in cellulose decomposition [100]. Mycorrhizal fungi that have an identity for their symbiotic partnership with host plants can also decompose organic matter. *Saccharomyces cerevisiae*, famously known as *yeast*, is used for the fermentation process to decompose sugars.

12.4.1.3 PROTOZOANS

A very few organisms of protozoans, that is, *amoebae*, *ciliates*, and *flagellates* perform decomposition living in water and soil. Additionally, they prey on bacteria to regulate their population. Protozoans like microcrustaceans and ciliates living in water bodies decompose aquatic plant litter, that is, leaves, and show support in decomposition to major decomposers, bacteria and fungi [101]. Conclusively, they participate actively in the ecosystem in energy and nutrient flow, becoming a part of the freshwater food web, but on the other hand, they can cause waterborne diseases to other organisms, that is, bacteria, and upon their death graze on them [102] (see Figure 12.3).

As these microbes succeed in breaking down organic matter, they *promote nutrient release* such as nitrogen, phosphorus, and potassium, which can be taken up by plants, ensuring a sustainable nutrient supply without any active efforts. This helps nutrient and carbon cycling as well in which they (microbes) breakdown organic matter, that is, dead organisms (plants and animals) into simpler substances releasing essential nutrients

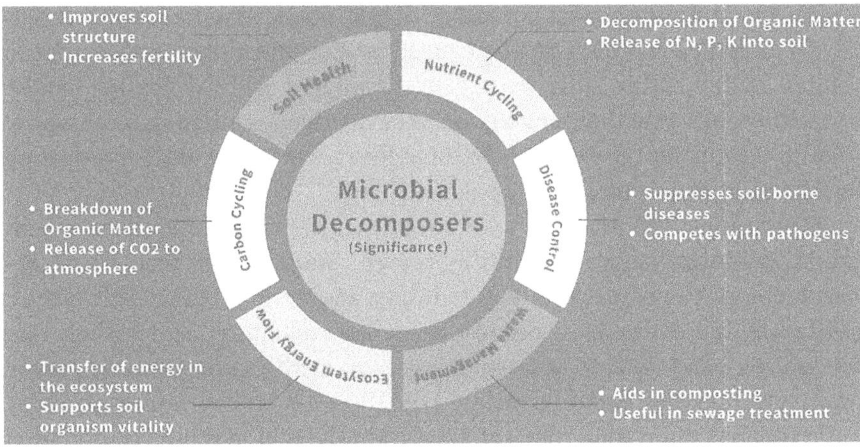

FIGURE 12.3 A visual representation of key roles of microbial decomposers.

(N, P, and K) into the humus content and liberate carbon dioxide (CO_2) to the surrounding, respectively. Nutrients released during nutrient cycling become available again to the growing plants; similarly, CO_2 released during carbon cycling becomes available to the plants for photosynthesis. However, litter decomposition in tropical forest regions may vary in terms of duration and other factors [103].

Decomposers help in *energy transfer within the ecosystem* by decomposing the intricate organic matter into fundamental constituents that can readily be consumed by the needy organisms. Without decomposers, nutrients will be stuck within the dead bodies, while the rest of the organisms may face severe deficiency, so decomposers make sure the smooth energy in the ecosystem. Often bacteria and fungi join hands to work mutually to decompose important matter that they separately find hard to break, that is, *macrophytes* of shallow lakes like *Zizania latifolia*, *Nymphoides peltate*, and so on, where bacteria can be temperature sensitive and fungi can be decomposing matter composition dependent [104].

The role of decomposers in *waste management* cannot be ignored as they help treat sewage, otherwise, it is an undesirable challenge to deal with waste. Some of the biodecomposers detoxify the harmful pollutants and contaminants by degrading them. Microbes aided with extracellular enzymes can also degrade biodegradable polymers that are very helpful for agroforestry for waste volume reduction [105].

We get *prevention from* a number of *soil-borne diseases*, by the courtesy of decomposers, that can be caused by the rotten dead bodies, putrid organic matter, feces, and pathogens generated from decaying matter. Some microbes, that is, *Sporanaerobacter acetigenes* uniquely decompose volatile organic matter and convert them into harmless products preventing health issues from volatile compounds [106].

Maintaining soil integrity over extended periods is crucial and essential, so microbes play a role in this regard. Some by-products of microbial decomposition serve as soil-building agents like humic substances, *enhancing soil structure*, promoting water retention, supporting root penetration, and overall guaranteeing soil health. For instance, some ammonia-oxidizing bacteria, that is, T-AOB-2, M-AOB-4, and so on decay cattle manure improving nutrient content in the soil making it cultivation fit [107]. Decomposers ensure *soil health* as they replenish the soil with useful nutrients by composing organic matter, making it best fit for cultivation, following the enhanced crop yield. Such microbes aided nutrient-rich soil are the idea for succession after an unwanted disaster [108].

Debate can be summarized as mycorrhizal fungi, NFB, and decomposers have importance in ecosystem balance, environmental energy flow, plants growth, nutrient cycling, soil fertility, stress tolerance, waste management, and so on, and all of these factors or phenomenon contribute toward the improvement in agroforestry needs. Agroforestry is very much concerned about the plant growth that is further associated with the yield of crops that is the need of the growing population, more crop production is linked with the standard environmental growing conditions, including fertile land. In this era where crop growth faces a number of environmental stress issues, such as nutrient deficiency, and so on, microbes resolve these issues to a great extent [48, 109, 110]. NFB and decomposers altogether increase soil fertility by fulfilling the deficiency of nutrients, and so on, but we conclude that soil structure and properties are greatly improved by mycorrhizal fungi. Because a report by Fall et al. [111] highlight the desired role of mycorrhizal fungi states that mycorrhizal fungi help greatly to boost the chemical, biological, and physical features of the soil which can highly be demanded in agriculture to deal with the environmental stress issues and climate change to gain high crop production through standard soil condition. Mycorrhizal fungi improve chemical properties of the soil by increasing nutrient content, that is, N, P, K, and so on that is associated with the biological properties of the soil which means that mycorrhizal fungi release secretions into the soil that become the reason for microorganisms and useful bacteria to gather forming communities ensuring the revamp of biological properties of the soil. These microbes bring about nutrient cycles and useful metabolic reactions resulting in accumulation of useful nutrients, which can improve the chemical structure of the soil. However, plants without MF association suffer regarding phosphorus uptake despite its availability in the soil [112], but MF offers a solution to this stress by making interaction with a phosphorus-solubilizing bacteria (PSB), which makes phosphorus available to the root of growing seedlings in the form of phosphates. Over and above this all, MF improves soil structure by producing stable soil aggregation glomalin, a glycoprotein that binds microaggregations to form stable macroaggregations improving biophysiological properties and soil structure. Dead hyphae of MF not only act as a binder for soil structure but also as a source of organic contents; so, overall, MF contribute exceptionally toward improving chemical, biological, and physical structure of the soil by reducing the chances of soil erosion, loss of organic mass, surface runoff, and so on (see Figure 12.4).

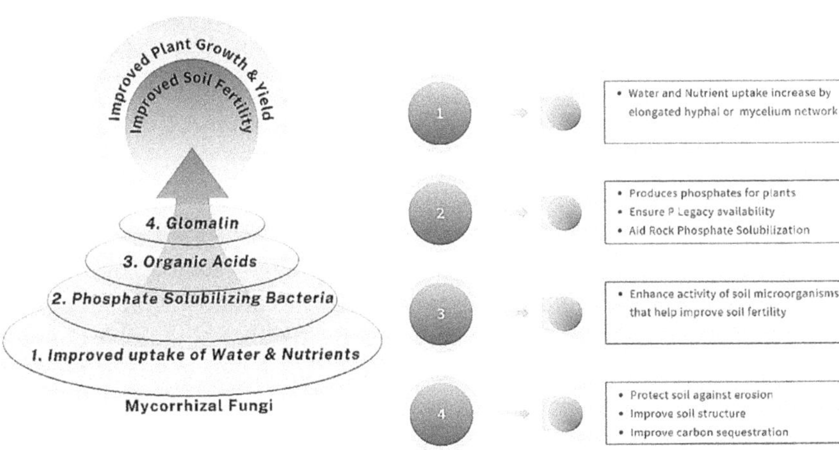

FIGURE 12.4 Contribution of mycorrhizal fungi in soil fertility.

Further research on the importance of mycorrhizal fungi, their associations (with bacteria for cellulose decomposition), working mechanism (importantly the nature of secretion produced by MF), and the role of protozoans as decomposers in the field of agroforestry is demanded in the future.

12.5 BIOLOGICAL CONTROL

The burgeoning world population will be demanding a food production increase of up to 70% by the year 2050. The cereal production is 2.1 billion tons now and will need to be raised by 3 billion tons, which means that crop production demands to be enhanced in order to meet the need of food for the world. But besides enhancing crop production, plant disease prevention and pest control are major threats to agroforestry. It is estimated that annually, US$220 billion are lost due to the diseases caused by pathogens [113]. The use of environment-friendly, biological agents for acquiring sustainable agriculture is a dire need at the time [114]. Soil health is directly related to the intensification of agriculture, and soil-borne diseases in plants are a major issue to be addressed in sustainable agroforestry [115]. With the advances in agriculture, the use of agrochemicals for the prevention of disease was enormously implemented, but the effects of these chemicals on the lives of plants as well as animals cannot be neglected [116]. Many types of harmful *genres* which act as plant pathogens are contained in the soil, such as bacteria (*Erwinia*, *Raltsonia*, *Rhizomonas*, *Agrobacterium*, *Streptomyces*, etc.), nematodes (*Meloidogyne*, *Heterodera*, *Longidorus*, *Paratrichodorus*, etc.), fungi (*Sclerotiumrolfsii*, *R. solani*, *Fusarium* spp., *Pythium*, *Phytophthora*, etc.) [19]. From the introduction of the concept of "friendly microorganisms" by Teruo Higa in the 1980s, which also highlights the concepts of positive and negative microorganisms, until now, many microorganisms are used practically in disease suppression in plants [117]. There are many approaches being used in the manipulation of microorganisms in soil-borne disease prevention such as soil microbiota (plant growth-promoting microorganisms, biological control agents, organic amendments [OAs], etc.). Whether the defense strategy is going to be successful or not mainly depends upon the *microbiota*. For instance, if the soil properties are being diminished or yield is constantly deteriorating, then the *microbiota* is involved in wiping off the soil-borne pathogens by enhancing natural soil suppressiveness [118].

12.5.1 THE ACTION MECHANISM OF PATHOGENS

Among the 1–10 million microorganisms that are present in 1 g of soil, many genres are disease-causing as discussed earlier. These pathogens are capable of causing diseases by completing their life cycle either residing in the soil or by occupying the plant itself. these pathogens are persistent because they have a very wide variety of hosts and can even survive when the normal host crop is absent [119]. The action of these pathogens is through the "disease triangle" mechanism, that is, the host, environment, and pathogens are interconnected. The environment is the most prominent factor in deciding the success of microbes to invade the plant roots to cause diseases [120]. It is known that the *rhizosphere* acts as the defense line and is the root–soil interface for the plants and the host–microbe or microbe–microbe interactions occur through it. So, it is very common for soil-borne pathogens to attack the plant through it [121].

12.5.2 ROLE OF MICROORGANISMS IN DISEASE CONTROL AND REDUCTION OF CHEMICAL INTERVENTIONS

Almost 25% of crops are destroyed because of different plant diseases. So, the management of plant diseases is the immediate need of the time. Traditionally, increased hygiene, crop rotation, use of more resistant varieties, and use of chemicals to prevent diseases were commonly implemented [122]. Many types of *fungicides* and chemical interventions have been used to combat the diseases since the 1960s, but the resistance of the pathogens to these chemicals, the presence of residues of these chemicals in food, and their effects on the environment have raised the need for their replacement with more useful and environment-friendly methods. For instance, the clubroot disease, caused by *Plasmodiophora brassicae*, infected its host plants in 75 countries and has been tried to be eliminated by chemical methods as well as by using other conventional methods, but the disease could not be completely eliminated [123, 124]. Many different methods, including the use of OAs, were being practiced, but their environmental pollution-causing effects cannot be overlooked. These conventional methods need to be eliminated or be replaced by up-to-date sustainable methods of disease management [125]. Considering all these problems, the use of biological control for disease prevention is obligatory.

Microorganisms are the true dominant part of the soil, not only because of their ability to suppress diseases, but also their ability to modify soil structure, outrun the nutrient cycle, aid the productivity of the soil, and change the ecosystem of the soil [126]. In soil-borne disease management, these microbes act as antagonists to pathogens and show their action through various modes: (1) producing antibiotics, (2) through lytic enzymes, (3) through competition for food, (4) through parasitizing of the pathogens directly. Due to the presence of these microorganisms, the soil often turns to be suppressive soil, that is, the soil where the disease-causing effect of the pathogens is mild or reduced in comparison of conducive soil [125, 127].

Different microbes have different modes of action to protect the plant from these diseases. PGPRs (plant growth promoting *rhizobacteria*) protect the plants through induced systemic resistance mechanism, through the phenomenon of *antibiosis*, via competing with the pathogens for the intake of iron, promoting the growth in plants and retarding the virulence of pathogens [128]. Likewise, *Trichoderma*, AFM, *Trichoderma harzianum*, and *B. subtilis* show the protective behavior against pathogens by different mechanisms. In addition to the sole action, a combination of different strains can be implied in order to attain positive results [129] (Figure 12.5).

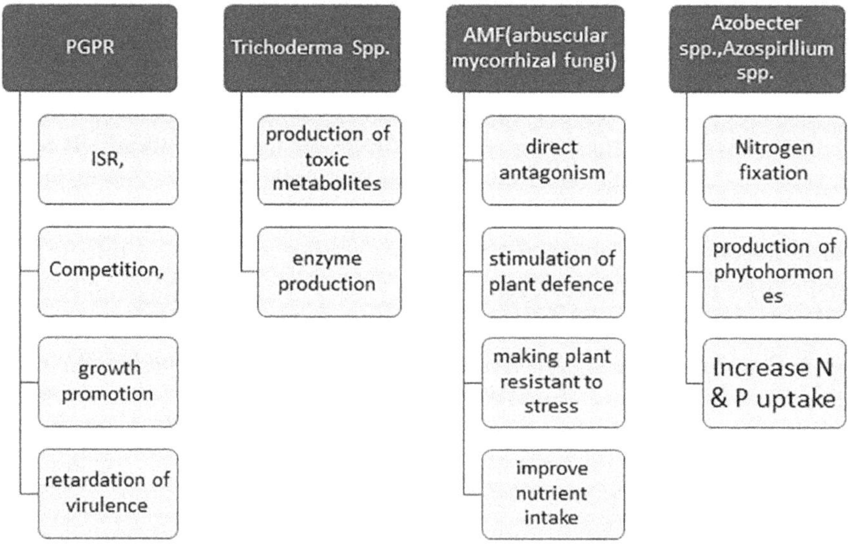

FIGURE 12.5 Some common examples of potential microbes involved in disease prevention.

Many microorganisms are being used in the control of disease management. As for the removal of caterpillars, *Bacillus thuringiensis* is used, and various bacterial strains are being used for the removal of the pests. Genus fungi also provides a prominent framework for combating the plant diseases, for instance, for pathogens such as *R. solani*, *Colletotrichum capsica*, *P. aphanidermatum*, *Colletotrichum acutatum*, *Pseudomonas* spp., and *C. gloeosporioides*, which are responsible for causing root disorders in different vegetables. The *Bacillus* strains are very effective for disease prevention [130] (Table 12.3).

12.6 MICROBIAL INOCULANTS

The biotic as well as abiotic factors are directly associated with maintaining soil health, which is ultimately balanced by all, chemical, biological, and physical characteristics of soil and their interaction with the environment. Factors like exhaustion of nutrients, depletion of organic assets of soil, erosion, pollution, and so on are gradually causing soil health deterioration and degradation of the soil. These gaps in the preservation of soil health and nutrients availability, and so on, need to be filled in order to achieve sustainable agriculture [140]. The use of agrochemicals for this purpose has been widely adopted, but the detrimental effects of these inorganic fertilizers to the environment cannot be neglected. According to statistics, 51 water bodies in the United States were detected to have deleterious elements, like arsenic, cadmium, mercury, lead, and so on, while the presence of numerous pesticides was detected in the air in inhabited areas in Japan. So, the need of eco-friendly and useful alternatives is the demand of the time [141]. Microbial inoculants also termed as biofertilizers are emerging as the most useful alternative to fill the nutrient gap, mitigate the environment stress, and maintain soil health. Microbial inoculation which is known to have its positive impact in agriculture for 120 years, is one of the best alternatives for the maintenance of soil health as well as mitigating the environmental stress. The first bioinoculant was "*Nitragin*" that contained nitrogen-fixing species of bacteria "*rhizobium*" [142]. These inoculants not only directly affect the plants or soil properties but the diminished properties of soil such as declined effects of beneficial microorganisms because of intense cultivation, excessive use of agricultural land, and so on, are regained by these microbial inoculants. The $1.34 billion market of these microbial inoculants is expected to be expanded to $3.15 billion in 2026, because of the use of these inoculants in developing

TABLE 12.3 List of Soil-Borne Diseases, Pathogens, Their Target Plants, and Potential Microbes for Disease Prevention

Soil-Borne Disease	Pathogens	Target Plant	Prevention	References
Common scab	*Streptomyces* spp.	Beet, mangel, turnip, rutabaga, radish, carrot, salsify, and parsnip	*S. diastatochromogenes* *B. velezensis* *P. mosselii*	[131]
Rhizoctonia root rot	*Rhizoctonia solani*	Members of the Poaceae, Fabaceae, etc.	*Trichoderma* *PGPR* *Bacillus bacterium*	[132]
Wilts	*Fusarium oxysporum* and *Verticillium* spp.	Aconitum, Dahlia, Salvia	*Pseudomonas fluorescens* PICF7 *Paenibacillus* *Streptomyces virginiae*	[133]
Phytophthora blight disease	*Phytophthora capsici*	Lavender, pepper, spurge	*Trichoderma* *Aspergillus flavipes* *Pseudomonas putida*, *Pseudomonas libanensis*, *Pseudomonas aeruginosa*, *Bacillus subtilis*, *Bacillus megaterium*, *Bacillus cereus*, etc.	[134]
Pythium damping-off disease	*Pythium debaryanum*, *P. ultimum*, *P. graminicola*, etc.	Green house crops, zucchini	*Trichoderma* cf. *harzianum*, *T.* cf. *asperellum*, *T.* cf. *virens*	[135]
Bacterial wilt disease	*Ralstonia solanaearum*	Tomatoes	*Trichoderma harzianum*, *Bacillus subtilis*	[136]
White mold disease (sclerotinia)	*Sclerotinia sclerotiorum*	Common bean tomato plants	*Trichoderma harzianum*	[137]
Lettuce drop disease	*Botrytis cinerea*, *Sclerotinia minor*	Lettuce	Mycostop (*Streptomycetes griseoviridis*)	[138]
Clubroot	*Plasmodiophora brassicae*	Radish	*Zhihengliuella aestuarii* *Lysobacter* *B. amyloliquefaciens* strain	[124]
Fusarium-wilt disease	*Fusarium oxysporum*	Watermelon	Arbuscular mycorrhizal fungi PGPR *Trichoderma* spp.	[129]
Root-rot disease	*P. myriotylum*	Cocoyam	*T. asperellum*	[139]

plant–microbial interaction, amplifying the accessibility of nutrients, declining the pest affliction, enhanced plant growth, and shielding plants against the environmental stress as well as intensifying the contention against encroaching species [143].

12.6.1 APPLICATION STRATEGIES

Microbial inoculants occupy a very crucial role in agroecosystem. Here is the list of some practical applications of microbial inoculants to enhance the AFS performance and mitigate the environmental stressors.

12.6.1.1 NITROGEN FIXATION

Nitrogen fixation is done via a number of microbial species, such as *Rhizobium*, *Bradyrhizobium*, *Frankia*, *Azospirillum*, and *Cyanobacteria* also termed as BGA (*Anabaena* and *Nostoc*), *Clostridium*, *Azobacter*, *Azoarcus*, and so on. The brief mechanisms and working of these species are given as follows:

12.6.1.1.1 Rhizobium Species

Stress (i.e., drought, pH, salinity, acidity, high temperature, nutrients deficiency)-tolerant *Rhizobium species* [144] form *Legume–rhizobium* symbiotic association to help fix nitrogen for partner plant, forming root nodules that reduce nitrogen (N_2) present in the air to ammonia (NH_3) which is readily available to the plant for use as per demand. This is done through catabolism of a plant-derived dicarboxylates allowing the reduction of N_2 to NH_3 (see Figure 12.8).

12.6.1.1.2 Frankia Species

Another important, nitrogen-fixing species *Frankia* found in the root nodules or living freely in the soil transforms the atmospheric nitrogen into organic compounds such as amino acid. This species has vesicles and spore-specific structures that are particular for nitrogen fixation [92].

This species forms actinorhizal symbiotic relationship with the host plant. Unlike *rhizobia*, *Frankia* has an oxygen preservation system, that is, it shields the oxygen-sensitive nitrogenase enzyme complex. This specification allows *Frankia* to let various host plants be involved in symbiotic relationship [145].

12.6.1.1.3 Cyanobacteria

Among the top-tier contributors in nitrogen fixation, *cyanobacteria* hold a principal place. The special type of cell known as heterocyst is the main site, while *Bryophyte–cyanobacterial* relationship is the chief cause of nitrogen fixation [146]. These are oxygenic photosynthetic bacteria, so the presence of oxygen-sensitive enzyme, *nitrogenase*, which is responsible for nitrogen fixation, and cause the *cyanobacteria* to evolve for doing their work well. So, the heterocyst lacks oxygenic photosystem and a glycolipid cell wall is present instead, which helps in keeping a lower concentration of oxygen in order to fix nitrogen [95].

12.6.1.1.4 Azospirillum

An important subclass of Proteobacteria, *Azospirillum* is recognized as one of the salient members of plant growth-promoting bacteria. They fix nitrogen in the roots of the plant, but without the formation of any nodules or new organs. It was reported that agricultural production was enhanced up to 25–30% because of these bacteria [147]. Some of the specialties of these bacteria include their efficiency in the highest amount of nitrogen fixation, and their forbearance to comparatively higher pH, salinity, and temperature [148].

12.6.1.1.5 Azobacteria

When it comes to *Azobecteria* for nitrogen fixation, the nitrogenase is protected by these bacteria by showing resistance to oxygen as well as there is uptake of hydrogenase for nitrogenase protection. An estimate says that these bacteria fix 20 kg N/ha/year [149].

12.6.1.2 PHOSPHATE SOLUBILIZATION

Phosphorus, the second-most important nutrient in plant life has limited availability to the agricultural soils because of its powerful binding ability to the soil. So, this problem can be solved by the implementation of microbial resources, which can improve the P cycle in plants and make the availability of phosphorus easier [150]. PSM are referred as phosphorus-solubilizing microorganisms and include some species of fungi and bacteria. PSB includes *Pseudomonas* spp., *B. circulans*, *Burkholderi*, *Erwinia*, *Enterobacter*, and so on, while among fungi *Aspergillus*, *Alternaria*, *Glomus*, *Penicillium*, and so on, are well known [151].

12.6.1.2.1 Pseudomonas

Inhibiting the plant tissue as well as *rhizosphere*, *Pseudomonas* have a significant role in attaining sustainable agriculture via playing a role in disease management along with phosphorus solubilization [152]. Recent studies show that *Pseudomonas plecoglossicida* solubilized 75.39 mg/L of phosphorus and also produced 38.89 ppm of indole acetic acid. These properties of *Pseudomonas* make it a special candidate of PSMs [153].

12.6.1.2.2 Bacillus Species

In addition to PGPR, PSB are used as biofertilizers for enhanced plant growth, availability of nutrients, and improved yield. *Bacillus* spp. is the most common type of PGPR, having endospores which enable it to survive and prevail in various environmental conditions. Among *Bacillus* spp., *Bacillus licheniformis* CKA1 are the one who showed the P solubilization and their potential is affected by the glucose dehydrogenase enzyme, which have a influence on gluconic acid production [154]. PSBs are responsible for the P solubilization and the mechanism behind this process is a reduction in pH of the soil because different organic acids or acid phosphatases are produced.

12.6.1.2.3 Burkholderi

A few *Burkholderia* are known to have the ability to dissolve phosphorus. Some strains of *Burkholderia*, which are isolated from white lupine are responsible for plant growth because they can dissolve phosphorus even If the concentration of phosphorus is not high. By *Phaseolus vulgaris*, *Burkholderia cepacia* promotes the phosphorus intake up to 44% more [155].

12.6.1.3 PLANT GROWTH-PROMOTING RHIZOBACTERIA

Besides being an effective, eco-friendly, and most effective alternative to the chemical fertilizers, PGPR are associated with biological nitrogen fixation, siderophore formation, phosphorus solubilization, phytohormone production, establishing symbiotic relationships between plant–microbes, and many more positive effects to sustainable agriculture [156]. These PGPR act as ecofriendly, efficient biofertilizers, and because of their properties, the global market of biofertilizers is expected to be raised up to $3.8 billion than $2 billion in 2019. They can also work as biocontrol agents, which provide protection to the plants by producing antibiotics in response to different disease-causing pathogens [157]. Since PGPR are associated with enhancing the availability of nutrients, the relationship between the PGPR and the host is quite important. So, two types of associations are there:

1. *Rhizospheric*: In this relationship, the PGPR are colonized and are attached to the plant via roots (sometimes peripheral intercellular spaces as well). But the colonization is affected by variations of properties of soil, which are the ultimate result of pH change, change in partial pressure of O_2, and so on.
2. *Endophytic*: This association develops when the PGPR are living inside the apoplastic spaces in plants and one of the best examples of this relationship is *legume–rhizobia* symbioses [158].

Moreover, PGPR influences plants via two approaches:

1. Direct approach where the PGPR directly affects the plant directly.
2. Indirect approach where the PGPR is not itself acting on the plant but having a influence via the production of some molecules, which have an impact on host plants [159] (Figure 12.6).

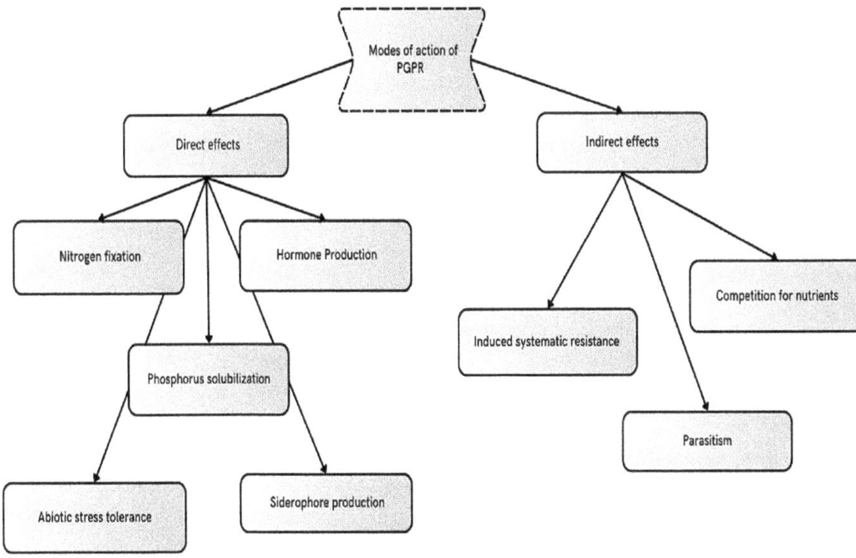

FIGURE 12.6 Illustration of modes of action of PGPR.

12.6.1.4 MYCORRHIZAL ASSOCIATION

Out of all seven mycorrhizal associations, the symbiotic relationship of plants with arbuscular mycorrhizae and ectomycorrhiza has utmost importance. High-yielding, firm, and substantial relationship between plants and mycorrhizae make the mycorrhizal inoculant technology as significantly favorable for sustainable agroforestry [160]. They play remarkable role by stimulating plant growth. There are various modes of actions for inducing stimulation in plants. These mechanisms are as follows [161] (Figure 12.7):

They also have a positive impact on agriculture by reducing the quantity of fertilizers being used. Besides, these mycorrhizal associations not only individually provide benefits to the plants but also work in coinoculation with other microbes, too. The most significant coinoculation of the mycorrhizal fungi is with bacteria [162]. AMF form a symbiotic relationship with endophytic fungi also. It was observed that these fungal inoculations enhanced leaf activity and also successfully maintained the concentration of hydrogen peroxide at a lower rate in comparison with the uninoculated plant leaves [163].

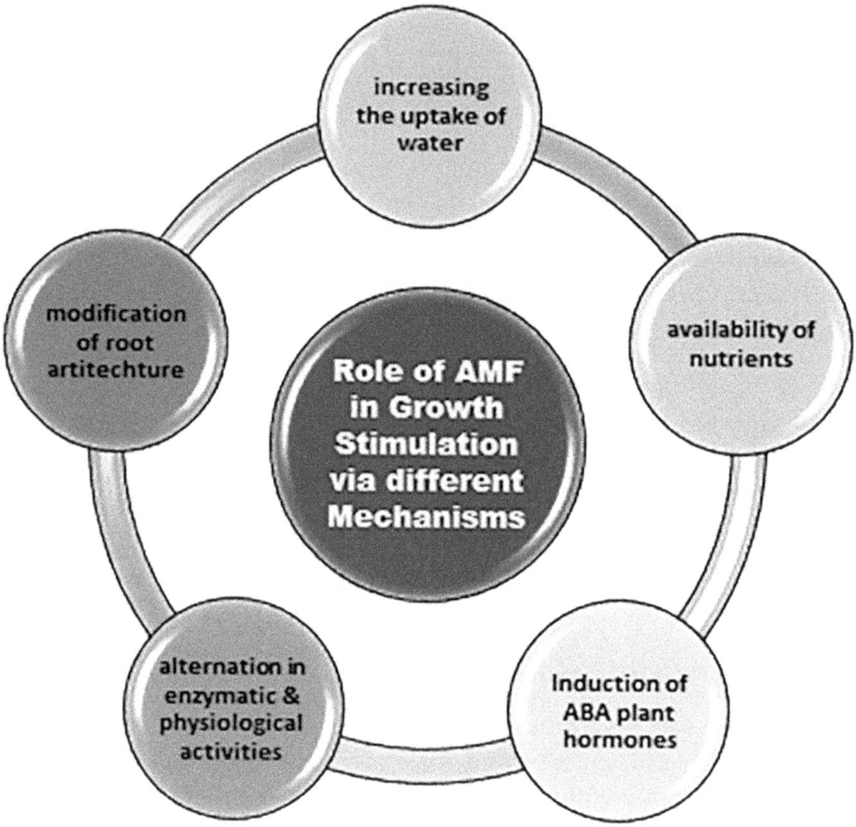

FIGURE 12.7 Different mechanisms used by AMF in growth stimulation.

12.6.1.5 STRESS ALLEVIATION

Production of various chemicals such phytohormones, extracellular polysaccharides, antibiotics, siderophores, and so on not only promote the growth of the plants or enhance nutrients availability but has prominent stress-reducing effects on the plants as well as soil health, which are a result of biotic and abiotic stresses. Thus, microbial inoculants act as mitigating elements to environmental stressors [164]. Soil salinity emerged as one of the crucial global apprehensions since it is a major global concern, causing disturbance to the soil quality and has deleterious effects on the agricultural productivity. Here, the halotolerant inoculants or PGPR act as the most effective salt stress regulator at molecular, and physiological level in addition to the biochemical level as well [165]. The role of microbial inoculants as mitigate environmental stressor cannot be negotiated since their use is far away than only promoting plant growth. For example, *Azospirillum* spp., *Pseudomonas syringae*, *P. fluorescens*, *Bacillus* spp., and *Trichoderma harzianum* are resistant to salinity and drought. In particular, drought tolerance comes from the production of volatile organic compounds along with the stress-responsive genes. Furthermore, many strains of AMF alleviate stress by increasing the uptake of nutrients [166, 167].

12.6.1.6 DISEASE RESISTANCE

ISR for disease control brings about some fruitful results in disease resistance by the production of different molecules by beneficial microorganisms, including some *Bascillus* spp., *Trichoderma* spp., and so on. ISR gets triggered by the production of antibiotics, siderophores, and different hormones and ultimately adopts a defensive approach to disease control [168]. ISR is activated in response to three mechanisms such as jasmonic acid, ethylene, and salicylic acid, known as "dependent-independent pathway." In addition to these, volatile organic compounds, phytohormones are beneficial in resistance to disease [168].

12.6.1.7 ROLE IN BIOREMEDIATION

Use of beneficial microbes for the removal of contaminants is regarded as bioremediation (a clean-up technology) and significant amount of these pollutants is removed by the action of microbial inoculants. They improve plant health as well as promote its growth. About 35,000 pollutants site as per reported in an EP survey, need to be cleaned, and soil pollution needs to be eradicated. Bioremediation is not eco-friendly but is also cost-effective which makes its implication even more useful [168]. These microbes either biodegrade, decompose, or bio-transform the contaminants, but their mechanism and action depend upon biotic as well as abiotic factors. Bioremediation involves the use of advanced technologies such as bioaugmentation, use of biofilms, bio-accumulation, and many more [169]. These microorganisms play this crucial role in bioremediation, where they either degrade or immobilize the pollutants while using the pollutants themselves as substrate, by releasing the enzymes that can degrade organic as well as inorganic pollutants, for example, polycyclic aromatic hydrocarbons, lead, chromium, and so on [170] (Table 12.4).

TABLE 12.4 Various Microbes and Their Role in Bioremediation

Microbial Inoculants	Bioremediation Application	Reference
Pseudomonas putida	Hydrocarbon degradation	[171]
Bacillus subtilis	Heavy metal removal	[172]
Clostridium acetobutylicum	Bioremediation of chlorinated solvents	[173]
Rhizobium leguminosarum	Soil fertility improvement	[174]
Rhodococcus erythropolis	Polycyclic aromatic hydrocarbons degradation	[175]

12.6.1.8 CARBON SEQUESTERING

As the highest contributor of 63% of total greenhouse gases, CO_2 concentration needs to be controlled in the environment in order to tackle the most attention-demanding problem globally, global warming. Carbon pool of the soil is the third largest pool, consists of mainly soil organic carbon and soil inorganic carbon, thus having an influential impact on carbon sequestering, and microbes living in soil can contribute well in managing the carbon sequestering because of different interactions [176]. For instance, electroactive bacteria play a role in sequestering by producing carbonates and bicarbonates from carbon dioxide. Likewise, in addition to producing secondary products, microbial electrosynthesis captures carbon. Furthermore, microalgae have quick action in capturing carbon at the expense of O_2 release [177]. In this regard, fungi and bacteria are the main contributors because they play a role in carbon sequestering by regulating various pathways as well as by owning those metabolic activities that are involved in capturing carbon, such as their capability of carbonates sedimentation, recalcitrance of tissues, and so on [178]. Similarly, PGPR not only increases the production of plant biomass but also provides climate change mitigation through carbon sequestering. In the soil, it enhances the carbon input up to 91% as compared to an average of 41% [179] (Figure 12.8).

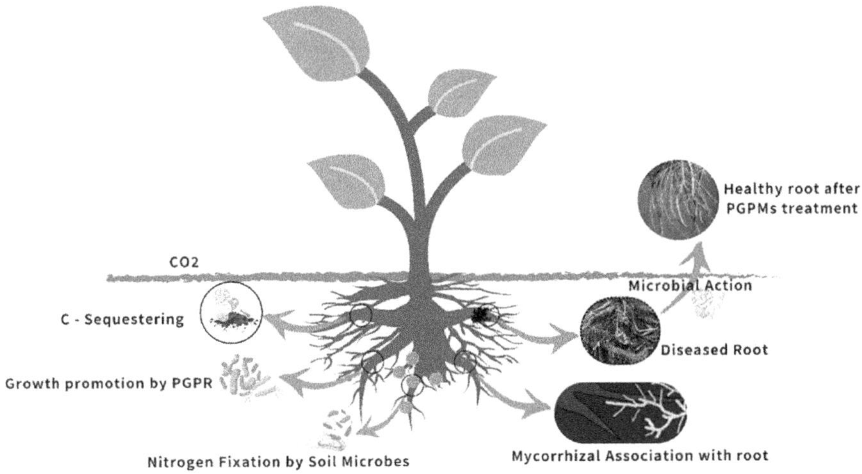

FIGURE 12.8 Different roles of microbial inoculants.

12.7 EMERGING RESEARCH DIRECTIONS

As discussed earlier, the action of microbes is interrelated in terms of environment, host, and microbes itself and a deep understanding of the relationship between these three parameters is crucial. The diverse applications of these microbes, such as making nutrients available to plant, protection from pathogens, stress alleviation, and so on need the microbe–plant relationship to be fairly comprehended in order to get maximum advantages from these microbes, because still a large number of microbes with potential beneficial use are unidentified and anonymous [180]. Some ongoing and future research directions that can advance our understanding for plant–microbe interaction can be illustrated as follows:

12.7.1 MULTIOSMIC APPROACH

The factors like climate change and the invasive species continuously affect plant health, and there is a need to apply a holistic approach in this regard. A multiosmic approach has a significant advantage in this regard since it can reveal a broad and clear picture of the plant–microbe interaction. So, multiosmic technologies can be proven as an effective approach for a better understanding. Some of these include:

1. *Genomics*: The study of genetic make-up of plant as well as microbes, such as fungi, bacteria, algae, and so on, prove to be an important tool in understanding the ecological questions in particular those, related to the change in the environment.
2. *Metagenomics*: The studies that involve the DNA investigation, which is directly influenced by the environment. This field particularly helps in realizing the functional role and variety of those microbes that are unculturable.
3. *Metabolomics*: The technique or approach that involves in a detailed comprehension of all molecules, even the smaller ones, in organisms. It is associated with the chemical analysis and provides a deep insight into the metabolites present in plants and microorganisms.

Other techniques include spectranomics and volatilomics, which are associated with the characterization of the response of plants to natural or environmental stress and comprehensive studies of the volatile compounds produced by organisms, respectively [181].

12.7.2 MICROBIAL ENGINEERING

Microbiome engineering can help in boosting plant growth and control of pathogens. There are two approaches to microbe engineering. One is an indirect approach that uses soil amendments and the use of selective substrates while the second approach is a direct approach that includes inoculation of plant with probiotic microbes or breeding or microbes as well as microbiome transplantation. The future of this approach lies in the fact that it is not always microbes that affect plants, but the plants also manipulate microorganisms, thus it is necessary as future perspective that we should focus on how we can maintain the right track for microbiome changes that occur in natural conditions [182].

12.7.3 CLIMATE CHANGE RESILIENCE

The increasing awareness about the climate change is imposing the need of plant–microbe interaction and the use of microbial capabilities to mitigate environmental changes. This can be done by either direct use of the microbes or their indirect manipulation to implement changes in the land and environmental practices. Since, these microbes are facing issues like arid soils, when the soil is wet again; these microbes have their microbial activity diminished. Drought stress, temperature increase, CO_2 gas emission, salinity, soil moisture change, increased precipitation, and floods are some of the main hurdles being tackled by the microbes and need to be addressed in the future. This can be done by connecting the details from landscape-scale resolution and microbiome studies and then use of these studies and advanced approaches to solve these problems [183].

12.7.4 MICROBIAL-PLANT-FAUNA INTERACTIONS

In studying and exploring the plant–microbial interactions, one of the most attention-seeking and important interactions, that is, the interaction of soil fauna is always neglected. Since most of the interactions occur through roots, so the fauna, inhabiting the roots needs consideration. Several interactions, such as bacteria–microfauna interactions, fungi–fauna interactions, mesofauna–fungi interactions, microfauna–fungi interactions, fauna–fauna interactions, are some important interactions that have a very strong impact on plant–microbe interactions. The mechanisms for these interactions are still unknown. Many multidisciplinary and interdisciplinary approaches need to be implemented in order to have a better understanding of these interactions as well as better future consequences [184].

12.7.5 PRECISION AGROFORESTRY

Nutrient availability, pests and disease management, improved soil health and resilience to climate change are demonstrated in AFS. Multifunctional land use system was first given by John Bene in 1970s, and it was then expanded worldwide. The effects of agroforestry include competitive landscape, microclimate, productivity, and many beneficial interactions. However, there is a huge lack of data-driven approaches that are needed for a better understanding of spatial interactions in real-time. Plant phenotype, wind effects, rain effects, light intensity, temperature, humidity, environmental measurements, soil nutrients availability, pH, texture, soil temperature are those perspective that need serious research approaches in order to get maximum advantages [185].

12.7.6 ECOLOGICAL RESTORATION

Interacting challenges such as climate change, decline in biodiversity, and social as well as ecological provocations demand a key strategy to respond to these challenges. The restoration of ecosystem is that key strategy and many initiatives on local as well as global scale are being taken. Ecosystem is going through a severe and rapid degradation and is disturbed due to unforeseeable social–ecological issue as well as other numerous reasons and explication of these issues for the restoration of the ecosystem is vital. But there are several challenges that need to be addressed. Three of the most important summons or challenges are climate change, instability in global politics and overutilization of resources. In addition to these challenges, spread of invasive species, fragmentation of habitat, pollution, and so on are also threatening challenges that seek attention. To solve these problems, certain measures need to be taken. Simply talking, the following two steps can be taken in order to restore our ecosystem.

- Initial assessment should be done and a consequential work frame and timeframe should be developed.
- Identification of possible responses. There should be formulation of specific activities that can combat these threats and support the key measures taken for restoration purpose.

However, while taking these steps the practical liability of these steps and likelihood should be considered [186].

12.8 CONCLUSION

This study concludes by highlighting the critical role that microbial partnerships play in optimizing AFSs—which combine crop farming and tree cultivation to address modern agricultural challenges. Plant–microbe interactions—especially those involving rhizobia, beneficial bacteria, and mycorrhizal fungi—are essential for improving plant growth, yield, and overall system resilience. These interactions are emphasized by the dynamic exchanges that the extensive root systems of agroforestry trees create with a variety of microbial communities. This thorough review examines the various plant–microbe synergies that occur in agroforestry and how mycorrhizal relationships enhance soil fertility, sequester carbon, and aid in nutrient uptake as well as climate change mitigation. The study also explores the benefits of beneficial bacteria in suppressing pathogenic organisms and the function of nitrogen-fixing rhizobia in enhancing soil. It is essential to comprehend and utilize these relationships between plants and microbes in order to maximize agroforestry techniques and promote sustainable agriculture. The study also discusses how important it is to control these interactions in order to lessen possible problems, such as competition for resources. In general, this study offers significant perspectives for professionals and decision-makers who aim to encourage the efficiency of agroforestry as a sustainable farming approach.

KEYWORDS

- **agroforestry**
- **plant–microbe interactions**
- **microbial partnership**
- **sustainable agriculture**
- **soil health**
- **nutrient cycling**
- **carbon sequestration**
- **ecosystem services**

REFERENCES

1. Li, Y., Zeng, C., Long, M. (2020). Variation of soil nutrients and bacterial community diversity of different land utilization types in Yangtze River Basin, Chongqing Municipality. *PeerJ*, 8, e9386.
2. Bastida, F., et al. (2021). Soil microbial diversity–biomass relationships are driven by soil carbon content across global biomes. *The ISME Journal*, 15(7), 2081–2091.
3. Guo, X., Zhou, Y. (2021). Effects of land use patterns on the bacterial community structure and diversity of wetland soils in the Sanjiang plain. *Journal of Soil Science and Plant Nutrition*, 21(1), 1–12.
4. Turley, N., et al. (2020). Agricultural land-use history and restoration impact soil microbial biodiversity. *Journal of Applied Ecology*, 57, 852–863.
5. Yao, H., et al. (2000). Microbial biomass and community structure in a sequence of soils with increasing fertility and changing land use. *Microbial Ecology*, 40(3), 223–237.

6. Chen, Q., et al. (2019). Comparison of soil organic matter transformation processes in different alpine ecosystems in the Qinghai-Tibet plateau. *Journal of Geophysical Research: Biogeosciences*, 124(1), 33–45.
7. Lamb, E. G., Kennedy, N., Siciliano, S. D. (2011). Effects of plant species richness and evenness on soil microbial community diversity and function. *Plant and Soil*, 338(1), 483–495.
8. Barberán, A., et al. (2012). Using network analysis to explore co-occurrence patterns in soil microbial communities. *The ISME Journal*, 6(2), 343–351.
9. Berthrong, S. T., Buckley, D. H., Drinkwater, L. E. (2013). Agricultural management and labile carbon additions affect soil microbial community structure and interact with carbon and nitrogen cycling. *Microbial Ecology*, 66(1), 158–170.
10. Lauber, C. L., et al. (2013). Temporal variability in soil microbial communities across land-use types. *The ISME Journal*, 7(8), 1641–1650.
11. Han, W., et al. (2023). Variations of soil bacterial community and denitrifier abundance with depth under different land-use types. *Journal of Soils and Sediments*, 23(4), 1889–1900.
12. Wagg, C., et al. (2019). Fungal-bacterial diversity and microbiome complexity predict ecosystem functioning. *Nature Communications*, 10, 4841.
13. Xu, M.-p., et al. (2022). Plant biomass and soil nutrients mainly explain the variation of soil microbial communities during secondary succession on the loess plateau. *Microbial Ecology*, 83(1), 114–126.
14. Lahlali, R., et al. (2021). High-throughput molecular technologies for unraveling the mystery of soil microbial community: Challenges and future prospects. *Heliyon*, 7(10), e08142.
15. Ming-yan, L., et al. (2020). Influence of 16S rRNA gene V4 and V3−V4 sequencing and sequencing depth on unraveling bacterial communities inhabiting oil reservoirs. *Microbiology China*, 47, 440–449.
16. Liu, S., et al. (2022). Composition and diversity of soil microbial community associated with land use types in the agro-pastoral area in the upper yellow river basin. *Frontiers in Plant Science*, 13, 819661.
17. Moonen, A.-C., Bàrberi, P. (2008). Functional biodiversity: An agroecosystem approach. *Agriculture, Ecosystems & Environment*, 127(1), 7–21.
18. Cacho, O. (2001). An analysis of externalities in agroforestry systems in the presence of land degradation. *Ecological Economics*, 39(1), 131–143.
19. Babafemi, O. P., et al. (2022). Advantages and potential threats of agrochemicals on biodiversity conservation. In: Chibueze Izah, S. (Ed.), *Biodiversity in Africa: Potentials, Threats and Conservation* (pp. 267–292), Springer Nature Singapore: Singapore.
20. Isobe, K., et al. (2019). Phylogenetic conservation of bacterial responses to soil nitrogen addition across continents. *Nature Communications*, 10(1), 2499.
21. Iizumi, T., Wagai, R. (2019). Leveraging drought risk reduction for sustainable food, soil and climate via soil organic carbon sequestration. *Scientific Reports*, 9(1), 19744.
22. Hati, K. M., et al. (2020). Soil organic carbon dynamics and carbon sequestration under conservation tillage in tropical vertisols. In: Ghosh, P. K. et al. (Eds.), *Carbon Management in Tropical and Sub-Tropical Terrestrial Systems* (pp. 201–212). Springer Singapore: Singapore.

23. Palma, J. H. N., et al. (2007). Modeling environmental benefits of silvoarable agroforestry in Europe. *Agriculture, Ecosystems & Environment*, 119(3), 320–334.
24. Manlay, R. J., Feller, C., Swift, M. J. (2007). Historical evolution of soil organic matter concepts and their relationships with the fertility and sustainability of cropping systems. *Agriculture, Ecosystems & Environment*, 119(3), 217–233.
25. Fonte, S. J., Barrios, E., Six, J. (2010). Earthworms, soil fertility and aggregate-associated soil organic matter dynamics in the Quesungual agroforestry system. *Geoderma*, 155(3), 320–328.
26. Roshetko, J., Lasco, R., Angeles, M. (2007). Smallholder agroforestry systems for carbon storage. *Mitigation and Adaptation Strategies for Global Change*, 12(2), 219–242.
27. Kirby, K. R., Potvin, C. (2007). Variation in carbon storage among tree species: Implications for the management of a small-scale carbon sink project. *Forest Ecology and Management*, 246(2), 208–221.
28. Ramachandran Nair, P. K., Mohan Kumar, B., Nair, V. D. (2009). Agroforestry as a strategy for carbon sequestration. *Journal of Plant Nutrition and Soil Science*, 172(1), 10–23.
29. Barreto, P. A. B., et al. (2011). Distribution of oxidizable organic C fractions in soils under cacao agroforestry systems in Southern Bahia, Brazil. *Agroforestry Systems*, 81(3), 213–220.
30. Udawatta, R. P., et al. (2008). Variations in soil aggregate stability and enzyme activities in a temperate agroforestry practice. *Applied Soil Ecology*, 39(2), 153–160.
31. Sørensen, J. T., Sessitsch, A. (2007). Plant-associated bacteria-lifestyle and molecular interactions. In *Modern Soil Microbiology*, Taylor & Francis.
32. Lovell, S. T., Sullivan, W. C. (2006). Environmental benefits of conservation buffers in the United States: Evidence, promise, and open questions. *Agriculture, Ecosystems & Environment*, 112(4), 249–260.
33. Araujo, A., et al. (2012). Microbiological process in agroforestry systems. A review. *Agronomy for Sustainable Development*, 32, 215–216.
34. Cardoso, I. M., Kuyper, T. W. (2006). Mycorrhizas and tropical soil fertility. *Agriculture, Ecosystems & Environment*, 116, 72–84.
35. Fahad, S., et al. (2022). Agroforestry Systems for Soil Health Improvement and Maintenance. *Sustainability*, 14, 14877.
36. Franco, S. D., et al. (2022). Assessing the presence of a monoculture: From definition to quantification. *Agriculture*, *12*(9), 1506.
37. Meena, V. S., Maurya, B. R., Verma, J. P. (2014). Does a rhizospheric microorganism enhance K+ availability in agricultural soils? *Microbiological Research*, 169(5), 337–347.
38. Singh, J. S. (2015). Plant–microbe interactions: A viable tool for agricultural sustainability. *Applied Soil Ecology*, 92, 45–46.
39. Sindhu, S. S., et al. (2016). *Potassium-Solubilizing Microorganisms (KSMs) and Its Effect on Plant Growth Improvement*. Springer: New Delhi.
40. Sahu, N., et al. (2017). *Strength of Microbes in Nutrient Cycling: A Key to Soil Health*. Springer: Singapore.

41. Howell, C. R. (2003). Mechanisms employed by trichoderma species in the biological control of plant diseases: The history and evolution of current concepts. *Plant Disease*, 87(1), 4–10.
42. Vinale, F., et al. (2009). Factors affecting the production of Trichoderma harzianum secondary metabolites during the interaction with different plant pathogens. *Letters in Applied Microbiology*, 48(6), 705–711.
43. Akhtar, M. S., Siddiqui, Z. A., Wiemken, A. (2011). Arbuscular mycorrhizal fungi and rhizobium to control plant fungal diseases. In: Lichtfouse, E. (Ed.) *Alternative Farming Systems, Biotechnology, Drought Stress and Ecological Fertilisation* (pp. 263–292). Springer Netherlands: Dordrecht.
44. Meena, M. K., et al. (2017). *Beneficial Microbes for Disease Suppression and Plant Growth Promotion*. Springer: Singapore.
45. Vidhyasekaran, P., et al. (2001). Induction of systemic resistance byPseudomonas fluorescens Pf1 against Xanthomonas oryzae pv. Oryzae in rice leaves. *Phytoparasitica*, 29, 155–166.
46. Syed Nabi, R. B., et al. (2021). Evaluation potential of PGPR to protect tomato against Fusarium wilt and promote plant growth. *PeerJ*, 9, e11194.
47. El-Aswad, A. F., et al. (2023). Efficacy evaluation of some fumigants against Fusarium oxysporum and enhancement of tomato growth as elicitor-induced defense responses. *Scientific Reports*, 13(1), 2479.
48. Saha, S., et al. (2016). Role of microbes in soil health improvement introduction. In: *State Agricultural Technologist Service Association-Annual Technical Issue* (pp. 53–62).
49. Fierer, N., Wood, S. A., Bueno de Mesquita, C. P. (2021). How microbes can, and cannot, be used to assess soil health. *Soil Biology and Biochemistry*, 153, 108111.
50. Kumar, R., Kumar, B., Ram, M., Kumawat, N., Kumar, R. (2014). Influence of fertilizers levels and bio fertilizers on growth and yield of linseed (Linum usitatissiumum L.) under rainfed condition of South Gujarat. *Madras Agricultural Journal*, 100(4–6), 403–406
51. Azarbad, H., et al. (2015). Microbial community composition and functions are resilient to metal pollution along two forest soil gradients. *FEMS Microbiology Ecology*, 91(1), 1–11.
52. Sarkar, S., et al. (2022). Role of soil microbes to assess soil health. In: Giri, B., et al. (Eds.), *Structure and Functions of Pedosphere* (pp. 339–363). Springer Nature Singapore: Singapore.
53. Morya, R., et al. (2023). Effect of different sources of nitrogen and phosphorus and application methods on yield and quality of hybrid pearl millet. *International Journal of Environment and Climate Change*, 13(12), 166–172.
54. Zhang, X., et al. (2020). Links between microbial biomass and necromass components in the top- and subsoils of temperate grasslands along an aridity gradient. *Geoderma*, 379, 114623.
55. Yadav, A. N., et al. (2018). Chapter 15—Microbiome in crops: Diversity, distribution, and potential role in crop improvement. In: Prasad, R., Gill, S. S., Tuteja, N. (Eds.), *Crop Improvement Through Microbial Biotechnology* (pp. 305–332). Elsevier.
56. Zhang, T. A., Ruan, H. (2018). Global negative effects of nitrogen deposition on soil microbes. *The ISME Journal*, 12, 1817–1825.

57. Schloter, M., et al. (2017). Microbial indicators for soil quality. *Biology and Fertility of Soils*, 54, 1–10.
58. Moghimian, N., et al. (2017). Impacts of changes in land use/cover on soil microbial and enzyme activities. *CATENA*, 157, 407–414.
59. Rilling, J. I., et al. (2019). Current opinion and perspectives on the methods for tracking and monitoring plant growth–Promoting bacteria. *Soil Biology and Biochemistry*, 130, 205–219.
60. Branco, S., et al. (2022). Mechanisms of stress tolerance and their effects on the ecology and evolution of mycorrhizal fungi. *New Phytologist*, 235(6), 2158–2175.
61. Solanki, M. K., et al. (2021). 6—Mycorrhizal fungi and its importance in plant health amelioration. In: Solanki, M. K., et al. (Eds.), *Microbiomes and Plant Health* (pp. 205–223). Academic Press.
62. Bowles, T. M., et al. (2020). Long-term evidence shows that crop-rotation diversification increases agricultural resilience to adverse growing conditions in North America. *One Earth*, 2(3), 284–293.
63. He, X., et al. (2022). Arbuscular mycorrhizal fungi enhance nutrient acquisition and reduce aluminum toxicity in Lespedeza formosa under acid rain. *Environmental Science and Pollution Research*, 29(20), 29904–29916.
64. Wang, Y., et al. (2021). Enhancement of photosynthetic parameters and growth of Zelkova serrata by arbuscular mycorrhizal fungi under simulated sulfuric acid rain. *Plant Ecology*, 222(12), 1361–1374.
65. Xia, L., et al. (2021). Improved tolerance of mycorrhizal torreya grandis seedlings to sulfuric acid rain related to phosphorus and zinc contents in shoots. *Journal of Fungi*, 7(4), 296.
66. Prisa, D. (2023). Mycorrhizal symbioses and plant interactions. *Karbala International Journal of Modern Science*, 9, 194–207.
67. Weng, W., et al. (2022). Roles of arbuscular mycorrhizal fungi as a biocontrol agent in the control of plant diseases. *Microorganisms*, 10(7), 1266.
68. Seleiman, M. F., Hardan, A. N. (2021). Importance of Mycorrhizae in Crop Productivity. In: Awaad, H., Abu-hashim, M., Negm, A. (Eds.), *Mitigating Environmental Stresses for Agricultural Sustainability in Egypt* (pp. 471–484). Springer International Publishing: Cham.
69. Birhane, E., et al. (2023). Arbuscular mycorrhizal fungi improve nutrient status of Commiphora myrrha seedlings under drought. *Journal of Arid Environments*, 209, 104877.
70. Rapparini, F., Peñuelas, J. (2014). Mycorrhizal fungi to alleviate drought stress on plant growth. In: Miransari, M. (Ed.), *Use of Microbes for the Alleviation of Soil Stresses, Volume 1* (pp. 21–42). Springer New York: New York, NY.
71. Tedersoo, L., Bahram, M., Zobel, M. (2020). How mycorrhizal associations drive plant population and community biology. *Science*, 367(6480).
72. Dierks, J., et al. (2021). Trees enhance abundance of arbuscular mycorrhizal fungi, soil structure, and nutrient retention in low-input maize cropping systems. Agriculture, *Ecosystems & Environment*, 318, 107487.
73. Hestrin, R., et al. (2019). Synergies between mycorrhizal fungi and soil microbial communities increase plant nitrogen acquisition. *Communications Biology*, 2(1), 233.

74. Bowles, T. M., Jackson, L. E., Cavagnaro, T. R. (2018). Mycorrhizal fungi enhance plant nutrient acquisition and modulate nitrogen loss with variable water regimes. *Global Change Biology*, 24(1), e171–e182.
75. Zhu, X., Song, F., Liu, F. (2017). Arbuscular mycorrhizal fungi and tolerance of temperature stress in plants. In: Wu, Q.-S. (Ed.), *Arbuscular Mycorrhizas and Stress Tolerance of Plants* (pp. 163–194). Springer Singapore: Singapore.
76. Pandey, D., et al. (2019). Mycorrhizal fungi: Biodiversity, ecological significance, and industrial applications. In: *Recent Advancement in White Biotechnology Through Fungi*.
77. Shi, Z., et al. (2021). Effects of acid rain on plant growth: A meta-analysis. *Journal of Environmental Management*, 297, 113213.
78. Alotaibi, M. O., et al. (2021). Arbuscular mycorrhizae mitigate aluminum toxicity and regulate proline metabolism in plants grown in acidic soil. *Journal of Fungi (Basel)*, 7(7), 531.
79. Riaz, M., et al. (2021). Arbuscular mycorrhizal fungi-induced mitigation of heavy metal phytotoxicity in metal contaminated soils: A critical review. *Journal of Hazardous Materials*, 402, 123919.
80. Chen, S., et al. (2020). Contrasting effects of long-term acid rain simulation on temperature sensitivity of soil respiration and enzymatic activities in a subtropical forest. *Journal of Soils and Sediments*, 20(1), 412–424.
81. Kuypers, M. M. M., Marchant, H. K., Kartal, B. (2018). The microbial nitrogen-cycling network. *Nature Reviews Microbiology*, 16(5), 263–276.
82. Wewalwela, J. J., Seneviratne, M., Shields-Menard, S. A. (2021). Role of nitrogen-fixing microorganisms for plant and soil health. In: Soni, R., et al. (Eds.), *Microbiological Activity for Soil and Plant Health Management* (pp. 161–179). Springer Singapore: Singapore.
83. Adeleke, B. S., Babalola, O. O. (2022). Roles of plant endosphere microbes in agriculture-a review. *Journal of Plant Growth Regulation*, 41(4), 1411–1428.
84. Doolotkelvieva, T., Bobusheva, S. (2016). Fire blight disease caused by erwinia amylovora on rosaceae plants in Kyrgyzstan and biological agents to control this disease. *Advances in Microbiology*, 06, 831–851.
85. Larran, S., et al. (2016). Endophytes from wheat as biocontrol agents against tan spot disease. *Biological Control*, 92, 17–23.
86. Ke, X., et al. (2019). Effect of inoculation with nitrogen-fixing bacterium Pseudomonas stutzeri A1501 on maize plant growth and the microbiome indigenous to the rhizosphere. *Systematic and Applied Microbiology*, 42(2), 248–260.
87. Hamayun, M., et al. (2017). Gibberellins producing endophytic fungus porostereum spadiceum AGH786 rescues growth of salt affected soybean. *Frontiers in Microbiology*, 8, 686.
88. Bergna, A., et al. (2018). Tomato seeds preferably transmit plant beneficial endophytes. *Phytobiomes Journal*, 2(4), 2471–2906.
89. Feng, N.-X., et al. (2017). Efficient phytoremediation of organic contaminants in soils using plant–endophyte partnerships. *Science of The Total Environment*, 583, 352–368.

90. Pajares, S., Bohannan, B. J. M. (2016). Ecology of nitrogen fixing, nitrifying, and denitrifying microorganisms in tropical forest soils. *Frontiers in Microbiology*, 7, 1045.
91. Gayathri, G., et al. (2023). Chapter 13—Improvement of soil fertility through plant microbial interaction. In: Swapnil, P., et al. (Eds.), *Plant-Microbe Interaction—Recent Advances in Molecular and Biochemical Approaches* (pp. 281–301). Academic Press.
92. Sharma, R., Salwan, R., Sharma, V. (2023). Chapter 20—Biology of nitrogen fixation in Frankia. In: Sharma, V., et al. (Eds.) *The Chemical Dialogue Between Plants and Beneficial Microorganisms* (pp. 271–281). Academic Press.
93. Cassán, F., et al. (2020). Everything you must know about Azospirillum and its impact on agriculture and beyond. *Biology and Fertility of Soils*, 56(4), 461–479.
94. Aasfar, A., et al. (2021). Nitrogen fixing azotobacter species as potential soil biological enhancers for crop nutrition and yield stability. *Frontiers in Microbiology*, 12, 628379.
95. Stal, L. J., Nitrogen fixation in cyanobacteria. In: *Encyclopedia of Life Sciences* (pp. 1–9).
96. Zilli, J. É., et al. (2020). The importance of denitrification performed by nitrogen-fixing bacteria used as inoculants in South America. *Plant and Soil*, 451(1), 5–24.
97. Debbarma, P., et al. (2023). Chapter 3—Role of quorum sensing in plant–microbe interactions. In: Chandra Pandey, S., et al. (Eds.), *Advanced Microbial Techniques in Agriculture, Environment, and Health Management* (pp. 57–66). Academic Press.
98. Ueno, A., et al. (2016). Anaerobic decomposition of humic substances by Clostridium from the deep subsurface. *Scientific Reports*, 6, 18990.
99. López-Mondéjar, R., et al. (2016). Cellulose and hemicellulose decomposition by forest soil bacteria proceeds by the action of structurally variable enzymatic systems. *Sci Rep*, 6, 25279.
100. Johnston, S. R., Boddy, L., Weightman, A. J. (2016). Bacteria in decomposing wood and their interactions with wood-decay fungi. FEMS Microbiology Ecology, 92(11), fiw179.
101. Reiss, J. (2021). The role of protozoans and microscopically small metazoans in aquatic plant litter decomposition. In: Swan, C. M., Boyero, L., Canhoto, C. (Eds.), *The Ecology of Plant Litter Decomposition in Stream Ecosystems* (pp. 217–233). Springer International Publishing: Cham.
102. Efstratiou, M. A. (2019). Microorganisms in beach sand: Health implications. In: Nriagu, J. (Ed.), *Encyclopedia of Environmental Health (Second Edition)* (pp. 385–389), Elsevier: Oxford.
103. Krishna, M. P., Mohan, M. (2017). Litter decomposition in forest ecosystems: A review. *Energy, Ecology and Environment*, 2(4), 236–249.
104. Zhao, B., Xing, P., Wu, Q. L. (2021). Interactions between bacteria and fungi in macrophyte leaf litter decomposition. *Environmental Microbiology*, 23(2), 1130–1144.
105. Roohi, et al. (2017). Microbial enzymatic degradation of biodegradable plastics. *Current Pharmaceutical Biotechnology*, 18(5), 429–440.
106. Shao, Y., et al. (2021). Composition and profiles of volatile organic compounds during waste decomposition by the anaerobic bacteria purified from landfill. *Waste Management*, 126, 466–475.

107. Xu, Z., et al. (2022). Cattle manure compost humification process by inoculation ammonia-oxidizing bacteria. *Bioresource Technology*, 344(Pt B), 126314.
108. Bani, A., et al. (2018). The role of microbial community in the decomposition of leaf litter and deadwood. *Applied Soil Ecology*, 126, 75–84.
109. Koskey, G., et al. (2021). Potential use of beneficial microorganisms for soil amelioration, phytopathogen biocontrol, and sustainable crop production in smallholder agroecosystems. *Frontiers in Sustainable Food Systems*, 5, 130.
110. Banerjee, S., van der Heijden, M. G. A. (2023). Soil microbiomes and one health. *Nature Reviews Microbiology*, 21(1), 6–20.
111. Fall, A. F., et al. (2022). Roles of arbuscular mycorrhizal fungi on soil fertility: Contribution in the improvement of physical, chemical, and biological properties of the soil. *Frontiers in Fungal Biology*, 3. https://doi.org/10.3389/ffunb.2022.723892
112. Rodrigues, M., et al. (2021). Tillage systems and cover crops affecting soil phosphorus bioavailability in Brazilian Cerrado Oxisols. *Soil & Tillage Research*, 205, 104770.
113. Singh, B. K., et al. (2023). Climate change impacts on plant pathogens, food security and paths forward. *Nature Reviews Microbiology*, 21(10), 640–656.
114. Sönmez, C., Mamay, M. (2018). Biological Control in Sustainable Agriculture, *1st International Gap Agriculture & Livestock Congress*, Turkey.
115. Sarvade, S. (2019). *Agroforestry and Soil Health: An Overview* (pp. 275–297).
116. Al-Ani, L. K. T., Mohammed, A. M. (2020). Chapter 12—Versatility of Trichoderma in plant disease management. In: Sharma, V., Salwan, R., Al-Ani, L. K. T. (Eds.), *Molecular Aspects of Plant Beneficial Microbes in Agriculture* (pp. 159–168). Academic Press.
117. Singh, S., Singh, V., Pal, K. (2017). *Importance of Micro Organisms in Agriculture*.
118. De Corato, U. (2020). Soil microbiota manipulation and its role in suppressing soil-borne plant pathogens in organic farming systems under the light of microbiome-assisted strategies. *Chemical and Biological Technologies in Agriculture*, 7(1), 17.
119. Lucas, P. (2006). *Diseases Caused by Soil-borne Pathogens* (pp. 373–386).
120. Chiaramonte, J. B., Mendes, L. W., Mendes, R. (2021). Rhizosphere microbiome and soil-borne diseases. In: Gupta, V. V. S. R., Sharma, A. K. (Eds.), *Rhizosphere Biology: Interactions Between Microbes and Plants* (pp. 155–168). Springer Singapore: Singapore.
121. Chepsergon, J., Moleleki, L. N. (2023). Rhizosphere bacterial interactions and impact on plant health. *Current Opinion in Microbiology*, 73, 102297.
122. O'Brien, P. A. (2017). Biological control of plant diseases. *Australasian Plant Pathology*, 46(4), 293–304.
123. Hirooka, T., Ishii, H. (2013). Chemical control of plant diseases. *Journal of General Plant Pathology*, 79.
124. Struck, C., Rüsch, S., Strehlow, B. (2022). Control strategies of clubroot disease caused by plasmodiophora brassicae. *Microorganisms*, 10(3), 620.
125. Akanmu, A. O., et al. (2021). Plant disease management: Leveraging on the plant-microbe-soil interface in the biorational use of organic amendments. *Frontiers in Plant Science*, 12, 700507.
126. Shah, K. K., et al. (2021). Role of soil microbes in sustainable crop production and soil health: A review. *Agricultural Science and Technology*, 13(2), 109–118.

127. Rashid, M. M., et al. (2022). Detection and diagnosis of important soil-borne diseases: an overview. In: Singh, U. B., et al. (Eds.), *Rhizosphere Microbes: Biotic Stress Management* (pp. 93–104). Springer Nature Singapore: Singapore.
128. Minchev, Z., et al. (2021). Microbial consortia for effective biocontrol of root and foliar diseases in tomato. *Frontiers in Plant Science*, 12, 756368.
129. Cai, X., et al. (2021). Effects and mechanisms of symbiotic microbial combination agents to control tomato fusarium crown and root rot disease. *Frontiers in Microbiology*, 12, 629793.
130. Ouhaibi-Ben Abdeljalil, N., et al. (2021). Combining potential oomycete and bacterial biocontrol agents as a tool to fight tomato Rhizoctonia root rot. *Biological Control*, 155, 104521.
131. Biessy, A., Filion, M. (2022). Biological control of potato common scab by plant-beneficial bacteria. *Biological Control*, 165, 104808.
132. El-Saadony, M. T., et al. (2022). Plant growth-promoting microorganisms as biocontrol agents of plant diseases: Mechanisms, challenges and future perspectives. *Frontiers in Plant Science*, 13, 923880.
133. Eljounaidi, K., Lee, S. K., Bae, H. (2016). Bacterial endophytes as potential biocontrol agents of vascular wilt diseases—Review and future prospects. *Biological Control*, 103, 62–68.
134. Avila-Quezada, G. D., Rai, M. (2023). Novel nanotechnological approaches for managing Phytophthora diseases of plants. *Trends in Plant Science*, 28(9), 1070–1080.
135. Khrieba, M. (2020). *Damping-off Caused by Pythium Species: Disease Profile and Management* (pp. 257–269).
136. Yan, L., Khan, R. A. A. (2021). Biological control of bacterial wilt in tomato through the metabolites produced by the biocontrol fungus, Trichoderma harzianum. *Egyptian Journal of Biological Pest Control*, 31(1), 5.
137. Costa Carvalho, D. D., et al. (2016). Biological control of white mold by Trichoderma harzianum in common bean under field conditions. *Pesquisa Agropecuaria Brasileira*, 50, 1220.
138. Chaurasia, A., et al. (2018). Actinomycetes: An unexplored microorganisms for plant growth promotion and biocontrol in vegetable crops. *World Journal of Microbiology and Biotechnology*, 34(9), 132.
139. Mbarga, J. B., et al. (2012). Trichoderma asperellum: A potential biocontrol agent for Pythium myriotylum, causal agent of cocoyam (*Xanthosoma sagittifolium*) root rot disease in Cameroon. *Crop Protection*, 36, 18–22.
140. Kumar, K. S. A., Karthika, K. S. (2020). Abiotic and biotic factors influencing soil health and/or soil degradation. In: Giri, B. Varma, A. (Eds.), *Soil Health* (pp. 145–161). Springer International Publishing: Cham.
141. Zhang, L., et al. (2018). The impact of agricultural chemical inputs on environment: Global evidence from informetrics analysis and visualization. *International Journal of Low-Carbon Technologies*, 13(4), 338–352.
142. Batista, B. D., Singh, B. K. (2021). Realities and hopes in the application of microbial tools in agriculture. *Microbial Biotechnology*, 14(4), 1258–1268.
143. Microbial Inoculents; by Barbara Bellows, Texas Institute for Applied Environmental Research; Mike Morris, NCAT; and Colin Mitchell, NCAT Published September

2020 ©NCAT IP604 Microbial Inoculants–ATTRA–Sustainable Agriculture (ncat.org)

144. Basu, S., Kumar, G. (2020). Nitrogen fixation in a legume-rhizobium symbiosis: The roots of a success story. In: Varma, A., Tripathi, S., Prasad, R. (Eds.), *Plant Microbe Symbiosis* (pp. 35–53), Springer International Publishing: Cham.
145. Van Nguyen, T., Pawlowski, K. (2017). Frankia and actinorhizal plants: Symbiotic nitrogen fixation. In: Mehnaz, S. (Ed.), *Rhizotrophs: Plant Growth Promotion to Bioremediation* (pp. 237–261). Springer Singapore: Singapore.
146. Mishra, A., et al. (2021). Role of cyanobacteria in rhizospheric nitrogen fixation. In: Cruz, C., et al. (Eds.), *Soil Nitrogen Ecology* (pp. 497–519). Springer International Publishing: Cham.
147. Pedraza, R., et al. (2020). *Azospirillum* (pp. 73–105).
148. Chennappa, G., et al. (2017). Azotobacter: A potential biofertilizer and bioinoculants for sustainable agriculture. In: Panpatte, D. G., et al. (Eds.), *Microorganisms for Green Revolution: Volume 1: Microbes for Sustainable Crop Production* (pp. 87–106). Springer Singapore: Singapore.
149. Sumbul, A., et al. (2020). Azotobacter: A potential bio-fertilizer for soil and plant health management. *Saudi Journal of Biological Sciences*, 27(12), 3634–3640.
150. Elhaissoufi, W., et al. (2022). Phosphate bacterial solubilization: A key rhizosphere driving force enabling higher P use efficiency and crop productivity. *Journal of Advanced Research*, 38, 13–28.
151. Alori, E. T., Glick, B. R., Babalola, O. O. (2017). Microbial phosphorus solubilization and its potential for use in sustainable agriculture. *Frontiers in Microbiology*, 8, 971.
152. Kumar, A., et al. (2017). Role of pseudomonas sp. in sustainable agriculture and disease management. In: Meena, V. S., et al. (Eds.), *Agriculturally Important Microbes for Sustainable Agriculture: Volume 2: Applications in Crop Production and Protection* (pp. 195–215), Springer Singapore: Singapore.
153. Rawat, P., et al. (2021). Phosphate-solubilizing microorganisms: Mechanism and their role in phosphate solubilization and uptake. *Journal of Soil Science and Plant Nutrition*, 21(1), 49–68.
154. Sharma, R., et al. (2016). Enhanced phosphorus solubilization by bacillus licheniformis CKA1 using central composite design and response surface methodology. *Journal of Pure and Applied Microbiology*, 9, 3131–3141.
155. Sun, X., et al. (2021). Plant growth-promoting effect of the chitosanolytic phosphate-solubilizing bacterium burkholderia gladioli MEL01 after fermentation with chitosan and fertilization with rock phosphate. *Journal of Plant Growth Regulation*, 40(4), 1674–1686.
156. Benaissa, A. (2019). Plant growth promoting rhizobacteria: A review. *Algerian Journal of Environmental Science and Technology*, 5, 873–880.
157. Riaz, U., et al. (2021). Plant Growth-Promoting Rhizobacteria (PGPR) as biofertilizers and biopesticides. In: Hakeem, K. R., et al. (Eds.), *Microbiota and Biofertilizers: A Sustainable Continuum for Plant and Soil Health* (pp. 181–196). Springer International Publishing: Cham.
158. Vessey, J. K. (2003). Plant growth promoting rhizobacteria as biofertilizers. *Plant and Soil*, 255(2), 571–586.

159. Nagargade, M., Tyagi, V., Singh, M. (2018). Plant growth-promoting rhizobacteria: a biological approach toward the production of sustainable agriculture. In: Meena, V. (Eds.), *Role of Rhizospheric Microbes in Soil* (pp. 205–223). Springer, Singapore.
160. Siddiqui, Z. A., Kataoka, R. (2011). Mycorrhizal inoculants: Progress in inoculant production technology. In: Ahmad, I., Ahmad, F., Pichtel, J. (Eds.), *Microbes and Microbial Technology: Agricultural and Environmental Applications* (pp. 489–506). Springer New York: New York, NY.
161. Sun, W., Shahrajabian, M. H. (2023). The application of arbuscular mycorrhizal fungi as microbial biostimulant, sustainable approaches in modern agriculture. *Plants*, 12(17), 3101.
162. Díaz-Urbano, M., et al. (2023). Development of agricultural bio-inoculants based on mycorrhizal fungi and endophytic filamentous fungi: Co-inoculants for improve plant-physiological responses in sustainable agriculture. *Biological Control*, 182, 105223.
163. Li, Q. S., et al. (2022). Arbuscular mycorrhizal fungi and endophytic fungi activate leaf antioxidant defense system of lane late navel orange. *Journal of Fungi (Basel)*, 8(3).
164. Kumar, P., et al. (2022). Bioinoculants as mitigators of multiple stresses: A ray of hope for agriculture in the darkness of climate change. *Heliyon*, 8(11), e11269.
165. Correa, P. A., et al. (2022). Chapter 6—Regulatory role of microbial inoculants to induce salt stress tolerance in horticulture crops. In: Seymen, M., et al. (Eds.), *Sustainable Horticulture* (pp. 125–155). Academic Press.
166. Chakraborty, U., et al. (2015). *Role of Microorganisms in Alleviation of Abiotic Stresses for Sustainable Agriculture* (pp. 232–253), CAB International.
167. Poudel, M., et al. (2021). The role of plant-associated bacteria, fungi, and viruses in drought stress mitigation. *Frontiers in Microbiology*, 12. https://doi.org/10.3389/fmicb.2021.743512
168. Chaudhary, T., Shukla, P. (2019). Bioinoculants for bioremediation applications and disease resistance: Innovative perspectives. *Indian Journal of Microbiology*, 59(2), 129–136.
169. Ganguly, D., et al. (2023). Role of microbes in bioremediation. In: Verma, P. (Ed.), *Industrial Microbiology and Biotechnology: Emerging concepts in Microbial Technology* (pp. 549–584), Springer Nature Singapore: Singapore.
170. Pandey, A., et al. (2023). Chapter 13—Role of microbial enzymes in bioremediation: Emerging opportunities and limitations. In: Sharma, V. K., et al. (Eds.), *Microbial Inoculants* (pp. 277–300). Academic Press.
171. Medić, A. B., Karadžić, I. M. (2022). Pseudomonas in environmental bioremediation of hydrocarbons and phenolic compounds—Key catabolic degradation enzymes and new analytical platforms for comprehensive investigation. *World Journal of Microbiology and Biotechnology*, 38(10), 165.
172. Ghosh, S., et al. (2022). Heavy metal removal by bacillus for sustainable agriculture. In: Islam, M. T., Rahman, M., Pandey, P. (Eds.), *Bacilli in Agrobiotechnology: Plant Stress Tolerance, Bioremediation, and Bioprospecting* (pp. 1–30). Springer International Publishing: Cham.

173. Hellal, J., et al. (2021). Chlorinated ethene biodegradation and associated bacterial taxa in multi-polluted groundwater: Insights from biomolecular markers and stable isotope analysis. *Science of The Total Environment*, 763, 142950.
174. Sindhu, S. S., et al. (2019). Soil fertility improvement by symbiotic rhizobia for sustainable agriculture. In: Panpatte, D. G., Jhala, Y. K. (Eds.), *Soil Fertility Management for Sustainable Development* (pp. 101–166), Springer Singapore: Singapore.
175. Lang, F. S., et al. (2016). Biodegradation of polycyclic aromatic hydrocarbons in mangrove sediments under different strategies: Natural attenuation, biostimulation, and bioaugmentation with rhodococcus erythropolis T902.1. *Water, Air, & Soil Pollution*, 227(9), 297.
176. Phonglosa, A. (2019). *Role of Microbes on Carbon Sequestration.*
177. Zhu, X., et al. (2022). The role of microbiome in carbon sequestration and environment security during wastewater treatment. *Science of The Total Environment*, 837, 155793.
178. Ahmed, A. A. Q., Odelade, K. A., Babalola, O. O. (2020). Microbial inoculants for improving carbon sequestration in agroecosystems to mitigate climate change. In: Leal Filho, W. (Ed.), *Handbook of Climate Change Resilience* (pp. 381–401), Springer International Publishing: Cham.
179. Nie, M., et al. (2015). Increased plant productivity and decreased microbial respiratory C loss by plant growth-promoting rhizobacteria under elevated CO_2. *Scientific Reports*, 5(1), 9212.
180. Gupta, R., et al. (2021). Plant–microbiome interactions for sustainable agriculture: A review. *Physiology and Molecular Biology of Plants*, 27(1), 165–179.
181. Crandall, S. G., et al. (2020). A multi-omics approach to solving problems in plant disease ecology. *PLoS One*, 15(9), e0237975.
182. Arif, I., Batool, M., Schenk, P. M. (2020). Plant microbiome engineering: Expected benefits for improved crop growth and resilience. *Trends in Biotechnology*, 38(12), 1385–1396.
183. Jansson, J. K., Hofmockel, K. S. (2020). Soil microbiomes and climate change. *Nature Reviews Microbiology*, 18(1), 35–46.
184. Bonkowski, M., et al. (2000). Microbial-faunal interactions in the rhizosphere and effects on plant growth. Paper presented at the 16th World Congress of Soil Science, 20–26 August 1998, Montpellier, France. *European Journal of Soil Biology*, 36(3), 135–147.
185. Ramil Brick, E. S., et al. (2022). A review of agroforestry, precision agriculture, and precision livestock farming—The case for a data-driven agroforestry strategy. *Frontiers in Sensors*, 3. https://doi.org/10.3389/fsens.2022.998928
186. Frietsch, M., et al. (2023). Future-proofing ecosystem restoration through enhancing adaptive capacity. *Communications Biology*, 6(1), 377.

Index

C

I

J

K

L

M

R

S

For Product Safety Concerns and Information please contact our EU
representative GPSR@taylorandfrancis.com
Taylor & Francis Verlag GmbH, Kaufingerstraße 24, 80331 München, Germany

www.ingramcontent.com/pod-product-compliance
Ingram Content Group UK Ltd.
Pitfield, Milton Keynes, MK11 3LW, UK
UKHW022207190926
475702UK00003B/246
* 9 7 8 1 7 7 9 6 4 1 3 8 0 *